To CHERYL & KONRAD ♡
MAY, 2017

I found this Book, good info!
Am not finished -yet
Kon & kids played with the Finlay
Let Them Eat Dirt   kid when we lived
in Sherwood Park.
ENJoy   LoVE
JoAnn & LEE

Let Them Eat Dirt

Saving Your Child from an Oversanitized World

B. BRETT FINLAY, PhD

MARIE-CLAIRE ARRIETA, PhD

GREYSTONE BOOKS

Vancouver/Berkeley

16 17 18 19 20   5 4 3 2 1

Greystone Books Ltd.
www.greystonebooks.com

Cataloguing data available from Library and Archives Canada
ISBN 978-1-77164-254-5 (pbk.)
ISBN 978-1-77164-255-2 (epub)

Cover design by Tim Green/Faceout Studio and Nayeli Jimenez
Text design by Steve Godwin
Printed and bound in Canada on
ancient-forest-friendly paper by Friesens

We gratefully acknowledge the support of the Canada Council
for the Arts, the British Columbia Arts Council, the Province of British
Columbia through the Book Publishing Tax Credit, and the
Government of Canada for our publishing activities.

To our kids, Jessica, Liam, Marisol, and Emiliano
for inspiring us to get the word out that kids need
more dirt in their lives.

Contents

Preface

We all want what is best for our kids. The problem is that there is no perfect handbook on how to raise them, nor is there any one best way, either. We read books and articles, talk to friends, and try to remember (or forget!) how our parents raised us. Both of us have children and have struggled and muddled through the parenting process the same way everyone does. We are also scientists who have worked with microbes for many years, and we couldn't help but consider how these ever-present microbes influence development as we raised our children. At first we studied microbes that cause disease, and we feared them just like anyone else. But more recently we began taking notice of all the other microbes that live in and on us—our "microbiota." As we continue to study the microbiota of humans, it is becoming clear that our exposure to microbes is most important when we're kids. At the same time, modern lifestyles have made childhood much cleaner than ever before in human history, and this is taking a huge toll on our microbiota—and our lifelong health.

The genesis of this book came from the realization that the studies in our lab—and the labs of several other researchers—prove that microbes really do impact a child's health. What shocked us most was how early this starts—the first one hundred days of life are critical. We knew microbes played a role in well-being, but we had no idea how soon this role began.

Several other factors converged to help convince us to write this book. Claire has young children, and all of her young parent friends were extremely interested in the concept of microbes and how they might affect their kids. Whenever we tell other parents about our work, the questions never cease—*Do I need to sterilize their bottles every time? What kind of soap should I use?* We realized that there are many questions out there about microbes . . . and a lot of wrong information.

Brett is married to a pediatric infectious disease specialist (Jane) who was constantly suggesting articles and findings about how microbes affect kids, which led us to realize that since this was such a new field, there was no one source parents could turn to if they wanted to learn more. Not to mention that scientific articles are usually dry, terse things with lots of jargon and, frankly, are terribly boring. However, this new area of research has a lot to offer to people raising children who are not likely to get this important information from dense scientific papers or from studies often misinterpreted by the press. There is a lot of information being produced by some of the best scientists in the world, which we consider extremely useful for the day-to-day decisions we make while raising our children, so we felt compelled to gather it all in one book and make it accessible to the everyday parent.

We start off by explaining a bit about microbes, and then explore what happens to a pregnant woman's body in terms of her microbiota and how it affects her child(ren) for life. We then discuss the delivery process, breastfeeding, solid foods, and the first years of life from a microbial perspective. In the middle of the book we cover lifestyle issues (*Should I get a pet? What do I do with a dropped pacifier?*) and the use of antibiotics. The latter part of the book features chapters dealing with specific diseases that are growing by leaps and bounds in our society, and the microbes that seem to affect them. These include obesity, asthma, diabetes, intestinal diseases, behavioral and mental health disorders such as autism, and a whole array of diseases in which, even five years ago, we had no clue microbes might be involved. Readers may want to skip over particular chapters if you feel that they are not applicable to you. However, each one is full of information that will educate you about the processes involved in these health issues. We think the section on the gut–brain connection (chapter 14) is particularly interesting in its exploration of how microbes might affect the brain and mental disorders. We finish the book with a discussion on vaccines and a futuristic view of what we can expect in terms of new therapies and medical interventions in the next few years. Each chapter ends with a few Dos and Don'ts—these are not meant to be comprehensive medical advice, but suggestions about things to do (or not do) that are based on current scientific evidence.

What we have learned in writing this book, and what we hope to convince readers of, is that microbes play a very large part in our children's lives. Even as scientists in the field, we were stunned to discover some of the profound roles these microscopic bugs have in normal childhood development. No doubt many of these findings, and many more to come, will have a major impact on how we think about raising our children.

—B. Brett Finlay and Marie-Claire Arrieta

Let Them Eat Dirt

PART ONE

We Are More Microbe Than Human

1: Children Are Microbe Magnets

Microbes: Kill Them All!

Microbes are the smallest forms of life on Earth. They encompass bacteria, viruses, protozoa, and other types of organisms that can be seen only with a microscope. Microbes are also the oldest and most successful forms of life on our planet, having evolved long before plants and animals (plants and animals actually evolved from bacteria). Although invisible to the naked eye, they play a major role in life on Earth. There are an astounding 5×10^{30} (that's 5 followed by 30 zeroes!) bacteria on Earth (for comparison, there are "only" 7×10^{21} stars in the universe). Collectively, these microbes weigh more than all the plants and animals on the entire planet combined. They can live in the harshest and most inhospitable environments, from the Dry Valleys of Antarctica to the boiling hydrothermal vents on the seafloor—they can even thrive in radioactive waste. Every form of life on Earth is covered in microbes in a complex yet usually harmonious relationship, making germophobia the most futile of phobias. Unless you live in a sterile bubble without any contact with the outside world (which is a time-limited proposition; see Bubble Boy, page 15), there is no escaping microbial life—we live in a world coated in a veneer of microbes. For every single human cell in our bodies, there are ten bacterial cells inhabiting us; for every gene in our cells, there are one-hundred fifty bacterial genes, begging the question: Do they inhabit us or is it really the other way around?

While in its mother's womb a baby is for the most part sterile, but at the moment of birth it receives a big load of microbes, mainly from its mother—a precious first birthday gift! Within seconds, the baby is covered in microbes from the very first surfaces it touches. Babies born vaginally encounter vaginal and fecal microbes, whereas babies born via C-section pick up microbes from the maternal skin instead. Similarly, babies born at home are exposed to very different microbes than if they are born in hospitals, and different homes (and hospitals) have different microbes present.

Why does all this matter? Well, until very recently hardly anyone thought it did. Until recently, whenever we thought of microbes—especially around babies—we considered them only as potential threats and were concerned with getting rid of them, and it's no surprise why. In the past century, we have experienced the benefits of medical advances that have reduced the number and the degree of infections we suffer throughout life. These advances include antibiotics, antivirals, vaccinations, chlorinated water, pasteurization, sterilization, pathogen-free food, and even good old-fashioned handwashing. The quest of the past hundred years has been to get rid of microbes—the saying was "the only good microbe is a dead one."

This strategy served us remarkably well; nowadays, dying from a microbial infection is a very rare event in developed countries, whereas only a hundred years ago, seventy-five million people died worldwide over a span of two years from the H1N1 influenza virus, also known as the Spanish flu. We have become so efficient at avoiding infections that the appearance of a dangerous strain of *Escherichia coli* (aka *E. coli*) in a beef shipment or *Listeria monocytogenes* in spinach leads to massive recalls and exportation bans, along with accompanying media hysteria. Microbes scare all of us, and rightly so since some of them are truly dangerous. As a result, with very few controlled exceptions such as yogurt or beer, we often think that the presence of microbes in something renders it undesirable for human use. The word antimicrobial is a sales feature in soaps, skin lotions, cleaning supplies, food preservatives, plastics, and even fabrics. However, only about one hundred species of microbes are known to actually cause diseases in humans; the vast majority of the thousands of species that inhabit us do not cause any problems, and, in fact, seem to come with serious benefits.

At first glance, our war on microbes, along with other medical advances, has truly paid off. In 1915 the average life span in the US was fifty-two years, about thirty years shorter than it is today. For better or for worse, there are almost four times more humans on this planet than there were just a hundred years ago, which translates to an incredibly accelerated growth in our historic timeline. Evolutionarily speaking, we've hit the jackpot. But at what price?

Revenge of the Microbes

The prevalence of infectious diseases declined sharply after the emergence of antibiotics, vaccines, and sterilization techniques. However, there has been an explosion in the prevalence of chronic noninfectious diseases and disorders in developed countries. One hears about these in the news all the time since they're very common in industrialized nations, where alterations to our immune system play an important role in their development. They include diabetes, allergies, asthma, inflammatory bowel diseases (IBDs), autoimmune diseases, autism, certain types of cancer, and even obesity. The incidence of some of these disorders is doubling every ten years, and they are starting to appear sooner in life, often in childhood. They are our new epidemics, our modern-day bubonic plague. (By contrast, these diseases have remained at much lower levels in developing countries, where infectious diseases and early childhood mortality are still the major problems.) Most of us know someone suffering from at least one of these chronic illnesses; due to this prevalence, researchers have focused their attention on identifying the factors that cause them. What we know now is that although all of these diseases have a genetic component to them, their increased pervasiveness cannot be explained by genetics alone. Our genes simply have not changed that much in just two generations—but our environment sure has.

About twenty-five years ago a short scientific article published by an epidemiologist from London attracted a lot of attention. Dr. David Strachan proposed that a lack of exposure to bacteria and parasites, specifically during childhood, may be the cause of the rapid increase in allergy cases, since it prevents proper development of the immune system. This concept was later termed the "hygiene hypothesis," and an increasing number of studies have explored whether the development of many diseases, not just allergies, can be explained by this hypothesis. There is now a large amount of very solid evidence, which we'll examine in the following chapters, supporting Dr. Strachan's proposal as generally correct. What remains less clear is what exact factors are responsible for this lack of microbial exposure. For his study on allergies, Dr. Strachan concluded that "declining family size, improvements in household amenities, and higher standards of personal cleanliness" contributed to this reduced contact with microbes. While this may be true, there are many other modern-life changes that have an even stronger impact on our exposure to microbes.

One of these changes can be attributed to the use, overuse, and abuse of antibiotics—chemicals that are designed to indiscriminately kill bacterial microbes. Definitely one of, if not *the* greatest discovery of the twentieth century, the emergence of antibiotics marked a watershed before-and-after moment in modern medicine. Prior to the advent of antibiotics, 90 percent of children would die if they contracted bacterial meningitis; now most cases fully recover, if treated early. Back then, a simple ear infection could spread to the brain, causing extensive damage or even death, and most modern surgeries would not even be possible to contemplate. The use of antibiotics, however, has become far too commonplace. Between the years 2000 and 2010 alone there was a 36 percent increase in the use of antibiotics worldwide, a phenomenon that appears to follow the economic growth trajectory in countries such as Russia, Brazil, India, and China. One troubling thing about these numbers is that the use of antibiotics peaks during influenza virus infections, even though they are not effective against viral infections (they are designed to kill bacteria, not viruses).

Antibiotics are also widely used as growth supplements in agriculture. Giving cattle, pigs, and other livestock low doses of antibiotics causes significant weight gain in the animals and, subsequently, an increase in the meat yield per animal. This practice is now banned in Europe, but is still legal in North America. It seems that antibiotic overuse in humans, especially in children, is inadvertently mimicking what occurs in farm animals: increased weight gain. A recent study of 65,000 children in the US showed that more than 70 percent of them had received antibiotics by age two, and that those children averaged eleven courses of antibiotics by age five. Disturbingly, children who received four or more courses of antibiotics in their first two years were at a 10 percent higher risk of becoming obese. In a separate study, epidemiologists from the Centers for Disease Control and Prevention (CDC) found that states in the US with higher rates of antibiotics use also have higher rates of obesity.

While these studies didn't prove that antibiotics directly cause obesity, the consistency in these correlations, as well as those observed in livestock, prompted scientists to have a closer look. What they found was astonishing. A simple transfer of intestinal bacteria from obese mice into sterile ("germ-free") mice made these mice obese, too! We've heard before that many factors lead to obesity: genetics, high-fat diets, high-carb diets, lack of exercise, etc. But bacteria—really? This raised skepticism among even the biggest fanatics in microbiology, those of us who tend to think that bacteria are the center of our world. However, these types of experiments have been repeated in several different ways and the evidence is very convincing: the presence and absence of certain bacteria early in life helps determine your weight later in life. Even more troubling is the additional research that shows that altering the bacterial communities that inhabit our bodies affects not just weight gain and obesity, but many other chronic diseases in which we previously had no clue that microbes might play a role.

Let's take asthma and allergies as an example. We are all witnesses to the rapid increase in the number of children suffering from these two related diseases. Just a generation ago it was rather unusual to see children with asthma inhalers in schools. Nowadays, 13 percent of Canadian children, 10 percent of US children, and 21 percent of Australian children suffer from asthma. Peanut allergies? That used to be incredibly rare, but is now so frequent and so serious that it has led to peanut-free schools and airplanes. As with the obesity research, it is now evident that receiving antibiotics during childhood is associated with an increased risk of asthma and allergies.

Our laboratory at the University of British Columbia became very interested in this concept and decided to do a simple experiment. As had been observed with humans, giving antibiotics to baby mice made them more susceptible to asthma, but what we observed next left us in awe. If the same antibiotics were given when the mice were weaned and no longer in the care of their mothers, there was no effect in susceptibility to asthma. There appeared to be a critical window of time, early in life, during which antibiotics had an effect on the development of asthma. When given orally, the antibiotic that we chose, vancomycin, kills only intestinal bacteria, and does not get absorbed into the blood, lungs, or other organs. This finding implied that the antibiotic-driven change in the intestinal bacteria caused the increase in the severity of asthma, a disease of the lungs! This experiment, as well as others from several different labs, came to the same conclusion: modifying the microbes that live within us at the beginning of our life can have drastic and detrimental health effects later in life. The discovery that this early period in life is so vulnerable and so important tells us that it's crucial to identify the environmental factors that are disturbing the microbial communities that inhabit us during childhood.

One of these factors has been observed by comparing children raised on rural farms to those raised in a city. Several studies have shown that exposure to a farming environment makes children less likely to develop asthma, even children from families with a history of asthma, and scientists are now beginning to learn why. Farm-raised children are exposed to more animals, more time outside, and a lot more dirt (and feces!), all things that are known to stimulate the immune system. A critical part of the training and development of the immune system occurs in the first years of life. Asthma, characterized by a hyperactive immune system, seems to have a higher chance of developing in a child with a limited exposure to these immune stimulants, because without them, the immune system does not have all the tools for proper development. By cleaning up our children's environments, we prevent their immune systems from maturing in the way they have for millions of years before us: with lots and lots of microbes. Life for our ancestors involved massive exposure to microbes from the environment, food, water, feces, and many other diverse sources. Compare that to our current way of life, where meat comes on sterile Styrofoam pans wrapped in plastic wrap, and our water is treated and processed until it's free of nearly all microbes.

Kids Will Be Kids

A friend, Julia, moved to a small free-range pig and poultry farm when her first child was a preschooler. She observed firsthand how differently a kid grows up in a city and on a farm. She has always been outdoorsy, so even when she was living in the city she would let Jedd, her oldest child, play outside a lot. They would go to parks and playgrounds, where she would encourage Jedd to get dirty, play in sandboxes and mud puddles—she even allowed him to put (safe-sized) objects in his mouth, like big rocks or leaves. Her outdoorsy nature, she thought, would make their transition to rural life easier, and it did in many ways. But nothing prepared her for the things she's seen her kids do on their farm. When her second baby was born, she would strap him on her back every morning so she could go to their chicken coop to pick up eggs. Jedd, timid with the animals at first, was now chasing and riding the chickens, tasting their feed and touching the fresh eggs. A couple of times she even caught him chewing on something he had picked up from the ground. Anyone who has stepped inside a chicken coop knows what's on the floor, so she's pretty sure Jedd has tasted chicken droppings at least a few times. Clearly, Julia freaked out at first, but it's hard to prevent a five-year-old boy from getting dirty when you're busy working and looking after a second child. After realizing that Jedd wasn't getting sick from his newly acquired tastes of the farm, Julia relaxed a bit. Jedd, now eight years old, is responsible for gathering the eggs every morning. Newly laid eggs are often soiled and he doesn't wear gloves. He washes his hands when he's done, but it's impossible that some of that stuff hasn't made it into his mouth.

Julia's second child, Jacob, was born and raised on the farm and, like his big brother, he was never the slightest bit hesitant to get dirty. He was once found playing knee-deep in a cesspool of pig waste. At fourteen months he swallowed a handful of fresh chicken droppings as Julia rushed towards him to prevent it. Her initial worry that her children were going to contract a disease from all this messiness dissipated as her kids remained healthy.

Nowadays, with her third baby strapped on her back, she doesn't even flinch at the sight of the two older boys doing what all farm kids do: getting very, very dirty. Every single day, they come home with dirt, poop, feathers, and who knows what else caked onto their skin and clothes. They try their best to keep their farm boots for outdoor use only, but it inevitably happens that dirty boots make it onto the living room carpet. Julia makes sure to wash their hands before they eat and they rarely miss a daily bath (the color of the bathwater is a constant reminder of why daily baths are mandatory in their house).

Even if they play outside a lot, most children growing up in urban environments rarely ever reach the level of dirtiness that Julia's kids experience on a daily basis. From this perspective, a farm kid (and his microbes) is very different from a city kid. We are by no means suggesting that we should all allow our kids to play with animal waste, as they could become sick from this. But farms in general provide a microbe-rich environment that has proven beneficial for the development of the immune system, and that really is akin to the way we used to live, which has been seriously altered only in the past few generations.

The vast majority of children have something in common with Jedd and Jacob, in that they all seek out dirt and enjoy getting messy and sucking on things. Why is that? Our natural behavior in the early years of life definitely tries to maximize our exposure to microbes: babies are in direct contact with maternal skin while breast-feeding, they are constantly putting their hands, feet, and every imaginable object in their mouths. Crawlers and early walkers have their hands all over the floor, and then in their mouths. It often seems that they're waiting for the few seconds that parents take their eyes off them to almost magically find and put the dirtiest thing they can reach in their drool-dripping mouths. It makes us wonder: Are kids instinctually drawn to microbes?

Older kids love digging in the dirt, picking up worms, rolling on the ground, catching frogs and snakes, etc. Perhaps this is actually natural behavior designed to populate kids with even more microbes. Children rarely hesitate to lick anything or anyone. As would be expected, children also suffer more infections than adults. Their vacuum-like behavior ensures that they taste the microbial world and subsequently train their immune system to react to it accordingly. If they encounter a disease-causing microbe, also known as a pathogen, their immune system detects it, reacts to it in the form of sickness, and then tries hard to remember it so that their body can prevent it from causing disease the next time this pathogen makes a visit. When the immune system encounters a harmless microbe—and the vast majority of microbes are harmless—it detects it and, through a series of mechanisms that science does not yet fully understand, decides to ignore or tolerate it. Thus, if children's lifestyles and behaviors dictate a limited exposure to these training events, their immune system will be partially immature and will not learn how to properly react to a pathogen or how to tolerate harmless microbes. The consequence of missing out on this early training appears to be that, later in life, the immune system may react too fiercely to these harmless microbes, which could trigger inflammatory responses in various organs of the body. This contributes to the appearance of "developed country diseases" (like asthma and obesity) that are becoming so prevalent today.

Microbes to the Rescue

Helping develop our immune systems is only part of what microbes do for us. They are in charge of digesting most of our food, including fiber and complex proteins, and chopping them into more digestible forms. They also supply the essential vitamins B and K by synthetizing them from scratch, something our own metabolism cannot do. Without the vitamin K from microbes, for example, our blood would not coagulate.

Good bacteria and other beneficial microbes also help us combat disease-causing microbes. Experiments in our lab have shown that infections from *Salmonella*, a diarrhea-causing bacterium, are far worse when antibiotics are given before the infection actually occurs. Similarly, many of us have experienced the side effects of a long bout of antibiotics: abdominal cramps and watery diarrhea. The microbes we harbor live in a balanced state that provides us with so many benefits, all in exchange for a portion of our daily calories and a warm, dark place to live with regular feeding and watering.

But changes in our modern lifestyles are altering this balance, especially during a critical window in early life. In many developed countries, about 30 percent of babies are born by cesarean section, antibiotic usage is a lot more frequent, and most children do not suffer serious infections thanks to vaccines. Far from suggesting that any of these things should be avoided, our aim is to educate parents, as well as parents-to-be, grandparents, and caregivers, about the potentially life-changing decisions we make on a daily basis by raising children in an environment that's much cleaner than ever before. As parents ourselves, we understand that most of us do the best we can with what we have, and it is not our intention to dictate how other people should raise their children. However, as microbiologists, we are becoming increasingly aware of the key role our resident microbes have in shaping our bodies' development. The microbial communities of babies and young children are being altered in ways that may make them sicker later in life, by the very same practices intended to keep them healthy. Talk about a double-edged sword!

The scientific community is just beginning to grasp this new knowledge, and the general public is just starting to hear about it in news articles of (often misinterpreted) studies. Preventing serious illnesses should always be one of our biggest concerns, but we can also do a great deal to try to distinguish between a necessary intervention, such as giving an antibiotic to fight a life-threatening bacterial infection, and an unnecessary and hyperhygienic practice, such as applying antimicrobial hand sanitizers every time a child plays outside. Not all children will or should be raised like Jedd or Jacob, but we can certainly change those unneeded aspects of our far-too-clean world.

In our classical training as microbiologists, we studied only the microbes that cause diseases and the ways to kill them. Now we acknowledge that we have, for many years, ignored the vast majority of microbes that keep us healthy. Our research labs are changing focus, and we are beginning to think it's time for everyone to become better hosts to our microbial guests.

BUBBLE BOY

David Vetter was born in 1971 in Houston, Texas, with a rare genetic disorder that left him without a working immune system. Any contact with a nonsterile world would mean certain death. Because of this, he was delivered by C-section and placed in a sterile bubble immediately after his birth. In a controversial medical decision, he lived in the hospital in a bubble that grew with him. His medical treatment included many courses of antibiotics to prevent any bacterial infection. Being devoid of bacteria meant that doctors also had to feed him a special diet, along with the essential vitamins K and B, which are normally produced by intestinal bacteria. David's story reflects the impossibility of living without an immune system in a world full of microbes, as well as a human's dependence on microbes and what they produce for us. Sadly, David died at the age of twelve from a viral infection a few months after a bone marrow transplant was finally performed.

2: A Newly Discovered Organ: The Human Microbiome

Invisible Life

The idea of humans being inhabited by countless microbes invisible to the naked eye is as old as the first microscope. Born in 1632 in the city of Delft, in what is now the Netherlands, Antoni van Leeuwenhoek was a tradesman with a special interest in lens making. His desire to see the intricacies of the cloths he marketed drove him to shape glass rods into spheres using a flame. These almost perfect spheres allowed him to magnify not just threads, but anything else he wanted to view in great detail. Although he wasn't formally trained as a scientist, he was one at heart and he soon began to put the oddest things under his rudimentary microscopes: water from a creek, blood, meat, coffee beans, sperm, etc. He methodically wrote everything down and sent his findings to the Royal Society of London, which began publishing his curiosities-filled letters.

One day in 1683, he decided to scrape the white residue between his teeth and place it under his lens, writing in his notes:

An unbelievably great company of living animalcules, a-swimming more nimbly than any I had ever seen up to this time. The biggest sort (whereof there were a great plenty) bent their body into curves in going forwards . . . Moreover, the other animalcules were in such enormous numbers, that all the water . . . seemed to be alive . . . All the people living in our United Netherlands are not as many as the living animals that I carry in my mouth this very day.

Naturally, Leeuwenhoek's observations of a never-before described world filled with microscopic "animalcules" were met with great skepticism and ridicule. It wasn't until other British scientists saw it with their own eyes that they began to acknowledge that Leeuwenhoek was not hallucinating. Leeuwenhoek had written many letters to the Society, but discovering microscopic life is what sealed his long-lasting fame. As a result of his many discoveries, Leeuwenhoek is considered the "Father of Microbiology."

Still, these findings remained nothing more than curiosities of the natural world, with no real connection to human biology until scientists discovered that those "animalcules" caused diseases. This revelation took place almost two hundred years later, when Robert Koch, Ferdinand Cohn, and Louis Pasteur each separately confirmed that diseases such as rabies and anthrax were caused by microbes. Pasteur's work also showed that microbes caused the spoilage of milk, and he thus designed the process known as pasteurization, in which microbes are killed with the use of high heat. Milk contamination led Pasteur to the idea that microbes could be prevented from entering the human body, and together with Joseph Lister, they developed the first antiseptic methods. These began to be widely adopted, with one of them still in use today: Listerine.

Avoiding Contagion at Any Cost

The work of Pasteur, Cohn, Koch, and others led to the widespread knowledge that diseases could be avoided by preventing contact with microbes, and by killing them, and so the quest to eradicate them began in earnest. Health departments opened in London, Paris, New York, and other big cities. Garbage, which had previously been left to pile high on sidewalks, was now collected and disposed of; drinking water was treated; rats and mice were hunted; sewer systems were built; and people with contagious diseases were often placed in isolation. It was through all this that the word "bacteria" gained its bad reputation and inherent connotation of disease, contagion, and plague. Germs were (and still are) entities to be feared, avoided, and fought.

Fast-forward another two hundred years and an equally astounding discovery is now in progress: in our quest to clean up our world, we have been killing more microbes than necessary and, ironically, this can make us sick. Why? Because our bodies know how to properly develop only in the presence of lots of microbes. This groundbreaking concept significantly expands on what science already knows about the nonharmful bacteria that inhabit our body: that they aid in the digestion of certain foods, and that they fabricate certain essential vitamins. However, only very recently have we begun to comprehend how profoundly necessary microbes are for our normal development and well-being.

Microbes: Partners in Evolution

The last twenty years of studying microbes has allowed us to understand that microbes aren't optional forms of life that live within us; they truly constitute part of who we are biologically. To get a better grasp on this, we must first understand that our partnership with microbes is as old as the first species of hominids (our ancestors), and that the evolutionary changes that hominids experienced were accompanied by changes in our microbiota, too. Throughout human history there have been only a few landmark evolutionary bursts (rapid evolutionary changes) that have marked the course of hominids. Interestingly, two of them can be clearly linked to changes in our intestinal physiology and thus with our microbiome.

As hunters and gatherers (a lifestyle that lasted about 2.5 million years), our ancestors had no permanent homes, living in temporary shelters with few possessions so they could easily move from one place to another. Depending on the geographic region they inhabited, early humans ate different mixtures of meats, roots, tubers, and fruits—whatever was in season. Then an extremely important event occurred that led to one of these evolutionary bursts: our ability to control fire and cook food. We completely take it for granted now, but cooking food made it safer to eat, as heat kills the disease-causing bacteria that thrive in decomposing meat. It also changes the chemistry of the food itself, making it much easier to digest and a lot richer in energy. This sudden increase in energy levels changed everything for humans. No longer did our ancestors have to spend hours chewing raw food in order to extract enough calories to sustain everyday life. Think of what our closest relatives in nature, apes, are almost always doing when see them in the zoo or on TV. If humans hadn't developed a way to cook food we, too, would have to spend six hours chewing five kilos of raw food every day to get enough energy to survive, just like our primate cousins do.

The fossil records of humans from this period consist of bones and teeth, making it impossible to determine what type of microbiota lived in the intestines of ancient hunters and gatherers. However, anthropologists have been able to show that the change in lifestyle and diet that resulted from the advent of cooking had anatomical consequences involving the intestines. As energy intake increased, the intestines of our human ancestors shortened and, amazingly, their brains grew, too, increasing in size by about 20 percent. Given what we know today about the link between gut microbes and brain development, it is very likely that intestinal microbiota had a part in this "sudden" brain growth. Brain enlargement improved our capacity to hunt, communicate, and socialize. In other words, cooking made us smarter—it made us human.

Another evolutionary landmark occurred about eleven thousand years ago. Certain groups of humans realized, probably by chance, that fallen grains from the wild wheat stalks they collected would give rise to more wheat if planted. When humans learned to domesticate plants for food, they tossed away their nomadic ways for a settled lifestyle. Having crops nearby meant that previously small tribes of a few dozen humans could grow to a few hundred, which in turn gave rise to basic traits of civilization, such as trade, written language, and math. If it weren't for farming, we would all still be picking berry after berry from bushes and walking miles every day. The emergence of agriculture coincides with the appearance of the first cities; inadvertently, agriculture built our modern social structures. This lifestyle change was so successful that farmers replaced foragers, and these days only a handful of people maintain a hunter-gatherer way of life.

As expected, the lifestyle associated with farming came with major dietary changes. Humans no longer ate small bites throughout the day with the occasional feast after a hunt since farmers had a steady and somewhat predictable supply of foods. So how did this affect our microbiota? By domesticating grains and consequently obtaining most of their daily calories from their new crops, the diet of farmers became less diverse. Based on what is currently known about the microbiota's response to diet, their microbiota likely became less diverse, too. In fact, comparing the intestinal microbiota of the Hazda people of Tanzania, one of the few contemporary tribes that relies on foraging, to a modern farmer is like comparing a rain forest to a desert, in terms of biodiversity. Less diversity in our microbiota is associated with a number of human diseases, many of which we cover in later chapters.

Although farming has been around for only eleven thousand years (just 0.004 percent of human history!), physiological changes have also been linked to the agricultural diet, and some of these changes involve our resident microbes. The new diet brought with it cavities and other periodontal diseases, mediated by bacteria rarely found in foragers. Our teeth, jaws, and faces have grown smaller, too, probably because chewing was reduced on such a diet. Some evolutionary biologists believe that we lived a healthier lifestyle as foragers, and that humans traded in that healthier lifestyle for food security and more babies (not a bad deal, actually!). Certain nutritionists have extrapolated from this a recommendation that, in order to promote health, all modern humans should eat the way hunters and gatherers did, but this has been debunked by top evolutionary biologists based on the fact that humans have adapted genetically to the challenges that farming generated (see the Caveman Diet, page 30).

What these two major events in human history teach us is that changes in lifestyle are accompanied by changes in our microbiota, and that these microbial changes might affect our health for better (e.g., cooking food and decreasing infections) or worse (e.g., agriculture and less microbial diversity). Whether we like it or not, we are married to microbes for life, in sickness and in health, for richer or for poorer.

Bugs "R" Us

Our microbes are part of what make us human, but our current way of living and eating, especially in the Western world, has exerted further changes in our microbiota and in our biology. In the past hundred years, and especially the last thirty years, humans have learned to process foods to make them tastier, more digestible, and more shelf-stable than ever before. On top of this, our push to clean up our world in order to fight infectious diseases, including the use of antibiotics, has further shifted the composition and diversity of our microbial communities. Double-punching our microbiota like this has induced huge changes in our intestinal environments and, as we will learn in the following chapters, on many other aspects of our bodies' normal functions.

In order to appreciate how the microbiota influences our health, it is important that we discuss certain basic biological concepts about our microbiota and the organ most of them call home, the human intestine. The human microbiota consists of bacteria, viruses, fungi, protozoa, and other forms of microscopic life. They inhabit our skin, oral and nasal cavities, eyes, lungs, urinary tract, and gastrointestinal tract—pretty much any surface that has exposure to the outside world. Another term that is frequently used is microbiome, which refers not only to the identity of all the microbes living within us, but also to what they do. A total of 10^{14} microbes are estimated to live in the human body and, as mentioned, the intestinal tract is the biggest reservoir of microbes, harboring approximately 10^{13} bacteria. It is this community that influences us, their host, the most. In fact, unless otherwise noted throughout this book, when we use the term microbiota, we are referring to the intestinal microbiota. Although bacteria are approximately twenty-five times smaller than human cells, they account for a significant amount of our weight. If we were to get rid of our microbiota we would lose around three pounds, or about the weight of our liver or brain! A single bowel movement is 60 percent bacteria numbering more than all the people on this globe, a deeply disturbing fact for germophobes.

For microbes, the gastrointestinal system is a fabulous place to live. It's moist, full of nutrients, and sticky (allowing microbes to adhere to it), and in many sections it completely lacks oxygen. Although it seems counterintuitive that any life-form would favor a place without oxygen, an enormous number of bacterial species either prefer or require such a place, as this world evolved for billions of years without oxygen. Microbes living without air are called anaerobes and our gut is packed with them.

About 500–1,500 species of bacteria live in the human gut; the types and numbers vary according to the different sections of the gastrointestinal system. Starting from the top down, the mouth harbors a diverse and complex microbiota—the tongue, cheeks, palate, and teeth are all covered in a dense layer of bacteria known as a biofilm. For example, the dental plaque that dentists remove from our mouths is one of these biofilms. The stomach, on the other hand, is not the best place for microbes, as it is as acidic as battery acid. Still, a few bacterial species have adapted to live under such conditions. Farther down are the small and large intestines, where the number of microbes continues to increase until we reach the very end of the large intestine. Oxygen follows the opposite pattern, as it gradually decreases towards the lower portions of the gut, allowing strict anaerobes (those that die when exposed to the slightest bit of oxygen) to flourish in the large intestine. The differences in living conditions within the small and large intestines determine the number and the types of bacteria that reside in each portion of the gut. For example, the slightly acidic and oxygenated environment in the upper small intestine allows for bacteria that are tolerant to these conditions, such as the bacteria we often eat in our yogurt, known as *Lactobacilli*. Unlike the upper small intestine, the large intestine, also known as the colon, moves or churns its contents very slowly and produces a lot of mucus, allowing for many more bacteria to grow, especially those that use mucus for food.

Another characteristic of the human microbiota is its variability between individuals. Although about one-third of bacterial species are shared between all humans, the rest of them are more specific, making our microbiome unique like a fingerprint. Similarities in microbiota are highly dependent on diet and lifestyle, and to a lesser extent, on our genes. For example, identical twins (who share all of their genes) can have very different microbiotas if one is a vegetarian and the other eats meat. Family members, including husbands and wives who are not genetically related, tend to have similar microbiotas due to a shared living environment and diet. Humans also have striking similarities with the microbiotas of several species of apes, but only those that are omnivores like us. Mountain gorillas, for example, have a microbiota much more closely related to pandas, because they both spend their days leisurely eating bamboo.

Once established in our intestine, microbial communities are very stable. Only drastic changes, such as adopting a vegan lifestyle or moving to a completely different part of the world, will significantly alter your microbiota. Going on antibiotics for a week to treat an infection will also affect your microbiota, but only temporarily in most cases. It will generally bounce back to something resembling its pre-antibiotic state after you finish the treatment and go about your old way of eating. However—and this is a big however—the microbiota takes about 3–5 years from the time we're born to become a fully established community, and during this period it's very unstable, especially during the first few months of life. Any drastic changes to it have a very high chance of altering the microbiota permanently. In fact, it is the early colonizers of the intestinal microbiota that have a major influence on the type of microbiome we have later in life. Thus, a short-lived event like a C-section may have long-lasting consequences, since a baby born this way starts with a very different microbiota than a baby born vaginally. The potential health outcomes and impact of this type of event during early life has major implications for later health and disease, as discussed in later chapters.

Immune Cell School

Given the strong associations between early-life alterations to the microbiota and immune diseases later in life, we might ask: What exactly are microbes doing to us when we're babies that is so important? As mentioned in the previous chapter, microbes help us use food that we can't digest properly, and they also fight off bacteria capable of causing us harm. We've known about these roles for decades, but they are just the tip of the iceberg. As soon as we're born and begin getting colonized with bacteria, bacteria kick-start a series of fundamental biological processes in our body. One of them is the maturation of the immune system, the network of cells and organs that defend us from diseases.

Before scientists started unraveling the role of the microbiota in immunity, every doctor and scientist was taught that we're born with an immature immune system that gets trained in a small organ called the thymus. Here, immune cells known as T cells—the strategists of our immune system—are taught who is a friend and who is a foe. This training boot camp lasts for a few years only, until the thymus disappears, and all our immune cells have acquired this knowledge. Immunologists deciphered a complex series of mechanisms showing exactly how this occurs, but they couldn't explain one big question: How does the thymus teach immune cells which kinds of bacteria are beneficial and which ones aren't? After all, since we're covered head to toe (also inside and out) with microbes, mostly good ones, how do immune cells know the difference? The thymus does not interact with bacteria, so where could it get this information? It turns out this very important aspect of the training doesn't occur in the thymus—it happens in our gut.

Before we're born, the lining of our gut is full of immature immune cells, and as soon as we come into the world and bacteria start moving into their new home, these immune cells "wake up" almost magically. They start multiplying, they change the type of activities they do, and they even move to other parts of the body to train other cells with the information they just received. Experiments with germ-free mice, which are mice that are born into and kept in a completely microbe-free environment, show that without microbes the immune system remains immature, sloppy, and unable to fight off diseases properly.

Scientists haven't figured out exactly how microbes do this at the molecular level, but it is known that most bacteria will teach these immune cells to tolerate them, whereas some bacteria—the pathogens that cause disease—have the opposite effect. This makes sense; if our immune cells started fighting off all bacteria indiscriminately, there would be an out-of-proportion inflammatory battle between the small quantity of immune cells and the vast numbers of bacteria right after we're born. In reality it's quite the opposite; despite the enormous amount of bacteria living in the intestine, it's a relatively controlled and harmonious place. The way this is achieved is by the microbiota modulating the immune system, allowing most microbes to be tolerated.

Many inflammatory diseases, such as asthma, allergies, and IBD, are characterized by an overreactive immune response. Knowing what we do now about the importance of microbiota in immune system development, it's not surprising that these diseases are being diagnosed in more and more children. They are, to a great extent, a consequence of the modern lifestyle changes that are altering the types of microbes that affect the immune system. There's a reason immune cells wait for microbes to come and train them right after we're born: because this is the way it has happened for millions of years and is the way it will always be. We need to find ways to modify our modern behavior so that immune cell school can function properly.

Feeding Our Microbes So They Can Feed Us

Another fundamental function of microbes is to aid in the regulation of our metabolism. Humans, just like any other living animal, obtain energy from food that is digested and absorbed in the intestines. Besides helping us digest certain foods that the intestines can't handle on their own, bacteria produce energy for us, and the amount they produce is noteworthy. Germ-free mice weigh significantly less than conventionally raised mice, but once bacteria begin to colonize them they have a 60 percent weight gain, despite not eating more food than regular mice. One of the mechanisms by which they accomplish this is a process known as fermentation. Think of the intestine as a bioreactor where bacteria ferment fiber, carbohydrates, and proteins that were not digested and absorbed in the small intestine. The end-products of this process are called short-chain fatty acids (SCFA), and three of them are very important to different aspects of human energy metabolism: acetate, butyrate, and propionate. Intestinal cells rapidly absorb SCFA and use them as an energy source to stay fueled. SCFA are also transported very rapidly to the liver, where they are transformed into critical compounds involved in energy expenditure and energy storage. SCFA help determine how and when we use the energy obtained from food, and, importantly, when to store it as fat. Thus, it's not surprising that alterations in the production of SCFA have been associated with obesity, both in mice and in humans.

SCFA are not exclusively produced by the microbiota. These compounds are too critical for our metabolism to rely entirely on bacteria for their production. Still, studies performed on patients genetically unable to produce propionate have shown that approximately 25 percent of the propionate in our body is derived from bacterial activity in the gut. The implications of this are significant, considering that treatment with many types of antibiotics severely alters intestinal SCFA production. If antibiotics are given during early childhood, especially in the first few months of life, the risk of experiencing long-lasting metabolic and immune alterations due to abrupt changes to the microbiota increases dramatically.

Scientists haven't yet figured out all the functions that our metabolism delegates to the microbiota. Immune training and metabolizing energy are two essential things that our microbes do for us, but it's clear that there are more. Brand-new research shows that the microbiota plays an important role in neurological development (discussed in chapter 15), and even in the health of our blood vessels. These types of discoveries have led scientists to call our microbiome a "new organ," perhaps the last human organ to be discovered by modern medicine. Although most of this knowledge has just recently emerged and many pieces of the puzzle remain unsolved, it is evident that protecting the initial developmental stages of our microbiota has a significant impact in human health.

In the next four chapters we discuss the life stages that are most influential in the development of the human microbiome, all of which occur during infancy and early childhood. We will explore how some of the actions parents take during pregnancy and birth, as well as through diet, can have profound implications in the communities of microbes that are part of our children's bodies. With scientific information parents have learned to make better choices when raising their kids, such as limiting sugar intake and even the amount of time spent in front of the TV. With our newfound awareness of how important the microbiome is, let's explore what we might do as parents to improve our children's health by caring for their microbes.

THE CAVEMAN DIET

The newest diet fad suggests that eating the way our Paleolithic ancestors did will make us be healthier and live longer. However, evolutionary biologists don't agree with this because it's not based on current scientific knowledge. Some assumptions of the "paleo diet" include:

- *Our ancestors ate mostly meat, and no legumes or grains.* Actually, our ancestors ate incredibly different diets depending on where they lived. One could expect this statement to be close to the truth in Arctic environments, but in more temperate weather this was not the case. Biochemical analysis of dental fossil records from this period show that foragers did eat grains and legumes. Also, the meat we consume today—from domesticated livestock—is completely different than the wild game our ancestors ate.
- *Our ancestors did not eat dairy.* While this is generally correct, modern humans from many regions of the world where dairy is consumed have genetically modified their metabolism to digest and absorb dairy products. In other words, we have evolved, in a somewhat short period of time, to digest foods that our ancestors didn't eat. Our genes have changed since we roamed the savannahs.

It is impossible for modern humans to eat the way our ancestors did because our foods today are completely different than before. Carrots, broccoli, and cauliflower did not exist back then, and neither did the leaves used to make salads. All of these are products of agriculture. What certainly is true is that the typical modern human diet has extremely low diversity and is heavily processed, compared to food consumed a hundred years ago.

In addition, only very recently have people stopped eating just what is in season and whole foods. These are the dietary changes that really have an impact on our health, in great part because of the effects on our microbiota. Yes, eating fewer refined carbohydrates and more vegetables will help you lose weight and feel better, but this does not reflect our Paleolithic past in the way "paleo" enthusiasts believe it does.

PART TWO

Raising Babies and Their Microbes

3: Pregnancy: Eating for Two? Try Eating for Trillions

The Pregnant Microbiota: Another Reason to Eat Well

Seeing that positive result on a pregnancy test changes everything for most women. All of a sudden they're going to the bathroom more times than they can count, forgetting where their keys are while they're holding them in their hands, falling asleep at work (at 10 a.m.!), feeling full right after a meal, only to feel famished ten minutes later. From differences in her skin and hair to buying pants in three sizes within one year, pregnancy is a time of major changes in a woman's body. In nine short months, a woman undergoes a series of drastic physiological transformations that nurture a single fertilized cell into a crying, hungry baby. Many of our organs alter their functions to facilitate these new biological needs of both the mother and her developing baby. For example, the liver produces 25–35 percent more fats in order to promote baby growth. Fats, also known as lipids, are formed as a way to store energy. By naturally adjusting liver metabolism to make more lipids, a pregnant mother's body ensures that there will be enough energy for the baby to grow, and for the future production of milk following delivery.

Like the liver, a pregnant woman's microbiota also responds to this new state. In fact, experts believe this change is a normal physiological adaptation to support the growth of the fetus. A recent study showed that the microbiota of a pregnant woman in her third trimester strikingly resembles the microbiota of an obese person (just what every pregnant women wants to hear . . .). Moreover, when the microbiota of a female mouse in late pregnancy was transferred into a germ-free mouse, the latter mouse gained a lot of weight, despite not increasing food intake or being pregnant. This study was carried out in the laboratory of Dr. Ruth Ley at Cornell University in New York, a scientist at the forefront of the microbiota field. She believes that late pregnancy is an energy-thirsty period, during which the body takes advantage of the energy-producing machinery of the microbiome to promote weight gain for the benefit of the mother and her baby. The timing for this large shift in microbiota couldn't be better, occurring towards the end of the pregnancy when babies start packing on the pounds and when women need to start preparing for the energy demands of breastfeeding.

This same study, which sampled ninety-one pregnant women (the largest to date), also showed that some species of bacteria that were more predominant in the third trimester of pregnancy were also found in their babies at one month of age. This suggests that another consequence of the big change in microbiota during pregnancy is to pass many of these bacteria on to the newborn. It's fascinating to think that a woman's body and her microbiota work together during pregnancy, likely because both benefit from having a new baby. From a genetic perspective, having babies is the only way to propagate our genes; from a microbial perspective, a newborn is brand-new real estate where microbial genes can also multiply and propagate.

Another recent study showed that the shifts to microbiota during pregnancy reflect the amount of weight women gain. According to the American Institute of Medicine, a woman of normal weight should gain 25–35 pounds during pregnancy, underweight women should gain 28–40 pounds, and overweight women should gain only 15–25 pounds. Women who gain more weight than what is considered standard have distinct changes in their microbiota. Given that a baby inherits many of its mother's microbes, and that some of these microbes actually promote weight gain, should we worry about passing obesity-associated microbes to our babies? Unfortunately, yes. Women need to watch their weight during pregnancy, especially during the last trimester. Obesity is a complex condition arising from both genetic and environmental (including microbial) factors (discussed in chapter 10), but it appears that even in cases in which obesity is considered genetic, microbes have a role in its development. This makes sense, as microbes are directly involved in the way we break down food and store fats. If you think no one is watching when you give in to that midnight snack craving, that's sadly not the case—microbes are watching what we eat at all times, since it affects them directly!

The good news is that, just as we can foster weight-gain microbes through a poor diet, we can promote the growth of beneficial microbes through a healthy diet. Although scientists haven't identified specific microbes associated with leanness yet, it has been shown that a varied diet that includes fruits, vegetables, and fiber promotes a diverse microbiota, a characteristic of lean (and healthy) individuals. Thus, you, and your microbiota, are what you eat—and there is probably no better time to watch your diet than when you're pregnant. Bad dietary choices during this stage of life will not only make women gain more weight than what is considered healthy, they also have the potential to influence a child's future ability to control weight. So, next time you walk by a candy machine, don't listen to your sugar-loving microbes, and nourish the trillions of microbes that are begging you to grab a piece of fruit instead.

The Vaginal Microbiota

During pregnancy, microbiota adaptation also occurs in the vagina, an organ that hosts millions of microbes. The composition of this microbiota influences vaginal health tremendously. Many women develop yeast infections after being on antibiotics or oral contraceptives (birth control pills alter the pH of the vagina). Bacterial vaginal infections, also known as vaginoses, are very common. These infections occur when yeast (often *Candida*) or bacteria overrun a beneficial group of microbes known as *Lactobacilli*, a type of lactic acid bacteria that is very common in the vagina. Lactic acid bacteria are also used in the dairy industry for the production of yogurt, kefir, cheese, and buttermilk. Many of them have health benefits and are used as probiotics.

During pregnancy, the number of vaginal *Lactobacillus* increases dramatically, which is thought to occur for two important reasons. First, by keeping the vagina acidic, the presence of *Lactobacillus* helps discourage disease-causing microbes such as *E. coli*, which do not like to grow in acidic conditions. There's probably no better time to arm the bacterial vaginal defenses than during pregnancy, when a pathogen could track up from the vagina, through the cervix, and into the uterus, where the baby is growing. In fact, it is known that certain vaginal infections during pregnancy are associated with pre-term and low-weight births. Second, *Lactobacilli* are great at digest-ing milk, as their name suggests (*lacto* is Latin for "of milk," and *bacillus* is the name given to rod-shaped bacteria). By ramping up the levels of *Lactobacillus* in vaginal secretions, more of these bacteria will reach the baby's gut (when born vaginally), and facilitate the digestion of the only food the baby will eat for months: her mother's milk. In this sense, *Lactobacilli* are probably a baby's first and best microbial friend.

The vaginal microbiota plays a very important role during preg-nancy and birth, as it is one of the sources (along with the gut micro-biota) of the first microbes to set up camp in a newborn. As soon as a baby is born vaginally, she gets covered in vaginal secretions and, yes, with fecal matter, too. Consequently, the composition of vaginal secretions is of utmost importance during pregnancy, and vaginal health should be taken very seriously during this period of time. Just as women should take care of their diet to promote a healthy in-testinal microbiota, they should look after their vaginal health, too.

To promote vaginal health, gynecologists recommend that preg-nant women wear cotton underwear, avoid vaginal douching (never recommended), avoid vaginal cleaning products, and use gentle, unscented soaps to clean the outside of the vagina only. The vagina is an organ that cleans itself through the production of secretions and needs little extra hygiene. In fact, cleaning the interior of the vagina is strongly associated with infections, as it alters the balance of the resident microbiota. In addition, it has been shown that the consumption of probiotics containing *Lactobacillus acidophilus* de-creases vaginal infections. Several clinical studies suggest that eating yogurt may help, too, although not to the same extent as probiotics alone. You can even get probiotic preparations in the form of vaginal suppositories, which are used to treat such infections. Safe sex is the best way to avoid sexually transmitted infections (STIs); it is a practice that should always be followed, and especially during pregnancy. An STI contracted during pregnancy can be more dangerous to the mother than an STI contracted at another time, as immune systems are weaker during pregnancy—a physiological adaptation meant to prevent a woman's immune system from reacting to the fetus. Unfortunately, this makes a mother-to-be more vulnerable to infection.

Stress, Your Baby, and Your Microbes

Another important measure to maintain a balanced microbiota during pregnancy is to avoid stress, which is always easier said than done. We've all felt it—stress is a condition that affects most people at some point or another. It can be helpful sometimes, like when it compels you to finish an assignment for work that's due the next day. The problems arise when stress becomes an everyday companion; this is when it affects our health. Stress can make you lose sleep, have headaches or stomachaches, overeat, or lose your appetite. While pregnancy is typically a very joyful time, it can also be difficult. Dealing with the physical discomforts such as nausea, exhaustion, and backaches may quickly add up. On top of that, hormonal changes affect mood and the ability to handle stress.

A moderate level of stress is unlikely to cause a major impact on the health of a mother or her baby. However, certain situations may lead to severe stress, which can have detrimental effects on the pregnancy and the health of the baby. Abrupt negative life events, such as divorce, serious illness, financial problems, partner abuse, depression, and the conflicting feelings surrounding an unplanned pregnancy—to name a handful—are all causes of long-lasting or severe forms of stress. Some women suffer severe stress and anxiety when faced with the idea of labor or parenting. Severe stress is associated with preterm and low-weight births, and with certain illnesses in children, including skin conditions, allergies, asthma, anxiety, and even attention-deficit hyperactivity disorder (ADHD; see chapter 14).

A recent study from the Behavioural Science Institute of Radboud University, in the Netherlands, suggests that the microbiota plays a leading role in the link between stress during pregnancy and the aforementioned disorders. This study, which recruited fifty-six pregnant mothers, found that women who experienced high and prolonged levels of stress had alterations in the vaginal microbiota that could also be detected in their babies' gut microbiota. Infants born to highly stressed mothers showed lower levels of beneficial microbes, such as lactic acid bacteria. In the same study, these changes to the microbiota were associated with more gastrointestinal issues and allergic reactions in babies. They also found that the negative effects of severe maternal stress could not be corrected by breastfeeding, even though it has been repeatedly shown to promote a healthy microbiome in infants.

A similar study aimed at exploring the link between maternal stress and the microbiota was recently performed in mice. The study showed that a reduction in vaginal lactic acid bacteria, caused by stress, is accompanied by decreased immune functions in the offspring. Furthermore, the changes in the baby mice were not limited to the types of bacteria growing in their guts; there were also important metabolic differences detected in their blood and their developing brains. It may well be that the vaginal microbiota is at the center of this, responding to maternal stress and transferring its imbalanced state to the newborn, where it can lead to lasting health consequences. Although a casual relationship remains to be established, it appears that lactic acid bacteria from vaginal secretions are not only involved in facilitating milk digestion in newborns, but also carry out important metabolic functions in the developing newborn—yet another reason to reduce stress as much as possible and to take daily probiotics during pregnancy.

Infections and Antibiotics: Can We Avoid Them?

Controlling your diet and your stress levels during pregnancy is an enormously challenging goal for most women, but it can be done. However, the microbiota of pregnant women can suffer a big blow through a situation that's out of their control: taking antibiotics to treat an infection. As mentioned before, pregnant women are more vulnerable to infections, and if they occur, they are likely to be more severe due to their compromised immune systems. This is why it's recommended that pregnant women wash their hands often, avoid caring for people with infections (good luck with that when you have other kids!), avoid gardening without gloves, cook meats thoroughly, avoid changing the cat litter box, and avoid deli meats, sushi, and unpasteurized milk. Pregnancy is definitely not the time to get dirty and eat dirt, as we will later suggest our kids should do (although some pregnant women have an urge to do so—see Care for a Spoonful of Soil? on page 51).

Despite best efforts to avoid them, infections during pregnancy are quite common, with urinary tract infections (UTIs) and bacterial vaginoses both affecting about 1 in 6 pregnant women in the United

States and about 1 in 10 pregnant women in Canada. Other commonly diagnosed infections during pregnancy are respiratory tract and skin infections. Fortunately, several antibiotic medications are safe to use during pregnancy, but they're being prescribed to a lot of women—very likely more than necessary. The most recent National Birth Defects Prevention Study in the US, which has been collecting data since 1997, showed that almost 30 percent of women receive at least one course of antibiotics during pregnancy. A population-based study (a term given to studies involving a very large number of people) in the UK showed that the same is true for British women, while 42 percent of French and 27 percent of German women take antibiotics while pregnant. There's no debate about the immense change that an antibiotic brings to the microbiota. After a course of antibiotics, the overall diversity of the microbiota is substantially reduced. Its effect can be compared to what happens when a lush rain forest gets chopped down, and only a few dominant species make a comeback. Fortunately, the adult microbiota is fairly stable, and after finishing a course of antibiotics, in a nonpregnant woman this microbial forest usually returns to normal. The concern during pregnancy is that the microbiota fluctuates considerably, which is a characteristic of unstable ecosystems that are more susceptible to abrupt changes and permanent damage. When expectant women take antibiotics, especially in the last two trimesters, their microbiota takes a major hit, and according to new research, so does the microbiota of their babies. What becomes even more concerning is that antibiotic use during pregnancy is now being associated with certain diseases seen later in children.

A study of more than seven hundred pregnant women from New York showed that children born to those who received antibiotics in their second and third trimesters had an 85 percent higher risk of childhood obesity by age seven. These results are very significant because they were obtained after correcting for other confounding variables of obesity, such as the weight of the mother, the birth weight of the child, and whether or not the infant was breastfed. All of these factors were previously shown to be associated with the risk of obesity, so it's important (for this and any other similar study) to remove these variables from the analysis. These findings are quite new (published in 2014) and they still need to be replicated, but if more studies show a similar trend, it suggests that childhood obesity may have roots in the very early stages of human development, and that antibiotic use during pregnancy has significantly more risk than is currently assumed in medical practice.

Antibiotic use during pregnancy has also been associated with asthma, eczema, and hay fever in infants. Two large studies from Finland, a country that has experienced a twelve-fold increase in asthma rates since the 1960s, showed that using antibiotics during pregnancy is a significant risk factor for early asthma in babies. Other epidemiological studies have found similar associations between antibiotic use during pregnancy and inflammatory bowel disease (IBD) and/or diabetes, each of which is discussed in detail in forthcoming chapters. What's very peculiar is that these diseases share common risk factors. They are all immune disorders that have become increasingly common in the past few decades, and they usually occur in individuals with certain known genetic predispositions. Recent research on humans and animals show that the risk factors associated with these diseases also involve the early microbiota. How early? According to the studies, these changes begin before we're born, through mechanisms that are just beginning to be understood.

As frequently occurs in science, the insights on the mechanisms that explain a disease come from animal experiments. In this case, neonatology researchers from the Children's Hospital of Philadelphia showed that baby mice born to mothers that received antibiotics during pregnancy had a reduced immunological response. Similarly, a separate study showed that mice predisposed to diabetes and born to females that were given antibiotics had persistent alterations in their immune cells. These same mice developed diabetes a lot sooner than mice born to females that did not receive antibiotics. While a lot more research is still needed to fully understand all of this, it's becoming evident that complex interactions between microbes, the immune system, and other aspects of human metabolism, occurring as early as in utero (before birth), influence the risk of disease later in life.

Getting Smart About Antibiotics

In light of all these findings it is crucial to understand that using antibiotics should not be discouraged when they're really needed, but the overuse or abuse of antibiotics should be prevented. So, when are antibiotics necessary during pregnancy? The answer is simple: antibiotics should be taken for serious bacterial infections, and only bacterial infections. However, this can be hard to put into practice, especially during pregnancy, when doctors want to prevent any possible complications that may arise from an infection. Because of this, many health providers are too quick to prescribe antibiotics, as a safety precaution, to expectant mothers for ailments that don't require antibiotics, like the flu. The flu is a viral disease that causes symptoms that many people confuse for a bacterial respiratory infection. Its onset is very sudden and people feel awful for about a week, until they start getting better. It's not hard to imagine a pregnant woman showing up at a doctor's office almost begging to get a prescription that will make her feel a little bit better. However, antibiotics should not be used for the flu, regardless of how bad a patient feels.

There are exceptions to this, though; the flu can lead to secondary bacterial infections that do require antibiotic treatment. This usually manifests a little bit differently: you feel truly awful, and after a week or so, you start to get better, but then you start feeling worse, with coughing and chest congestion, which can lead to pneumonia. This is the classic example of a secondary bacterial infection following the flu, which should be treated with antibiotics.

However, the key concept here is to *prevent* infections from occurring in the first place if possible. As such, it is currently recommended that pregnant women get a flu shot. Fortunately we have an effective vaccine that is completely safe to use during pregnancy, which significantly decreases the chances of getting the flu and a secondary respiratory bacterial infection during flu season.

Despite the precautions you can take, infections do happen during pregnancy and antibiotics are prescribed. So what then? Based on the current research, it seems that the period at which antibiotics are taken is important, with microbial changes in the later stages of pregnancy being the most influential. If antibiotics must be used in the second and especially the third trimester, one should start or continue microbial supplementation with probiotics and a diet rich in fiber and vegetables. It's important to choose a probiotic that contains several species of *Lactobacillus* and *Bifidobacterium*, both known to be important early members of an infant's microbiota. As with any supplement or medication taken during pregnancy, we recommend discussing this with your health care provider.

Heading Off Group B Strep

During the births of her first two children, Neve had been given antibiotics, an increasingly common occurrence nowadays, with 1 in 3 women receiving antibiotics during labor. Neve knew how frequent antibiotic use is during delivery because she had tested positive for a type of bacteria known as Group B streptococcus, or GBS for her first two births. (Other very common circumstances that require antibiotics during labor are scheduled C-sections, which will be discussed extensively in chapter 4.) In many countries, all women between 35–37 weeks of gestation get tested for GBS. These bacteria commonly reside in 15–40 percent of all pregnant women, yet they rarely cause any symptoms. However, between 40–70 percent of GBS-positive women will pass it on to their babies during natural birth, and a small but very significant number of babies (1–2 percent) will develop a GBS infection (for further discussion of GBS infections, see chapter 4). Fortunately, if a pregnant woman who tests positive for GBS is treated with antibiotics during labor, the risk of her baby developing a GBS infection is reduced by 80 percent, making GBS prevention a pertinent use of antibiotics.

However, recent studies have shown that receiving antibiotics during labor alters the microbiota of the newborn, even if they are administered only an hour before birth. Reading about these studies made Neve, pregnant with her third child, feel uneasy. She knew that GBS could potentially be very serious and she understood the need for antibiotics during labor, but she wondered if anything could be done to *prevent* testing positive for GBS. Her second child has asthma and although it's impossible to know whether his exposure to antibiotics during birth is to blame, she's left wondering if it contributed. More importantly, Neve wanted to do whatever she could to decrease the risk of her new baby developing asthma, too. She hoped to help by testing negative for GBS, but how could she do something about that?

It turned out that she might actually have some say in the matter. GBS are bacteria that will expand in numbers only if they're given the chance. Normally other members of the microbiota keep them in check, usually our bacteria superstars, the *Lactobacilli* in the gut and the vagina. In fact, if you grow *Lactobacilli* and GBS together in the lab, the *Lactobacilli* make it very hard for GBS to multiply; they beat them easily. Furthermore, a small number of studies suggest that applying probiotics directly to the vagina increases *Lactobacilli* and decreases the number of GBS. This finding was shown in healthy nonpregnant women and remains to be supported in bigger studies, but given how safe it is to administer probiotics to pregnant women, Neve was open to trying this approach and her midwife supported this prophylactic treatment.

Neve ended up testing negative for GBS at her 36-week visit, and she is expecting to have an antibiotic-free birth very soon. However, it's important to mention that it remains to be proven in a randomized clinical trial that the prophylactic use of vaginal probiotics prevents or reduces the chance of a GBS-positive test during pregnancy. The use of vaginal probiotic suppositories, as with any treatments during pregnancy, should always be discussed with a health practitioner.

Can Bacteria Influence Us Before Birth?

So far we have discussed different ways to take care of the maternal microbiota during pregnancy in order to prepare the best kind of microbes that a mother can give to her baby at birth. This is when babies get soaked in microbes, during their trip down the vaginal canal. But very recent research shows that microbes may pay a visit to babies even before birth. For many years it has been widely accepted that humans are germ-free immediately before birth and that the presence of bacteria in utero is considered infectious and dangerous. Often this is true—bacteria growing in the placenta or the amniotic fluid can be a sign of infection and a cause of premature birth or even stillbirth. But what we're just now beginning to learn is that there may be very low numbers of bacteria that commonly reach the baby in the uterus without causing any harm. We still don't know how they get there and, more importantly, what they do, but in two separate studies bacteria were detected in the amniotic fluid and placentas of healthy babies. Although some scientists (including us) remain skeptical about these findings, the authors of these studies speculate that these bacteria are involved in immune stimulation of the fetus. Additional studies are needed before we can explain why this occurs, or if it even does.

Another more likely exposure to microbes before birth may occur in the form of bacterial metabolites, which are very small substances produced by the enormous amount of bacteria in our guts. Bacterial metabolites are known to travel in the bloodstream at all times, and are involved in biochemical reactions in just about every human organ, influencing many aspects of our metabolism. Thus, even if very few bacteria actually reach the fetus during pregnancy, the metabolites may reach the growing baby through the bloodstream and potentially affect fetal growth and development. Much-awaited studies are under way to explore the impact these microbes might have in human development before birth.

Dos and Don'ts

◆ **Do—** eat for your microbes, not just your cravings. Make vegetables, fruits, and fiber staples of your diet, along with the other food groups, and reduce sugary foods. A varied diet is a healthy diet for you, your baby, and your microbiota.

◆ **Do—** add daily probiotics, yogurt, or kefir (a fermented milk drink) to your diet. Increasing the growth of beneficial bacteria in your vagina will promote their passage to the newborn, where they carry out very important functions.

◆ **Do—** prevent infections if possible. Not only will you avoid feeling awful while pregnant, but it also reduces the chances of having to take antibiotics. Wash your hands often, avoid being in close contact with sick people, and follow the current recommendations of foods that pregnant women should avoid. If antibiotics are necessary, start or continue taking probiotics.

◆ **Don't—** sweat the small stuff, and do try to control stress as much as possible. Severe stress is associated with a number of disorders in children and also with alterations to the microbiome. If stress is becoming a big part of your life, reach out for help through your health practitioner. Even if your stress is moderate, incorporating exercise, yoga, or meditation into your routine can help keep the edge off.

◆ **Do—** consider vaginal probiotic suppositories in your third trimester in order to reduce the chances of testing positive for GBS. A negative GBS test will make an antibiotic-free birth more likely.

CARE FOR A SPOONFUL OF SOIL?

Perhaps the most bizarre of pregnancy cravings is the urge to eat dirt—a form of pica, a term used to describe an intense craving for nonfoods. Some suggest that dirt pica is the body's attempt to consume minerals and that it may be linked to iron deficiency, which occurs in many expectant women. Still, it is not known for certain what drives some mothers-to-be to eat dirt.

The rates of dirt pica vary depending on culture and socioeconomic status. In Kenya, it is so common that people see it as a sign of pregnancy, with 56 percent of pregnant women following this practice. Even in the US, 38 percent of low-income women from southern Mississippi claim to crave dirt or clay. Dirt pica is common enough that you can order dirt online to satisfy your craving! However, pregnant women are also more vulnerable to infectious diseases, and eating dirt may prove dangerous. Dirt is a known source of pathogens, toxins, and even lead, making it a bad option for those hard-to-curb cravings.

4: Birth: Welcome to the World of Microbes

The Best Laid Plans

At 3:50 a.m. a week before her due date, Elsa realized she was in labor. She was sleeping (sleeping should really have a different name in late pregnancy, as it is just not the same thing) when her water broke, alerting her and her startled husband that it was time. Soaking wet, they nervously laughed at the realization that they were going to meet their baby boy soon. They had a hospital delivery plan written down—labor in a bathtub, "laughing gas" for pain management, clear communication about interventions—and then, when the contractions became closer together, they would calmly put on comfortable clothes, gather their already-packed hospital bag (which included magazines, an iPad to serve as a music player and video camera, a massage device, and a heating pad), gather snacks and energy drinks, phone the grandparents, and drive to the hospital. The infant car seat had been installed in their car for about a month, and they had even practiced driving the route they were going to take. They already knew the best place to park in the hospital parking lot and the exact location of the maternity ward. Elsa and her husband had it all covered . . . or so they thought!

The first thing that kiboshed their perfect plan was having her water break before feeling contractions, also known as PROM (premature rupture of membranes). Elsa wanted to labor at home, but she knew that she had to go to the hospital right then. When the water breaks, the bag full of amniotic fluid, which keeps the baby protected, ruptures. It's not unusual for it to occur before labor, with 1 in 10 women experiencing that, but babies need to be monitored when this happens due to an increased risk in complications, such as an umbilical cord prolapse or an infection.

Within fifteen minutes they were out the door. They got dressed, grabbed the bag, forgot the snacks (oops), and decided to call their parents on the way to the hospital. It took Elsa another ten minutes to find a not-too-uncomfortable position to sit in the car, and just then, she started to feel her first real contraction. It was overwhelmingly strong. "If this is early labor," she thought, "I won't be able to deal with the pain." Elsa's husband, Paul, had previously volunteered to monitor her contractions. He had an app in his phone that would time contractions, and allow them to give each one an intensity score from one to five. As soon as Paul noticed the first contraction he reached for his phone and started to record its duration. Excited, he then asked Elsa: "How would you rate that contraction, babe?" With her gaze and voice lost, Elsa slowly opened her hand and showed him five fingers. "A five?" Paul said, "That can't be, we just got started!" And with the look that so many husbands have experienced during their wives' labor, Elsa just said, "Drive!"

By the time they reached the hospital, Elsa was already dilated five centimeters (halfway there) and in intense labor. "Forget the *&#^$ plan!!" she yelled. "I WANT AN EPIDURAL NOW!!" The nurse strapped a monitor to Elsa's belly to measure the baby's heart rate and Elsa's blood pressure. On the next contraction (they were coming three minutes apart now) the nurse noticed that the baby's heartbeat had dropped, not a lot, but enough to bring the obstetrician in to have a look. Then, just as the nurse was about to put an IV in Elsa's arm, the baby started squirming around, causing Elsa even more pain. Worse yet, the baby's heart rate dropped significantly. The obstetrician monitored the baby during the next sets of contractions and surmised that the baby must be pinching the umbilical cord. "We have to get him out *now*," the doctor said.

In what felt like hours but was only a few minutes, Elsa was rushed to the operating room and given spinal anesthesia for the C-section, after which they allowed Paul in the room. Elsa and Paul were both terrified.

However, very soon thereafter they heard the sweetest sound of their baby boy, Elijah, crying. A pediatrician and nurses quickly took Elijah to make sure he was all right (he was). After weighing and measuring him, they brought him to his parents, who were crying with relief, excitement, and love. "So much for the best laid plans," said Paul. Their cries turned into laughs as they realized that nothing had gone according to plan. It didn't matter . . . their baby was here and everyone was okay. Paul pulled out his phone, took the first picture of Elsa and Elijah, and sent it to the proud new grandparents, just over two hours after Elsa's water had broken, back in their bedroom.

Cesarean Epidemic

Although births come in different circumstances, durations, and outcomes, they have two things in common. First, just like with Elsa and Paul's experience, they seldom go as planned; births are unpredictable. Second, no one ever forgets when, how, and what it feels like to give birth. No other event in life compares in intensity and emotional impact. Biologically speaking, having a baby is the pinnacle of our existence, yet the human birth experience is very painful and often risky. In fact, compared to apes, human birth is longer and more perilous. Elsa's labor was unusually short at only two hours, but most first births average ten hours, and many are even longer. In addition, about 1 in 250 mothers carry a baby with a head too big to fit through the birth canal, requiring a cesarean section (C-section). One would think evolution would have favored easy deliveries, yet our bodies have not greatly improved on the process. Before the development of modern obstetrical medicine, there were about 70 deaths per 1,000 births. Those statistics have improved, but still, to this day, 500,000 women die annually worldwide from complications during childbirth. Why is human birth such hard and hazardous work?

Scientists believe that our births are more complicated because of the "human condition": we walk on two legs and have very big brains. Walking on two legs was truly advantageous to our human ancestors; they had their arms free to reach for fruit and other foods, they could carry items (babies included), they could hunt and craft tools, and they could look above the vegetation by standing upright. However, this advantage came with the anatomical price of narrower hips in order to achieve better balance and support the body's weight on two legs. Another aspect that makes humans unique is the large size of our brains. Thanks to our developed brains, humans can do math, build skyscrapers, and read books. Big brains (and, consequently, big heads) plus narrow hips? Any human can do this math: this causes the level five painful contractions Elsa was feeling and the medical need for C-sections.

C-sections are a medical miracle in terms of their ability to save the lives of so many mothers and babies. Try to imagine how much scarier Elsa's birth would have been had a C-section not been an option. Elijah's umbilical cord had twisted, preventing him from getting enough oxygen and blood flow. Elijah could have suffered a serious brain injury or even died from asphyxia if a trained doctor hadn't been able to pull him out surgically. A hundred years ago, dying during birth was a lot more common for both mothers and babies and modern C-sections played a pivotal role in changing this.

The history of when and where the first C-sections took place is a bit murky, but there are accounts of C-sections dating as far back as Ancient Greece. It is commonly believed that the name of this surgical procedure originates from the birth of the Roman emperor Julius Caesar. Regardless of whether this is true or not, Roman law decreed that all dying or dead birthing mothers had to be cut open in an attempt to save the child. Unfortunately, mothers rarely survived these early medical procedures and they were performed only as a last resort. Once anesthetic and antiseptic practices became the norm, C-sections became a much safer procedure and were used to save many lives. At the beginning of the twentieth century, for every 1,000 births, 9 women and 70 babies would die during childbirth, compared to 0.1 women and 7.2 babies today. That's more than a 90 percent reduction in mortality, a true triumph for modern medicine.

Still, for many decades C-sections were performed only when it was medically necessary: if the lives or health of the mother and/or the baby were at risk. However, towards the last quarter of the twentieth century, C-section rates skyrocketed. In 1970 the C-section rate was 5 percent in the US, rising to almost 25 percent by 1990 and to 33 percent in 2013. It has gone from a rate of 1 in 20 babies to 1 in 3 babies in the span of forty years. Canada's C-section rate is slightly lower at 27 percent, but it has still experienced a 45 percent increase since 1998.

Unlike the initial decrease in mother and infant mortality, the surge in C-section rates experienced in the past thirty-five years did not bring an improvement in mortality or morbidity (disease) rates. On the contrary, a C-section performed without a medical indication, also known as an elective C-section, is riskier than a vaginal birth. A C-section is a major surgical procedure that poses an increased risk of blood loss and infection for the mother. Also, any mother that has birthed via C-section can attest that healing takes much longer than a vaginal birth, not to mention the limited mobility of the new mother, who must let the incision to her abdomen heal; it's harder to hold the baby, to get up to change diapers (wait—maybe this is a plus), and sometimes even to breastfeed. Since 1985, the World Health Organization (WHO) has determined that the ideal rate for C-sections should be between 10–15 percent. Newer studies show that the number is likely closer to 10 percent. When C-sections rates approach 10 percent in a population, mortality surrounding birth decreases. But when the rates rise above 10 percent, mortality does not improve.

There are many explanations for this unnecessary but widespread increase in C-sections, and discussing them and their complexities are probably the subject for an entirely separate book. Suffice it to say, C-section rates are still increasing, and they are becoming epidemic and an emerging global health issue. Many experts disagree with this view and support the current rate of C-sections, because even if they are riskier than natural births, they are still very safe procedures. Modern obstetricians are extremely skilled in this surgery, and most complications that result from it, which are rare, can be treated with good outcomes in a hospital setting. There are maternal advantages associated with an elective C-section as well, including a reduction in urinary incontinence (loss of bladder control), avoidance of labor pain, reduction of fear and anxiety related to labor, and the overall convenience of planning the timing of birth. To some, the idea of a planned, painless birth is a dream come true.

On the baby front, C-section supporters claim that the health complications for babies born by elective C-section are rare and usually treatable. Babies born via C-section do look a bit different than babies born through the vaginal canal (their heads don't get squished), but after a few days they all look the same. However, while C-section advocates may be correct that severe birth complications, such as a stillbirth, are very rare in elective C-sections, we are now learning that there are significant health concerns associated with C-sections, including an increased risk of chronic disorders later in life, such as asthma, allergies, obesity, autism, IBD, and celiac disease. The elevated rates of these issues associated with C-sections hover around 20 percent for most of them. This is tremendously worrisome, considering that many countries have a C-section rate well above what the WHO recommends. Approximately 6.2 million unnecessary C-sections are performed around the world, with Brazil, China, the United States, Mexico, and Iran accounting for 75 percent of them. Brazil and China have an outright C-section epidemic; many hospitals in those countries deliver more than 85 percent of their babies surgically. The situation in Brazil has reached critical levels, as many women there have to give birth by C-section without the medical need for it, simply because of the shortage of hospital beds allotted for vaginal deliveries (see Brazilians Love C-sections, page 69).

The good news (kind of) is that it isn't the procedure itself that causes these disorders. Rather, it's something extremely important that *does not* occur during the few minutes it takes for a doctor to surgically remove a baby from the womb: the baby does not come in contact with his mother's microbe-rich vagina and feces.

A Dirty Birth Is a Good Birth

A baby's very first encounter with microbes most likely happens when his head comes out through his mother's vagina. As previously mentioned, the vagina contains an extremely high number of microbes, so the seconds (or minutes) it takes for a child to exit the birth canal are enough to impregnate a newborn's mouth, nose, eyes, and skin with many of them. It's also very common for women to defecate during birth, especially during the pushing stage. Babies usually exit the birth canal with their mouths facing their mom's anus, and it is now proposed that this position allows for additional exposure to maternal fecal microbes.

It makes total sense. The world is full of microbes, and all babies are going to get soaked with them immediately after birth, regardless of how they are born. Why not make sure that a baby gets coated in the microbes from which she will benefit most? Nature sees to it that the type of microbes first encountered by babies born vaginally are the ones that are going to aid in the digestion of milk, as well as contribute to the development of a baby's immature immune system, and even protect them against infections. Vaginal secretions are packed with *Lactobacillus*, whereas another milk-digesting bacteria known as *Bifidobacterium* come from feces. You've probably heard these two types of bacteria mentioned in yogurt advertisements. It's no coincidence that these bacteria are used in the dairy industry, as they're experts at digesting or fermenting milk and are also associated with health benefits. Unknowingly, every mother seeds her baby with a special custom package of microbes that will best suit her baby's needs. Babies instinctively seek their mother's breast shortly after birth, and breast milk is exactly what these microbes need to flourish in the baby's gut. This wonderful synchrony of biological events is a fine lesson in how nature works.

However, not every birth ensures the passage of beneficial microbes to newborns. As discussed in chapter 3, if the vaginal microbiota is unbalanced (low amounts of *Lactobacilli* in vaginal secretions), or if a woman has tested positive for Group B streptococcus (GBS), a baby will not get the same kind of microbial bath from her mom. Given how important it is to receive those beneficial microbes at birth, it's critical that women pay special attention to their vaginal microbiota in the weeks preceding birth. If there are any signs of a vaginal infection (itchiness, burning sensation during urination, or abnormal discharge), it's recommended that the mother consult a doctor and follow treatment with oral and vaginal probiotics as appropriate. In fact, given the proven safety of probiotics during pregnancy, all expectant mothers should consider including probiotics in their diet, especially in the weeks preceding birth (see additional recommendations in chapter 3).

If one could view birth through a microscope, a C-section is drastically different than a vaginal delivery: their microbiota is remarkably dissimilar. Studies comparing the gut microbiota of newborns in the days and weeks following birth consistently show that babies born by C-section have lower numbers of *Lactobacillus* and *Bifidobacterium*, as well as divergences in several other bacteria. These babies are colonized by microbes often found on skin, soil, and other external surfaces, instead of vaginal and fecal microbes. Even more worrisome, some of these differences persist and can still be detected when children are seven years old, according to a 2014 Dutch study.

To better understand how different a C-section is in the context of microbes, lets trace a baby's possible route of microbial exposure following a C-section. The brand-new bundle of joy goes from the doctor's sterile gloved hands to a table or a scale where he's touched with medical utensils and cloths. He may also brush someone's lab coat or hand in the process. If all is well, minutes later the baby is brought to his parents, and they can finally touch and kiss him, providing skin and mouth contact. Very often the baby is not allowed to breastfeed until his mother has started to recover from the anesthesia, which takes hours in most cases (although a few hospitals are now allowing this right after delivery). During this period, the baby will likely be wiped clean, warmly bundled in a clean hospital blanket, and placed in a cot, heated by a lamp, where he is offered warm (sterile) formula. During all this, the baby is exposed to the air, which has many microbes, but they are very different from mom's microbes, the ones humans are adapted to get exposed to at birth. It can take up to two hours before the baby is returned to his mother, when he can finally try breastfeeding for the first time.

Seeding Hope for the Future

Clearly, a baby born via C-section surely misses out on something crucial: that first splash of mom's microbes. But rather than judging mothers who have decided to give birth this way, whether by choice or due to medical necessity, we need to look at what can be done to make C-sections a more microbiota-friendly choice.

How can one restore a baby's microbiota following a C-section?

If you think about it, the way vaginally born babies are exposed to microbes is very simple: they come in contact with vaginal secretions. Why not inoculate a baby born by C-section with mom's vaginal secretions shortly after birth? Such procedures, called "seeding," are currently being used and tested in several hospitals around the world, and have been gaining an increasing amount of attention.

Veronica, a thirty-three-year-old mom from Edmonton, Canada, had to schedule a C-section some weeks prior to her due date because her baby was in breech position. However, she was aware of the importance of imparting her microbiota to her baby during vaginal birth and decided to talk to her midwife about this. Her midwife came up with a plan. She inserted a piece of sterile gauze into Veronica's vagina while she was waiting to be taken to the operating room. Minutes before her C-section, her midwife removed the gauze and placed it in a sterile glass container. Right after their baby girl was born, Veronica's husband took the gauze with gloved hands and swabbed it inside the baby's mouth and on her skin. Veronica also swabbed her own nipples, with the hope that the infant would take in even more vaginal microbes while breastfeeding.

As far-fetched as this method may sound, Veronica is part of a growing trend of moms and health practitioners who are trying it. Not only does it make scientific sense, but there's also scientific evidence backing up its effectiveness. Dr. Maria Dominguez-Bello, a scientist at NYU and a leading expert in the field of microbiota studies, has focused her attention on the development of early microbiota. She recently conducted a study involving eighteen births, in which babies born by C-section were "seeded" with mom's vaginal secretions and placed on mom's chest. Her team found that this process resulted in the microbiota of "seeded" babies becoming much more similar to that of a baby born vaginally. "While not equivalent to a baby born vaginally, there is some important restoration happening," she says. It's still unknown whether this simple procedure will reduce a baby's risk of suffering a chronic illness later in life. Her research group will follow up with these children in the years to come. Additionally, her group is working on conducting a much larger study that can provide sufficient evidence in terms of the safety of this practice. In the meantime, there's a compelling argument that women planning to have a C-section should discuss this option with their doctor or midwife.

Antibiotics During Birth

Antibiotics are routinely administered in conjunction with a C-section, given intravenously as a precaution against infection. As one can imagine, with the surge in C-sections, there has been a similar increase in the use of antibiotics during birth. In this instance, the antibiotics are truly necessary, as 10–15 percent of women that undergo C-sections will develop an infection. But it's up for debate whether the antibiotics have to be administered before surgery, or if it can wait until after the baby has been delivered. If given before the C-section, the baby will likely be exposed to the antibiotics, further compromising her microbiota at birth. If given after, the mother will still get the treatment she needs to prevent an infection and the baby will not be directly exposed to the antibiotic.

This was the case for Carley, now the mom of a healthy three-month-old daughter. During a doctor visit early in her third trimester, Carley learned she would have to deliver her baby via C-section (an umbilical cord abnormality made a vaginal birth too risky). As a naturopathic doctor herself, Carley had hoped for a vaginal birth, but she was aware of the need for a C-section for the safety of both her and her baby in this case. At the same time, Carley was aware that C-section babies have an increased risk of developing allergies, asthma, and obesity, with current research showing that a difference in microbial exposure influenced this risk. She had been taking daily probiotics throughout her pregnancy, but knowing that she would receive antibiotics before her birth, she was concerned that her baby would not received the optimal amount and type of microbes during birth. Carley explained her concerns to her obstetrician, who agreed to administer the antibiotics after her baby was born. They also agreed to "seed" her baby with her vaginal secretions after birth. Carley's C-section went smoothly and she recovered very well from it. She continued to take probiotics and to eat a healthy and varied diet to help restore her microbiota afterwards.

As in Carley's case, doctors are getting an increasing number of requests to administer antibiotics to the mother only after the baby is delivered, and even to forego antibiotic treatment altogether. While delaying the administration of antibiotics is a reasonable proposition, eliminating antibiotics during a major surgical procedure puts the mother at a very significant risk of infection. Like all medical decisions, the risks must not outweigh a patient's benefits. In this case, the desire to protect the mother's microbiota is outweighed by the increased risk of a severe infection acquired during surgery.

Another common use of antibiotics at birth is the application of antibiotic ointment (erythromycin) in the eyes of newborns. This is routine in the US and Canada, aimed at preventing the development of eye infections from the bacteria that cause gonorrhea and blindness caused by chlamydia. Because the possible outcome of these infections in a newborn is so severe, it is a medical indication in all births, although countries such as Australia, the UK, Norway, and Sweden forego the practice. In the US, thirty-two states are required by law to administer this treatment, regardless of whether the mother has chlamydia or gonorrhea, or whether the baby was born vaginally or via C-section (the infection can occur only during a vaginal birth). Recently, the Canadian Paediatric Society stopped recommending routine eye prophylaxis; however, this has not yet filtered down to common practice and many children still receive this treatment.

All pregnant women should be tested for sexually transmitted infections (STIs), including chlamydia and gonorrhea, and in fact most of them already are. But considering that the majority of women test negative for these diseases, and that many of them are part of a monogamous relationship, it seems reasonable to recommend eliminating the use of topical erythromycin after birth, at least in places where the law allows for a parent's right for an informed refusal of treatment. Although a small amount of antibiotic in the eyes will not have the same effect as an antibiotic administered intravenously, it could certainly affect the microbiota on the skin. Additionally, indiscriminate use of antibiotics aids in the development of antibiotic resistance, a larger public health issue.

Several other circumstances require the use of antibiotics during birth, including premature water breaking, labor lasting for more than twenty-four hours, signs of infection (e.g., fever) in the mother or the newborn, or if the mother has a known infection (such as a urinary tract infection). After birth, if the baby shows any symptoms that could indicate infection, the baby is tested. This usually takes 24–48 hours, during which it is assumed that an infection is taking place and the baby is administered antibiotics while awaiting laboratory results. In the vast majority of cases, these tests turn out to be negative, meaning many babies are given antibiotics unnecessarily. However, the consequences of undertreating an infant that is indeed suffering from a life-threatening infection can be disastrous. Clearly, there's a real need for better and faster methods to diagnose newborn infections, but until then antibiotic use in these circumstances is medically necessary. Newborns are especially susceptible to diseases, given how immature their immune systems are at birth, and the outcome of an infection can be very severe. Antibiotics certainly have a place during and after birth, and they have saved many lives, but considering how strongly they affect a baby's microbiota, their use should be limited to medical necessities.

Premature Babies

Some pregnancy complications can lead to the delivery of a baby well before she is ready to survive outside the womb. Medical treatments have advanced enormously in this field and premature babies can sometimes survive when born as early as twenty-three weeks (barely five months of pregnancy!). These babies often face major difficulties, like the inability to breathe or eat on their own, and much of their development has to occur in a hospital incubator, under the vigilant care of doctors and nurses in neonatal intensive care units (NICUs). Despite best efforts, this just isn't the same as a dark, warm, wet womb. Premature babies are born with immature intestines, making them vulnerable to the development of an extremely serious intestinal disease known as necrotizing enterocolitis, or NEC. Around 7 percent of low birth weight infants suffer from NEC in the US and Canada, and up to 30 percent of these infants die as a result. Naturally, there has been a big push to determine what causes NEC and how to prevent it.

NEC usually occurs a week or more after birth, and pathogenic bacteria (bacteria that cause disease) are often found in premature babies that suffer from NEC. This has led to the suspicion that NEC is caused by an infectious agent. Recent microbiota surveys from these babies have shown that NEC is very likely caused by an imbalance of the immature microbiota in their young bodies. Researchers found that in healthy premature babies that do not develop NEC, their microbiota becomes similar to a full-term baby at around six weeks of age. In contrast, premature babies that develop NEC have an abnormal growth of harmful bacteria days or even weeks before the symptoms of NEC appear.

The microbiota of premature babies have a lot going against them. Since these children are frequently delivered by C-section and given antibiotics after birth even when they're not showing any signs of infections, it's no surprise that a balanced bacterial intestinal community cannot develop properly in their bodies. However, recent studies have shown that the incidence of NEC can be significantly reduced if these babies start probiotic therapy shortly after birth. In a large study of three hundred premature babies in the NICU of a large hospital in Montreal, the administration of a mixture of *Bifidobacterium* and *Lactobacillus* led to a 50 percent reduction in NEC. That's a pretty astonishing figure! The authors of the study estimated that about twenty-five hundred cases of NEC in North America could be prevented each year by this very simple treatment.

Dos and Don'ts

◆ **Do—** your homework regarding your chosen mode of delivery. There are a number of studies that can inform a pregnant woman about the risks and benefits of vaginal birth and C-sections. Include what is known about the microbiota during birth as part of your decision. Also remember that births usually don't go as planned, so make yourself knowledgeable in case you need to make a quick but necessary change in plans.

◆ **Do—** look after your vaginal and gut microbiota during the weeks preceding birth (see chapter 3). A high number of *Lactobacillus* and *Bifidobacterium* will help establish these microbes in your baby, and help him adjust to life outside the womb. Take probiotics containing combinations of several of these microbes to boost their levels in your body.

◆ **Do—** discuss with your midwife or physician the procedure known as "seeding," to help restore your baby's microbiota if she is delivered by C-section.

◆ **Do—** consider sharing your concerns regarding the use of antibiotics during birth with your midwife or physician. If medically unnecessary, antibiotics should not be given as a routine measure to either the mother or the baby.

◆ **Don't—** question the need for antibiotics if medically necessary. If mother and/or baby are deemed at risk of infection during or after birth, antibiotics are the only effective treatment to prevent a severe outcome. Trust your health care practitioner. If a baby has been given antibiotics shortly after birth, ask your doctor if he can be given infant probiotics soon after the treatment stops.

◆ **Do—** talk to the pediatrician about the use of probiotic treatment, if you have a premature baby who has to spend time in a NICU. Probiotic use significantly reduces the risk of developing a very severe disease known as necrotizing enterocolitis (NEC).

BRAZILIANS LOVE C-SECTIONS

No other country in the world beats Brazil for number of C-sections. A large proportion of Brazil's growing middle class use private medical insurance, which means that certain private hospitals look after these patients exclusively. In many of these hospitals, obstetricians rarely attempt natural births, and the rates of C-sections reach 90 percent. Because of this, a C-section birth is now the norm for millions of middle class and affluent Brazilians, and has become a type of status symbol.

Even public hospitals have succumbed to this trend, with almost 50 percent of all births in Brazil delivered by C-section. With so many women booked for C-sections in advance, it has become difficult to find available hospital beds for vaginal births, which take much longer than scheduled C-sections. Some physicians are known to ask for extra payment if a patient wants to deliver naturally, as it involves many more hours of medical care (and often occurs in the middle of the night).

This issue has become so critical that the Ministry of Health recently approved new rules aimed at reducing Brazil's alarming C-section rate. Doctors are now mandated to inform women about the risks of C-sections and ask them to give their signed consent before the procedure. Additionally, doctors have to medically justify the need for a C-section based on a complete record of labor and birth. It is not yet known whether these rules will improve a Brazilian woman's chance to give birth vaginally if she wants to.

5: Breast Milk: Liquid Gold

Born Too Young

After minutes (or hours!) of intense pushing or skillful surgical maneuvers, a brand-new life meets the world. What a beautiful moment! Finally, after months of pregnancy and the intensity of the actual birth, all the hard work is over and parents can enjoy touching and smelling their new child. Ha! As if! For most parents, it becomes evident very quickly that the hard work is just beginning: sleepless nights coupled with the near-constant needs of a newborn.

Babies demand a lot of care during those initial few months. Compared to most other mammals, humans are born in a very immature and fragile state. While in utero, all babies grow to a point in their development in which they can withstand being outside of the womb. For example, a calf or a foal will shakily stand up and take its first steps mere hours after birth. In contrast, a human mother needs to deliver her baby before his head becomes too big to fit through the birth canal, which necessitates babies being born somewhat immaturely. For humans, it takes an additional 5–12 months for a baby to move on its own and eat anything other than milk.

During this somewhat fragile early stage of life, a baby's developing intestines and other organs demand constant nutrition and a lot of rest. This translates into feeding every 2–4 hours and sleeping around the clock, hence the unmistakable zombie-like look most parents have during the first months after birth.

A baby's intestine in particular goes through a lot after birth. Not only does the newborn gut have to mature and develop all the mechanisms necessary to digest and absorb food, it also has to withstand the onslaught of trillions of foreign microbial cells rapidly colonizing its entire surface. On one hand, a baby's gut has to allow the entry of the much-needed nutrients from milk, but it also has to prevent microbes from entering the rest of the body. This is an important concept to grasp: while trillions of microbes inhabit our intestines, they live in the inner intestinal space, known as the intestinal lumen. Think of the gut as a tube (which it is; see chapter 12), with the mouth and the anus at either end of that tube, and the rest of our body surrounding it. The intestinal lumen is the inner part of the tube, and although trillions of bugs live and perform all sorts of good tasks for us in the lumen, they aren't supposed to breach the borders of that tube. Our internal organs, such as the heart or the kidneys, do not benefit from having microbes floating around them. Quite the contrary—the entry of microbes into our body is a signal for our immune system to react, attack, and clear the intruders, and a strong immune response at this stage of life carries too much risk in such a wee little body. Thus, keeping microbes at bay within the intestinal lumen is a good strategy to prevent infection and a strong immune reaction. As mentioned in chapter 2, an overly active immune system can create havoc, so a calm and tolerant immune state is favored in the presence of so many microbes.

With all of this going on, the developing gut of a newborn is a somewhat chaotic place for a few months. It's no surprise that babies seem uncomfortable right after they eat, needing to be burped and sometimes shocking parents with the loud and messy sounds that come out of both ends. This is one of the main reasons why we can't go ahead and feed pasta with meatballs to a baby soon after he's born. It takes about 4–6 months for his intestines and immune system to be ready for all the nutritious foods you want to give him. Yet as usually happens in nature, millions of years of evolution have designed a perfect food for this challenging developmental process: human milk.

Feeding Trillions

Health organizations around the world agree that human breast milk is an amazing liquid and that it's the healthiest food for babies. Its benefits are most noticeable when babies drink it exclusively during the first 4–6 months and combined with solid food until age two. Scientists have studied breast milk for a long time, yet we're still discovering fascinating facts hidden in its biochemical composition, which is extremely complex and actually changes depending on many factors, such as how far along in the pregnancy birth occurs, the baby's age, and the woman's diet.

The nutritional components of breast milk include proteins (~10 percent), fat (~30 percent), and carbohydrates (~60 percent), plus many vitamins and other small molecules, encompassing all the necessary nutrients a baby needs in order to grow. As anyone who has raised a baby knows, they grow extremely quickly. It's not your imagination when you notice that your child grew overnight; babies grow every single day, hence the constant need for food and a mother's constant breast engorgement and accompanying exhaustion.

In addition to the nutritional components in breast milk, there is an ever-expanding list of ingredients found in it that are extremely beneficial for a baby's development. For example, maternal antibodies help fight potential pathogens (disease-causing microbes) that may enter a baby's immature gut; *lactoferrin*, a protein that binds iron, steals it from iron-loving bacteria (often pathogens) and prevents them from thriving; *lysozyme*, a very potent enzyme, aids in food digestion; and *growth factors*, which are potent immunity-enhancing substances, promote intestinal development and at the same time keep the immune system tolerant. All of these are present in breast milk (although not in formula), and they help the intestine mature while protecting it from the overwhelming load of microbes rapidly setting up house.

Interestingly, not all nutrients in breast milk can be digested by the newborn's intestinal system. A significant amount of the breast milk carbohydrates—known as *oligosaccharides*—go right through a baby's stomach and small intestine without being digested by a baby's digestive enzymes. Ready for the fun and fascinating part? It turns out that oligosaccharides, which comprise about 10 percent of human milk content, are *only* digested by bacteria present in the baby's large intestine. Hence, a nursing mom is not just feeding her baby's cells, she is devoting about 10 percent of her breast milk to feed the trillions of bacteria that have colonized her baby's gut.

It takes a lot of calories to make breast milk. Why would a mother's body invest so much time and energy to produce and pack into breast milk a certain nutrient exclusively to feed her baby's bacteria?

The only explanation is that the baby benefits immensely from hosting these bacteria and mom wants to make sure they're well watered and fed. These bacteria do a lot of things in those early weeks and months that affect the infant. It's been shown that the metabolic activity of a baby's microbiota (that is, breaking down stuff and producing energy) is even higher than that of the liver.

Recent research also shows that many of the benefits historically attributed to breast milk are actually mediated by the baby's microbiota. For example, the presence of a specific *Bifidobacteria* species enhances the effect of a key growth factor present in breast milk, known as the transforming growth factor, which keeps the immune system tolerant, thereby preventing an excessive inflammatory response in the gut. Administering *Bifidobacterium breve* to preterm infants made them more responsive to the effects of this growth factor, thus showing that the presence of this bacteria is an important component in early development.

Yet there's more: Feeding a baby's microbiota is not the only way that breast milk promotes the growth of beneficial microbes. Cutting-edge research has recently shown that breast milk itself comes with its own microbiota. These bacteria, as we've discovered, come not only from the mother's skin (when a child breastfeeds, a lot of skin bacteria make it into his tummy), but from the milk itself. At first these findings puzzled some scientists, who doubted the results and assumed that the breast milk samples were contaminated during the sampling process. It took many experiments to show that breast milk indeed has its own bacterial residents. Before these findings came to light, breast milk was thought to be a sterile fluid, and the presence of bacteria was a sign of a painful breast infection known as mastitis.

In one of the experiments, pregnant mice were orally given specific

Lactobacillus bacteria, which were labeled in a way that would distinguish them from other bacteria and that could also be detected later. Lo and behold, the scientists found that the exact bacteria they had fed to the pregnant mice turned up both in the mice's breast milk and in the baby mice's tummies. The research group, from Complutense University of Madrid, later went on to show that the same was true in humans. Somehow, a mother's body manages to add her own bacteria to her breast milk in order to further promote the growth of beneficial bacteria in her baby, just like packing a special treat in a lunch box.

How does this happen? Where are these bacteria coming from? The truth is, we're still not sure how this happens, but recent experiments have shown that these bacteria may come from the mother's gut. Specialized immune cells that live in the intestines protrude out of the intestinal wall, reaching into the intestinal lumen to "swallow" bacteria. Some scientists hypothesize that these cells then take a long ride from the intestines to the mammary glands, where they're passed to the nursing baby, along with their bacterial hostages. Others propose that the bacteria themselves sneak out of the intestinal lumen and make their own way to the breast, without the need of an immune cell as an escort. This process marks yet another way breast milk shapes the bacterial ecosystem of a baby's gut.

Regardless of how they get there, bacteria inhabit breast milk—and there are a lot of them! A detailed analysis of the type of bacteria that live in breast milk found that they're present in surprisingly large amounts. A single feeding can provide a baby with up to 100,000 bacteria. This analysis, performed in Dr. Mark McGuire's lab at the University of Idaho, also showed that the breast milk microbiota doesn't contain just a few species of bacteria, but is quite a diverse community. Even colostrum (the first watery fluid produced by the breast right after delivery) has hundreds of types of bacteria. What's even more interesting is that the type of bacteria changes depending on several factors, such as the age of the infant and even the mode of delivery. For example, the type of microbiota in colostrum is very different from the microbiota in breast milk at one or six months after birth. Also, the breast milk microbiota in mothers that gave birth vaginally is different than the breast milk microbiota in women following a C-section. It's not known why this is, but it's speculated that the physiological stress and hormonal changes that happen during vaginal birth influence the transmission of microbes into the breast. Clearly, breastfeeding is not the sterile practice many people might think it is. Every single time a baby reaches for a breast to feed, he's not only getting calories, he's also acquiring mouthfuls of beneficial bacteria and the right food to keep these bacteria fed.

Breastfeeding: Not as Easy as It Sounds

Pregnant women and new mothers around the world hear time and again that breastfeeding is the best food for the baby, that it promotes emotional attachment between mother and child, that it provides all the nutrition a baby needs, not to mention that it also helps burn all those extra pounds gained during pregnancy. It seems that every time a mother-to-be sits down in a doctor's office there's a brochure featuring a picture of a blissful-looking mother nurturing her baby from her breast. *Breast is best* is the word on the street, and it truly is, but it's rare to hear the truth about how hard and exhausting breastfeeding actually is, especially in the beginning.

Jacky, the mom of six-year-old Steph, vividly remembers her agonizing first few weeks of motherhood. It didn't matter how hard she tried, Steph wouldn't latch properly. Steph would also only remain calm or fall asleep when she was on her mom's breast, always waking up the second Jacky tried to pull her off. By the time Steph was a week old, Jacky couldn't handle it anymore. Every time Steph would latch, it was unbearably painful; the delicate skin covering her nipples was split open in many places, and she would bleed during every feeding, a truly torturous experience Jacky will never forget. The advice she was getting from her mother and aunt was not what she wanted to hear: "Give her formula!" they kept saying. "You were fed formula when you were a baby and you were a healthy kid." They were right, she was a superhealthy kid, but she really struggled with the idea of not providing breast milk to her baby given everything she knew about its benefits.

Jacky was very close to giving up on breastfeeding when even the weight of her clothes on her breasts was so painful that she would break into tears. Instead, she decided to listen to a friend, who recommended a lactation consultant, and the following day, Jacky, her husband, and Steph found themselves in the overcrowded waiting room of this particular professional. The room was filled with overwhelmed and exhausted moms, who, just like her, were desperate to figure out how to make breastfeeding work. Yet there they were again—the stupid brochures with a picture of a beautiful and surprisingly rested mom blissfully breastfeeding her child. Jacky wanted to burn every single one of them. Instead she waited patiently, hoping that help was on its way. It took quite a bit of work and several visits to the consultant, but eventually Steph learned to latch correctly and breastfeeding became a lot easier and painless.

Still, Jacky couldn't help but wonder why no one had mentioned how hard breastfeeding was going to be. Can the proven benefits of breastfeeding really outweigh the risks of parenting while emotionally spent? There's no doubt that breastfeeding is worth the effort a new mother puts into it, but the process is not always as intuitive or easy as it may look, and a very large number of women don't receive proper advice on how to do it, or are unable to breast-feed for a variety of reasons.

The statistics don't lie: Six out of ten moms experience difficulty breastfeeding during the first six months, and three out of ten suffer mastitis, a painful and debilitating infection caused by pathogenic bacteria that make their way into the breast. These bacteria infect one or more mammary glands, causing redness, swelling, fever, and pain—quite a bit of it. Women are instructed to continue breast-feeding or to express their milk to help wash out the bacteria, and they are often also prescribed antibiotics. Although a necessary mea-sure, breastfeeding while taking antibiotics to treat a breast infec-tion will undoubtedly result in the baby drinking antibiotic-laden milk—not at all the cocktail the baby's microbiota wants to drink. Antibiotics will also likely change the microbiota of the breast milk, further altering the type of bacteria that a mom is passing on to her baby. In this situation, it's very advisable to administer a high dose of probiotics to the mom, and also to give pediatric probiotics to the baby to replenish at least some of the key beneficial species of bacte-ria (these probiotics should contain a mixture of different *Lactobacilli* and *Bifidobacteria* species).

A serendipitous discovery made by a group of scientists research-ing the microbiota in breast milk suggests that probiotics can be used as a treatment for mastitis as well. They found that when they gave a mixture of *Lactobacilli* species to nursing moms, those who were suffering from mastitis experienced a noticeable improvement. In a different study, 352 moms with mastitis were treated orally either with antibiotics or with two types of *Lactobacilli* probiotics. Within three weeks of receiving treatment, the women given probiotics experienced much lower pain scores, and 88 percent of them fully recovered, compared to 29 percent of the women given antibiotics. Recurrence of mastitis was also much lower in the probiotic group, with only 10 percent of those women developing mastitis again, compared to 30 percent of the antibiotics group. This was a pretty surprising finding, until one realizes that certain *Lactobacilli* species are quite adept at fighting off disease-causing bacteria. Based on these studies, women experiencing symptoms of mastitis might consider using probiotics as a treatment. However, it's critically important to recognize that mastitis can be a very dangerous infection and that medical advice should be sought out and followed as well.

When Breast Milk Is Not an Option

Millions of babies around the world drink formula instead of breast milk for much of their infant and toddler lives. Baby formula is a societal need, because the realities surrounding the decision or the length of time a woman breastfeeds is contingent on many factors, including cultural and religious practices, as well as socioeconomic reasons.

Formula is a helpful and often necessary solution for so many women, and it does contain the nutrients that babies need to grow and thrive. A significant number of women in the world don't produce enough milk (if not expressed consistently, breast milk production can dwindle, even more so during the early postpartum period). Other women may pass on an infection (e.g., HIV) or a drug to their babies, making formula a safer option. Some women live and work in countries where there are limited or even nonexistent maternity work benefits and they simply cannot afford the time to breastfeed. Others opt for formula because of the demands of their careers. Regardless of the reason, the decision to breastfeed a child is entirely personal, and ideally, an informed one. Not breastfeeding a child does not equal inferior parenting; it's simply a different approach to nourishing a child.

Certainly, breastfeeding should be promoted and societal changes are needed to make breast milk available to more children, but until that happens, science should also look at improving baby formulas to mimic the incredibly complex composition of breast milk. Baby formulas have seen significant improvement over the years, but we still lack the scientific and technical knowledge to make formula exactly the same as breast milk—that will take much more time, if it ever happens. Until then, one way formula can come closer to breast milk is by adding microbes to it in the form of probiotics.

Only very recently has the scientific community realized the role of the microbiota within breast milk. One constant in studies on the topic is that babies fed with formula have a very different microbiota than babies who are breastfed. How different? It's been reported that breastfed infants tend to have a more uniform population in the gut. There are also differences in the type of *Bifidobacteria* that live in the gut, as well as in many of the other less abundant species. There are also distinctions in the type of bacterial metabolites present in breast milk and formula, suggesting that the various bacteria in breast milk do very different things. Also, formulas contain cow-derived oligosaccharides, not human oligosaccharides, further altering the type of food available for the microbiota of breastfed children.

This area of research is incredibly new, and very little is known about the changes that these differences create at the molecular level. However, the epidemiological data that link formula feeding with a variety of diseases is, in some cases, very strong. Formula feeding is associated with a higher risk of developing infectious diseases, obesity, and asthma, and the evident changes in the microbiota hint at the involvement of early-life microbes in these diseases.

What's more important is that a number of studies have shown that adding probiotics to formula makes a formula-fed infant's microbiota more similar to that of breastfed babies. Other ways that probiotic administration brings formula closer to breast milk is by improving stool consistency and frequency. Because most studies suggest beneficial clinical effects, and because we know that probiotics and prebiotics (foods that promote the growth of probiotic species; see chapter 16) are very safe, it makes sense that formulas should include them, and many formula brands already do. In addition to choosing a formula with probiotics, parents may further supplement with probiotic drops.

Breastfeeding After a C-section

Whether scheduled or not, a C-section makes breastfeeding an even bigger challenge. Melanie, mom to three babies born this way, remembers well how difficult it was for her to breastfeed after her first C-section, despite her best intentions to do so. She opted for a C-section upon advice from her doctor, since her baby remained in breech position towards the end of her pregnancy. Disappointed at first, Melanie became comfortable with the idea of a surgical birth and was even relieved knowing that she would avoid labor and pain (who wouldn't?).

A few hours after she was admitted for her C-section, Melanie began to realize that foregoing labor did not mean her birth was going to be easy. In fact, she was incredibly anxious during the procedure, as well as uneasy from feeling the forceful (though mostly painless) pushing and pulling movements inside her abdomen. Things definitely got trickier after birth. Recovering from the effects of the epidural anesthesia proved to be really hard for her. She was dizzy and nauseous and her legs and arms were extremely numb; even three hours after delivery she wasn't comfortable holding her baby out of fear that she would drop her. For days afterwards Melanie felt like she had been run over by a train, and she struggled to cope with the early days of parenting, and with breastfeeding. Melanie's abdomen remained sore for weeks, and she still couldn't move well or hold her baby comfortably the first few days after birth.

Breastfeeding was difficult from the get-go for Melanie. She had a brief skin-to-skin moment with her child while she was being sutured, but this lasted only a couple of minutes and it was really more of a face-to-face, as she was unable to actually hold her baby. For the next few hours, her infant remained in the nursery, where she was given warm formula to keep her nourished and comfortable. By the time Melanie was reunited with her little girl, she was peacefully sleeping in her cot and continued to sleep for a few more hours. The first time Melanie was able to offer her breast was almost six hours after birth, and it was with the help of her husband, as she didn't feel safe holding her baby yet. Throughout her stay at the hospital, Melanie continued to breastfeed through the nausea, dizziness, and pain. She needed assistance, as she couldn't stand up for the first twenty-four hours, so it was her husband or a nurse who brought her baby to her and held her while she nursed.

Once they got home it became even harder. Although Melanie was taking pain medication, her incision caused her a lot of discomfort, and getting up to pick up and feed her baby was excruciating.

This was taking a toll on her husband as well, who hadn't had much sleep for days and was very frustrated by not being able to ease his wife's pain or his baby's constant hunger. They gave their infant a few more formula feeds at night to allow Melanie to rest and recover. By the fourth day, Melanie's milk came and with it a new challenge: more pain and constant engorgement. Her little girl wanted to breastfed every two hours, as most newborns do, but it was incredibly difficult for Melanie to find a comfortable position; it just hurt too much. By the end of the first week, Melanie was completely exhausted, so they decided to switch to formula at nights to give Melanie much needed rest. It worked; her baby started sleeping for 3–4 hours at night and that felt like heaven for Melanie. However, her baby began rejecting her breast and preferring her bottle, a common occurrence known as "nipple confusion." It's easier for a baby to drink from a bottle than from a breast, and many of them choose the bottle when given the option. By the time Melanie's baby was a month old, she was feeding only once or twice from the breast and taking the rest of her nutrition from formula. At three months old, she was taking only formula; Melanie's milk had dried out and her baby rarely accepted it anyway.

Melanie's first baby had very few issues with formula. She grew beautifully, gaining weight as she was supposed to, and overall was a happy and healthy child. Still, when Melanie got pregnant again, she decided to give breastfeeding another chance based on all the things she read about breast milk being the best nourishment for babies. Aware that she would have another C-section (following a C-section with a vaginal birth is not impossible, but most health care practitioners advise against it due to an increased chance of uterine rupture during labor), she knew that breastfeeding might be challenging again. She talked to her doctor and explained that she wanted to breastfeed her baby right after birth. So instead of offering formula, she enrolled the help of a nurse, who brought her baby to her to breastfeed while she was recovering from the anesthesia. Then her mother-in-law stayed with her during the night to help handle the baby during feedings. Once they got home, friends provided support—looking after their older child (now two years old) and doing some household chores. Melanie knew that if she wanted to succeed, she needed the time to spend with her baby, as well as the time to recover from a major surgery. It wasn't easy, but Melanie managed to exclusively breastfeed her two younger children. Her advice for mothers who want to breastfeed after a C-section? *"Be determined and get help!"* (See further recommendations in Best Bets for Breastfeeding, page 86).

Dos and Don'ts

◆ **Do—** get informed about the benefits of breastfeeding. Breast milk not only provides the best type of nutrition to babies, it also contains the right type of nutrients for a baby's microbiota. Breastfeeding has been repeatedly associated with better health outcomes, including protection against infections, asthma, and obesity.

◆ **Don't—** expect breastfeeding to be easy, because often it is not, at least during the first few weeks. Prepare for it as you prepare for birth by attending a prenatal clinic or getting advice from a nurse, doula, or lactation consultant. Be patient, as breastfeeding correctly is not always as intuitive as it looks and it takes time to do it correctly. If you experience difficulties breastfeeding, look for help as soon as possible.

The first few weeks after birth are both overwhelming and very important to establish successful breastfeeding, so this is the right time to ask for professional help.

◆ **Do—** consider taking probiotics if mastitis or other infections develop, especially if they require antibiotics. Probiotics will help replenish the microbes that inhabit breast milk, which may have been lost due to antibiotic treatment. Probiotics have also been shown to be effective in treating mastitis, so consider using them as an alternative form of treatment for this condition—however, always consult with your health care practitioner first.

◆ **Don't—** think that providing formula to your baby is an inferior form of parenting. Formula is a societal need and the decision to breastfeed or not ultimately depends on the mother. If you decide to use formula, complement it with pediatric probiotics. One of the biggest differences between breast milk and formula is the type of microbiota that result from them, so probiotics will help make formula more similar to breast milk. Consult your doctor or nutritionist for advice on the best options in the market.

◆ **Do—** look for extra help after a C-section. Breastfeeding is hard as it is, without the added physical and mental recovery from major abdominal surgery. Establishing successful breastfeeding will depend on preparing for it through a hands-on consultation with a nurse or lactation consultant, and getting support at home. The efforts to breastfeed after a C-section are worth it, but it will require extra work.

BEST BETS FOR BREASTFEEDING

Below are a series of recommendations that will increase the chances of success in breastfeeding after a C-section.

1. Choose a hospital that is supportive of breastfeeding right after a C-section. In advance of your procedure, communicate with the nurses and doctors about your decision to breastfeed during recovery. This will require assistance, as you may not be able to handle the baby safely.

2. If the mother and child must be separated during the first few hours after birth, request a hospital breast pump to stimulate milk supply. The collected colostrum can be given to the baby.

3. Recruit help with home chores after the surgery, whether it's friends, family, paid help, or a postpartum doula.

4. Breastfeed often, as frequently as every two hours, to increase milk supply and prevent breast engorgement, which can lead to mastitis.

5. Breastfeed lying down, holding the baby on your side, to avoid added pressure and discomfort on the recovering abdominal area.

6. Try to avoid giving bottles for the first few weeks to prevent nipple confusion.

7. Be determined to stick to breastfeeding and don't hesitate to seek help from a lactation professional if it's not working.

6: Solid Foods:
A Growing Diet for Microbes

New Food Means New Microbes to Eat It

The early days and weeks of babyhood seem to go by so slowly, yet they also go by so fast. In the blink of an eye, the child who ate and slept all day begins to stay awake for longer periods of times, enjoying his surroundings and delighting mommy and daddy with cooing sounds and gummy smiles. Soon after, that same smile becomes very drooly, leaving a trail of slobber on every clothing item, bag, shoe, and piece of furniture you own. It's around this time that he starts staring at you while you're eating. After months of happily drinking only milk, he starts scrutinizing that weird, solid stuff that his parents put in their mouths, until one day he reaches for it, opening his mouth to get a bite of whatever it is that suddenly smells so good. This usually happens somewhere between four and six months of age, the time when solids foods should be introduced. Before this, babies obtain all the calories and nourishment they need from milk (including formula). Plus, as we discussed in the previous chapter, the intestine needs to mature for quite some time before it's ready for solid food.

The first time a baby eats solid foods marks an event that is usually photographed, recorded in baby diaries, and, nowadays, also shared on social media for the entire world to see. Who doesn't want a picture of their baby with goop all over his face, chair, hands, hair, and everywhere but his mouth? It's one of those wonderful "firsts" that no one wants to miss. Another event that fortunately *does not* get immortalized through pictures or diary entries is what happens in the hours or days after introducing babies to solid foods—those new changes in diaper content. It takes only one or two diaper changes to realize that things in your baby's tummy are changing and that diaper duty just got a lot more unpleasant.

Nothing influences the type of microbiota we harbor more than diet, so the introduction of solid foods marks a big change in the microbial community that calls a baby's gut home. With those first few spoonfuls of solid foods, the bacteria that specialize in digesting milk start being replaced by other species, and within a few months the infant microbiota begins resembling the microbial communities found in adults (and smelling like it, too!). The ecological changes in the gut microbiota after the introduction of solid foods vary greatly between individuals, but one common feature is an increase in microbial diversity.

The Boon of Diversity

Ecological diversity, also referred to as biodiversity, is determined by the number of different species in a particular habitat or place. In this case, introducing solid foods kick-starts a process that allows several new microbial species to feast on the new foods. Research from our lab and others has shown that by the time a child is one year of age, she will have approximately 60 percent more bacterial species in her gut than she did just seven months earlier—a huge jump in biodiversity. These new microbes have the ability to metabolize more complex sources of nutrients, as shown in a recent study led by Dr. Fredrik Bäckhed, a scientist at the University of Gothenburg in Sweden.

Just like many other ecosystems, microbial diversity in the intestine has been repeatedly proven to be a marker of good health. The same is true for a lake, a forest, an ocean, or other natural habitat—low biodiversity can be sign of an unhealthy and unstable ecosystem. Our intestine is no different; diseases like obesity, type 2 diabetes, and gastrointestinal disorders all share low microbial diversity as a common feature. Although more research is needed to clarify the extent to which microbial diversity promotes health, the message is clear: the typical Western lifestyle does not promote diversity.

The so-called Western diet—high in fats, sugars, and highly refined grains—is very strongly associated with a number of human diseases, and also with a less diverse microbiota. The reason for this low diversity is likely because most of the food eaten in these societies (and increasingly in developing countries) comes from very few species of plants and animals. Seventy-five percent of the world's food comes from only twelve plant species and five animal species. Amazingly, just three species—rice, corn, and wheat—account for 60 percent of the calories that humans obtain from plants, a shocking and sudden change in human practices when one considers that in the early 1900s there was 75 percent more genetic diversity in plants used for crops.

These days, everyone is eating the same stuff, and a lot of it, except in regions where a lack of economic development has kept people's farming and dietary practices more similar to those of a century ago. The effect this has on the gut was shown in a study comparing the microbiota of children living a rural lifestyle in Burkina Faso in West Africa to the microbiota of urban, city-dwelling children in Italy. The African children ate a high-fiber diet of vegetables, grains, and legumes, with an absence of processed foods, whereas the diet of the European children was saturated in sugars, animal fats, and refined grains, which also have more calories. The microbiota composition of the children from Burkina Faso was very different from—and much more diverse than—that of the Italian kids. Now, it's hard to argue that children from Burkina Faso have a healthier lifestyle than Italian children—they're more likely to suffer severe infections and malnutrition, and unfortunately have a lower life expectancy than a child born in Western Europe. However, they also have a decreased risk of suffering from the immune diseases that are becoming almost epidemic in the Western world. Coincidence? Not according to a mounting body of evidence suggesting that the early microbiota plays a very important role in the development of these disorders.

In an ideal world, children would harbor a rich and diverse community of microbes without the threat of severe infectious diseases, yet our current societal practices only address half of this equation (decreased infections). Given how well bacteria respond to diet, eating a variety of foods is most likely the best way to increase microbial diversity. Furthermore, there's no better time to establish a diverse microbiota than during the first 2–3 years of life. Remember, our microbial communities remain almost unchanged after early childhood, making it the optimal time to promote a diverse microbiota through diet. For example, don't feed a baby only rice cereal for weeks until the package is finished—offer them a variety of grains, including oats, rice, barley, quinoa, etc. It's also important to offer whole grains instead of refined ones. The Western diet is extremely low in fiber, and refined grains contain very little of it. Protein-rich legumes, such as lentils, beans, and peas, have an abundance of fiber and can be easily mashed for babies. Add more fiber by including vegetables in all or most meals and by offering nontraditional starchy vegetables such as sweet potatoes, parsnips, or cassava (tapioca), rather than just sticking to low-fiber veggies such as potatoes.

Understandably, the majority of people in developed societies won't crave these foods the same way they crave the texture of macaroni and cheese or the like, but the infant stage is the best time to introduce good dietary practices. For kids, eating healthy foods becomes a habit the same way cleaning their room does: by doing it frequently. When considering how to feed your growing baby and toddler, think not only about her, but also about her microbiota. Feed *them* a wholesome, varied diet rich in fiber and low in fats and sugar; both your child and her microbiota will thank you for it.

When, What, and How Much?

Knowing when to introduce solid foods and what foods to first offer your child can be confusing because the rules keep changing. For example, Claire experienced a change in guidelines between her two children, who are not even two years apart. With her daughter, Marisol, she was instructed to introduce solid foods at six months of age, starting with cereals and slowly moving into vegetables, fruits, and meats. Just over a year and a half later, with her second child, their family doctor recommended that she give him solids between 4–6 months, upon him showing signs of readiness. Then last year, a friend of Claire's had a baby and the guidelines had changed again, with the advice to start at 4–6 months and give meats first. Surely, certain things still need to be figured out when it comes to the timing of solid food introduction, although one thing is clear: there's wiggle room for starting at the four-, five-, or six-month mark, depending on when the baby shows interest and readiness.

Before 4–6 months of age, babies don't need any other nourishment than milk, but approaching six months of age their levels of iron start to decrease. Iron is an incredibly important mineral—it carries oxygen from the lungs into the rest of the body, among several other important functions. Having low levels of iron in the blood is known as iron-deficiency anemia, a condition that can be prevented by offering iron-rich foods to babies at around six months. Iron is also important for brain development and an iron deficiency is associated with a lower IQ. Current recommendations are to offer foods such as meat, meat alternatives (eggs, tofu, or legumes), or iron-fortified cereal. However, getting the full daily requirement of iron would take half a cup of iron-enriched cereal, which is a huge amount for a baby just starting solids, and it's not as well absorbed as natural sources like meat, anyway (which is unfortunate news for vegetarians, since iron from nonanimal food sources is also not absorbed as easily as iron from meat). Note that babies given formula should receive 2.5 ounces of iron-fortified formula per pound of body weight each day, gradually decreasing this amount as the baby gets older and eats more solid food.

Although the eventual goal, as mentioned previously, is a diverse diet, there are compelling reasons to nurture a baby's palate slowly. Foods should be introduced in small quantities and, ideally, one at a time. Initially, the baby will have just a taste, then move up to a teaspoon of food, then a whole tablespoon, and so on. This gradual increase should follow the baby's cues: if he wants more, offer a bit more; if he turns away, hold off for now. Offering foods one a time allows the baby's digestive system to get used to one ingredient for a couple of days before it tries a different one. It also allows a parent to detect if one particular food does not sit well with the baby or if it causes an allergic reaction (more on food allergies in the next section).

Within a few weeks, a baby will have tasted lots of foods that can start to be mixed and matched. This is when parents should think about variety as the central theme of their child's diet, offering options for every food group (meats, grains, fruits, and vegetables) and choosing foods rich in fiber as much as possible. If a baby doesn't like a food, don't force it, but definitely try again later. Some babies may need to be given a new food as many as 10–15 times before they'll actually eat it. A quick look at what babies around the world eat is proof that babies do eat whatever they're offered, even if it needs to be offered many times (see First Foods Around the World, page 100).

Another important thing to keep in mind is that the introduction of solid foods doesn't mean that babies should be weaned from milk. On the contrary, it has been proven beneficial to continue to breastfeed on demand until the baby is two years old, or even longer if the mother and child still want to. A recent study from the laboratory of Dr. Fredrik Bäckhed showed that the microbiota of babies who are weaned early shifted more quickly to an adultlike microbiota than babies that continued to breastfeed. Given that the presence of milk-loving bacteria has such strong and beneficial immune effects, it's thought that delaying the maturation of the microbiota is beneficial to young children. As such, it's recommended to continue to breastfeed or to provide a formula with probiotics to babies eating solid food.

Another way to boost your baby's levels of probiotic species is to introduce them to fermented foods such as yogurt or, even better, kefir. Kefir is a drink very similar to yogurt, except it has a more watery texture and is slightly more sour (and has a ton of probiotics in it; its advantage over yogurt is that it has a lot more different species). For either yogurt or kefir, it's optimal to choose a brand low in sugar and without artificial sweeteners. The fewer refined sugars your baby drinks the better, both for the baby and for the trillions of microbes feasting on everything that ends up in that tummy.

Dangerous Eats

Except for poisonous foods and choking hazards, food is safe, right? Ideally, yes . . . but it turns out that certain foods agitate the immune system and elicit an allergic reaction in some people. Many foods are known allergens—the usual suspects are wheat, eggs, milk, peanuts, tree-nuts, fish, shellfish, strawberries, sesame, and soy. These allergic reactions tend to run in families, so if a parent or a sibling has a specific food allergy, there's an increased risk that the new baby will develop one, too. However, the genetic basis for food allergies does not explain the huge increase in cases in the past generation. According to a study by the CDC, food allergies in children increased 50 percent between 1997 and 2011, an enormous jump in incidence that has one in every twenty kids suffering from a food allergy. Twenty years ago it would have made no sense to see a note on a package of gummy bears warning that it may contain traces of peanuts, whereas nowadays millions of parents have to scrutinize safety signs on packaged food, because without one, it can't go in the lunch box.

Food allergies have a large spectrum of symptoms, from mild ones such as an itchy mouth, to severe ones like anaphylaxis—a potentially deadly reaction. Few things are scarier than imagining your child suffering a serious allergic reaction, yet a child visits an emergency room for this reason every three minutes in the US. Malcolm and Jeannie are the proud parents of three teenage boys, and one of them, fourteen-year-old Callum, has a severe allergy to certain nuts. Jeannie found this out when Callum was about fifteen months old, after she offered him a cashew. Callum quickly became fussy and would not stop crying until he fell asleep. A few minutes later he woke up irritated and they noticed he was covered in hives. They phoned his pediatrician right away, who urged them to call 911 and wait for an ambulance. Once the first responders arrived, they assessed Callum and took him to the ER, where they treated him and gave him a referral to an allergist. A few weeks later, through a series of tests, they found out that Callum was not only allergic to cashews, but also to pecans, almonds, and peanuts. Malcolm and Jeannie were given the same (and only) advice that thousands of parents receive after their child is diagnosed with a food allergy: avoid the problem food completely and carry an EpiPen in case accidental exposure occurs.

It's not clear why there's been such a jump in the number of food allergy cases, but it's likely that exposure (or lack thereof) to early-life microbes plays a role. For example, babies born and raised on farms have a lower chance of developing food allergies, and breastfeeding for the first six months reduces the risk and severity of certain food allergies. In addition, epidemiological studies are showing that the time at which these foods are introduced is very important. In Israel, where peanut allergies are ten times less prevalent than they are in the UK Jewish population, babies are given a popular healthy snack called Bamba as soon as they can safely eat it. Nothing seems out of the ordinary until one realizes that 50 percent of the ingredients in this snack are peanuts. As any parent that has raised a child in the past twenty years can confirm, giving peanuts to a baby under one year of age is a huge no-no. Or is it? According to the latest research, in trying to protect our children from food allergies, it appears that we may have been doing the opposite.

If navigating the recommendations about when to start solid foods is confusing, deciding on when to introduce foods that are known to induce allergies is even harder. It's also hard for pediatricians and other health practitioners to keep up with the current lines of thought and offer the right advice (sometimes they have to offer different advice to the same parent for each individual child!).

When food allergies started becoming more common around the 1990s, experts agreed that delaying the introduction of these foods would reduce the likelihood of developing an allergy. In the year 2000, the American Academy of Pediatrics issued guidelines that infants should wait until age one to have cow's milk, until age two to have eggs, and until age three to have shellfish, fish, peanuts, and tree nuts. Eight years later, the guidelines were revised and the accompanying statement cited little evidence that delaying the introduction of these foods was beneficial in preventing food allergies, yet it didn't include new recommendations, leaving parents and doctors in limbo. "As these guidelines were implemented we've seen a paradoxical increase in foods allergies in young children, especially with peanut allergies," said Dr. Anna Nowak-Węgrzyn, a professor of pediatrics at the Icahn School of Medicine at Mount Sinai in New

York. It's taken a few years to gather enough evidence to show that delaying the introduction of these foods is not only ineffective, but it may be making things worse.

A recent study published in the *New England Journal of Medicine*, one of the most prestigious medical science journals, revealed that children who received a delayed introduction to peanuts had an increased risk of developing peanut allergy, compared to children that encounter them early. Just how early? Very—between four and seven months of age. Because of this landmark study, as well as others, the new recommendations issued by the American Academy of Asthma, Allergy and Immunology (AAAAI), Canadian Paediatric Society (CPS), and Canadian Society of Allergy and Clinical Immunology (CSACI) state that allergenic foods should be introduced in the same way as other foods: slowly and gradually, starting at 4–6 months. The AAAAI recommends that once an infant has been given a few nonallergenic foods (meat, vegetables, etc.) the allergenic ones can be offered without delay, ideally before seven months of age.

These new guidelines not only follow what the evidence says, but they also make sense based on what's currently known regarding the immune system of young infants. During the first months of life, exposure of foods to the gastrointestinal immune system encourages immune tolerance. In addition, once an allergenic food is introduced, it appears to be equally important to maintain frequent, regular ingestion early on in order to maintain tolerance and truly prevent a food allergy.

It's important to take into account that one of the factors keeping the infant's immune system in this tolerant state is the presence of a bunch of microbes that specifically work to promote immune tolerance. Thus, breastfeeding throughout this stage or supplementing formula with infant probiotics will help maintain higher levels of these hardworking microbes. The same guidelines issued by the AAAAI recommend breastfeeding at least until four months of age, although it could be argued that, based on the current understanding of how allergies develop, extending breastfeeding beyond four months is probably beneficial, too. Another concept that has changed with these guidelines is the restriction on pregnant mothers eating allergenic foods. New evidence shows that such a restriction does not prevent the development of allergies in babies.

Dos and Don'ts

♦ **Do—** look for signs of readiness for your baby to start eating solids: he can sit up without support; he doesn't automatically push solids out of his mouth with his tongue; and he seems interested in food and watching *you* eat. Try offering solid foods between 4–6 months of age once you notice your baby is ready.

♦ **Do—** start slowly, always following your baby's cues, and increasing amounts gradually. Begin with one food at a time, trying the same food for 2–3 days. This will both help your baby's digestive system get used to a particular food and help detect a possible food allergy.

♦ **Do—** diversify your baby's microbiota. After single-course solids have been eaten for a few weeks, try varied choices from all food groups. For meats, offer different meat sources as well as meat alternatives such as eggs, tofu, or legumes (beans, peas, and lentils). For grains, choose wheat, rice, oats, corn, barley, rye, quinoa, etc. Pick whole grains and whole grain flours, as they add substantial amounts of fiber. Provide various types of vegetables and fruits in every meal and consider serving nontraditional starchy vegetables, such as sweet potatoes, parsnips, or cassava.

◆ **Do—** keep sugary foods to a minimum, especially juice. A baby with a sweet tooth will likely become a toddler with a sweet tooth.

◆ **Don't—** stop breastfeeding once your baby starts eating solids (if possible). A baby continues to benefit from breast milk until she's two, so the longer you're able to breastfeed, the better. If formula feeding, look for a brand that contains probiotics, or administer them separately in the form of pediatric probiotic drops. This will help delay the process of your baby's microbiome becoming adultlike too early.

◆ **Don't—** delay the introduction of allergenic foods. Offer peanuts, soy, shellfish, etc., after less allergenic foods have been tolerated, between 4–7 months of age. Do this slowly and using the "one at a time" rule, just like with any other food.

◆ **Do—** keep up with the most current medical information regarding solid food introduction and ways to prevent food allergies. The guidelines will likely continue to change as more studies get published and revised by medical associations.

FIRST FOODS AROUND THE WORLD

For almost half a century the first food staple for North American babies was processed single-grain cereal. Fortunately, the recommendations have changed and babies are now encouraged to eat veggies, meats (or meat alternatives), and iron-fortified cereals. A look at the foods that babies around the world taste for the first time makes rice cereal seem terribly boring—and it proves that babies will eat anything.

Chinese babies get lots of rice in their first bites, but it's often mixed with tofu, fish, and vegetables. Japanese babies start on solids similar to Chinese babies, but also enjoy seaweed. Most Japanese babies have had raw fish before they turn two! Filipino babies often begin solids with rice simmered in broth, chicken bits, onions, and garlic. East Indian babies start eating lentils and rice, spiced with coriander, mint, cinnamon, and turmeric as early as six months of age. Mexican and Central American children munch on corn tortillas, mashed beans, and vegetables as first foods. In Mexico, babies eat chilies before they turn one, and candy is often mixed or sprinkled with chilies, lemon, and salt.

So the next time your child refuses to eat something you offer, remind her that she should be thankful she's not in Tibet, where babies eat barley flour mixed with yak butter tea!

7: Antibiotics: Carpet Bombing the Microbiota

The Antibiotic Paradox

Pam was excited about her upcoming delivery. The pregnancy had gone well, but then, when the labor finally came on, things began to go wrong. Pam was fully dilated, but when she started to push, the baby wasn't coming down the birth canal properly. The obstetrician tried forceps, but the baby was really stuck. Pam then underwent a C-section, followed by antibiotics, just to be safe, to prevent any infections that might result from the procedure.

Pam is one of the few women who experience a C-section on top of full length of labor, a huge ordeal. Unfortunately, Pam's troubles weren't over yet. After the difficult birth, she returned home from the hospital with her beautiful newborn daughter, but then had trouble breastfeeding. Her nipples were cracked and she developed mastitis, followed by diarrhea. The diarrhea became so severe that Pam had to be hospitalized for dehydration and high fever. She tested positive for *Clostridium difficile* (also known as *C. diff*), a severe intestinal infection that occasionally follows antibiotic treatment, which landed her in isolation at the hospital (to help prevent the spread of *C. diff*). She was given another antibiotic (vancomycin) to treat the *C. diff*, and eventually things began to return to normal.

Pam's story highlights both the pros and cons of antibiotics. They're great drugs for controlling microbial infections, but we now realize they can also cause problems and side effects that we didn't fully appreciate before, which is making us rethink their applications.

Wonder Drugs of the Twentieth Century

Arguably, antibiotics have had the greatest effect on improving human health and longevity of any class of drugs used in the twentieth century. Their invention was truly magical, easily treating diseases that previously could lead to death. Before antibiotics, 90 percent of children with bacterial pneumonia would die. Children with strep throat were placed on bed rest to try to avoid the dreaded complication of rheumatic fever. Talk to your grandparents—or anyone who grew up in the pre-antibiotic era (prior to 1945)—and you'll realize how scary a simple infection could be.

The word *antibiotic* comes from Greek, meaning "against life," and we use the term to describe drugs that work against microbial life. This includes chemicals called antibacterials, antivirals, anti-parasitics, and antifungals. However, we tend to use antibiotic and antibacterial interchangeably, and antibiotics usually refer to drugs that target bacteria as opposed to other microbes such as viruses or fungi. Some antibiotics are broad spectrum (meaning they target many types of bacteria; see Going Bananas with Amoxicillin, page 115); others are narrow spectrum (they target fewer microbes, but still don't specifically target a single microbial species).

The discovery of antibiotics dates back to the early 1900s. A chemist named Paul Ehrlich had the idea of a "magic bullet" that could target a disease-causing microbe without harming the human host. He tenaciously tested many compounds and synthetic dyes, screening them to find a compound that would target the bacterium that caused syphilis. After Herculean efforts, he discovered a compound that could be used to treat this infection. Although the drug did have serious side effects, such as rashes, liver damage, and "risk of life and limb," the cure was better than the disease. Ehrlich's work laid the foundation for the discovery process that has given us so many new antibiotics.

By far the most famous antibiotic discovery was made by the British scientist, Sir Alexander Fleming. Apparently, Fleming wasn't the most organized of scientists and in 1928 he came back to his lab after a vacation to find lots of petri dishes that he had left out covered with bacteria. He began to sort through all this mess when he noticed something out of the ordinary in a petri dish that contained *Staphylococcus aureus* (the bacterium that causes skin boils and abscesses). One area on the petri dish was free of bacteria, but it had a blob of mold instead. This mold, which was later identified as the fungus *Penicillium*, produced a substance that inhibited growth of the bacterial pathogen. Penicillin is a complex molecule, and it took almost two decades for chemists to figure out how to synthesize it. With the outbreak of World War II, there was a huge need to control bacterial infections from the battlefield. In the pre-antibiotic era, wars were terrible for infections (in the American Civil War, more people died from infections than bullets). So during World War II, with infections raging, penicillin was invaluable. Initially, it was so precious that it was kept for military use only, and it was recovered and re-isolated from patients' urine in order to reuse it.

However, penicillin production was rapidly scaled up, and by 1945 it was mass-produced and distributed to the general population, saving countless lives.

This breakthrough changed the world of infectious diseases and how they were treated, opening up therapies for diseases that were previously considered untreatable and often fatal. Several new classes of antibiotics were discovered in the 1950s–1970s, mainly from soil organisms. Soil organisms produce antimicrobials to kill their neighbors since they're in competition for scarce nutrients. Microbiologists collected soil samples from the far corners of the world and tested them for antimicrobial activity. When promising compounds were found, chemists modified and tweaked them to enhance their activity or uptake into microbes. These semisynthetic compounds formed the backbone of the antibiotic industry, providing many potent new antimicrobial drugs that could be used to treat a variety of infectious diseases.

Like all good things, we just couldn't get enough. There are at least 150 million antibiotic prescriptions written in the US each year (that's one for every two people.) In 2010, the estimated global consumption of antibiotics was 63,000 tons, which will increase by 67 percent in 2030 to over 100,000 tons. Much of this is due to their increased use as growth supplements in livestock (80 percent of the US's antimicrobial usage is in livestock). It turns out that low doses of antibiotics enhance the weight gain of farm animals. Europe has stopped this practice, but unfortunately Canada and the US have not, and developing countries are just beginning to use antibiotics in livestock management. Some of the consequences of this are discussed in chapter 10, where we look at how this practice affects childhood obesity.

So what effect does dumping into the world thousands of tons of chemicals that kill microbes have on the microbiota? As we shall see, the microbes have responded, and it's casting serious shadows on these wonder drugs.

Resistance Is Futile

Well, maybe not if you're a microbe. As mentioned previously, most antibiotics come from compounds made by soil microbes to give them a competitive advantage. However, microbes must have a way to resist killing themselves with these toxic molecules. As a result, microbes have developed resistance mechanisms to accompany antibiotic production. We call this antimicrobial resistance, and microbes have been doing this for as long as they've been producing antibiotics. For example, antimicrobial-resistant genes have been found in human remains that were frozen in permafrost, obviously millennia before antibiotics were discovered. Resistance genes have also been found in environments where there has been no human contact, such as underground lakes below sheets of ice. As soon as penicillin was discovered, it was also realized that microbes exist that can resist its killing effects.

The other thing we need to remember about microbes is that, unlike us, they regularly exchange DNA with one another, allowing them to rapidly adapt and evolve during their lifetime (which can be as fast as twenty minutes, but is often a few short hours). Think of the microbes in your gut as being hooked up to a genetic Internet—they exchange genes much like we download songs and apps. Having an "antimicrobial resistance app" would be a lifesaving feature if you were getting hit over the head with a lethal antibiotic. In response to our massive use of antibiotics, microbes have spread resistance genes like wildfire to other microbes (their "must-have" apps!), with antibiotics providing a strong selection pressure to get that app (live or die). We now see massive microbial resistance to all major antimicrobials that are used extensively, with resistance arising within a year or two, often making the drug obsolete within 3–5 years.

We are also discovering more and more microbes that are resistant to most, if not all, antibiotics—we call them "superbugs." They cause infections that we could previously treat, and they're wreaking havoc in our hospitals and health care systems worldwide. They go by lovely acronyms such as MRSA, XDR TB, MDR E. coli, and VRE, and are causing a major rethink in how we use antibiotics.

Since 2009 only two new antibiotics have been approved, and the number in the pipeline continues to shrink. Most pharmaceutical companies have either drastically downsized or closed their antibiotic discovery divisions. This is creating the perfect storm: no new drugs, and the ones we have no longer work. The World Health Organization recently summed it up nicely, saying that antimicrobial resistance is a "serious threat [that] is no longer a prediction for the future, it is happening right now in every region of the world and has the potential to affect anyone, of any age, in any country." These strong words suggest that we're headed for a post-antibiotic world, taking us back to the fears of mortal infections common to our great-grandparents.

"Mommy, My Ear Hurts!"

These words strike fear into any parent's heart, knowing it usually means a sleepless night, if not a trip to the ER in pajamas. Ear infections, called otitis media, are quite common in young children, and are usually treated with antibiotics. However, it isn't always clear that antibiotics are warranted, which is confusing to parents (and physicians). Take the story of Jack, a two-year-old who had developmental difficulties and was also prone to recurrent ear infections. After a particularly sleepless night, Jack's mother was convinced her son had yet another ear infection and took him kicking and screaming to see his pediatrician. Diagnosing otitis media usually means visually observing the eardrum (called the tympanic membrane) to see whether the eardrum is bulging, which indicates fluid in the middle ear, and whether it looks red, which suggests inflammation. Redness of the eardrum can also result from a child crying and does not always mean an infection is present. The pediatrician had difficulty assessing Jack's eardrums, as he was a combative child, and he also had wax in his ears. She thought the ear looked red, and considering the mother's description of the symptoms, prescribed a common antibiotic used to treat ear infections. Unfortunately, this resulted in Jack subsequently getting *C. diff* diarrhea, which required him to go on metronidazole (a strong antibiotic also called by the brand name Flagyl). The first course of metronidazole did not work. The second time, his mother also wisely gave him a yeast probiotic (see Probiotics with Antibiotics—an Oxymoron?, page 112), and he finally got better. After that, an ear, nose, and throat surgeon placed drainage tubes in Jack's eardrums in order to decrease the number of ear infections, rather than use multiple courses of antibiotics to treat them.

Treating otitis media with antibiotics is controversial. First of all, it's often overdiagnosed. The eardrum can be red from crying or due to a viral infection, so antibiotics wouldn't work in those instances. Anyway, most cases of ear infections caused by bacteria resolve on their own. Studies in the Netherlands indicate that seven kids with otitis media would have to be treated with antibiotics in order for one to benefit from antibiotic therapy. There is a very small risk that otitis media, if left untreated, will progress to more serious illnesses such as mastoiditis, a dangerous infection of a bone that sits behind the ear. However, this is uncommon. The Dutch study calculated that only 1 in 2,500 children develop a complication from otitis media; if they were all treated, it would result in a lot of kids receiving unnecessary antibiotics. Seventy-five percent of children have at least one episode of acute otitis media before their first birthday, so this is certainly an infection every parent will encounter at some point. It's usually preceded by an upper respiratory infection (more often than not a viral cold), which plugs up the Eustachian tubes that drain the middle ear, which then fills up with fluids, making a perfect broth in which bacteria happily grow.

Most pediatric organizations have developed guidelines for treating otitis media, along with adopting a cautious use of antibiotics. They generally suggest a "watch and wait approach," especially if the child is older than six months, is otherwise healthy, and has mild symptoms (no major fever, etc.). Doctors will also ensure that parents have access to painkillers to help the child ride things out. Generally, the recommendation is to wait 48–72 hours, and then follow up if the infection has not resolved. This results in approximately only one-third of children getting antibiotics, which is much more reasonable than administering them to every single child who shows early symptoms.

The other current practice is how pediatricians aim to "hit hard and fast" if they're going to use antibiotics. Five days of antimicrobial treatment are at least as effective as ten days of antibiotics in children older than two years of age. A longer span of treatment may be needed in complicated cases, but the general idea is to expose your child to fewer antibiotics, if medically reasonable.

Breastfeeding has been shown to decrease ear infections, probably because the maternal antibodies in breast milk help protect against infections. Bottle-feeding causes a child to suck hard on the negative pressure inside the bottle, which then causes negative pressure inside the ear, which may draw fluid and microbes into it (fully ventilated bottles can avoid this problem). Similarly, extensive use of a pacifier in children younger than three years old increases the risk of otitis media by 25 percent, presumably for the same reasons. Since ear infections usually follow respiratory infections, minimizing the time young children (particularly those younger than twelve months) spend in day care, and thus minimizing their inevitable exposure to colds, also decreases the risk of ear infections. Infants are routinely given a pneumonia vaccine (Prevnar), which may also help make a child less susceptible to ear infections. Influenza vaccination can help as well. Maternal smoking in the first year of life is a significant risk factor, too, especially in children that had a low birth weight.

Wonder Drugs That Aren't So Wonderful

In the title of this chapter we used the expression "carpet bombing," which is a military term that entails bombing a defined area with indiscriminate destruction and lots of collateral damage, in hopes of destroying the desired military targets. Unfortunately, this description can also be applied to antibiotics and their effects on the microbiome. We now know that antibiotics cause massive disruption to the microbiota (naturally, as they're designed to kill as many microbes as possible). This indiscriminate killing of many bystander microbes, in addition to the desired infectious agent, has unintended consequences. We're also starting to realize that we may be wiping out microbes from our society before we even realize that they're beneficial. As detailed in Dr. Martin Blaser's book *Missing Microbes*, previous generations had much more diverse microbiota, and our quest to kill them all may have serious consequences for future generations. Who would have thought that we might have to put microbes on the endangered species list?

As discussed more in chapter 13, multiple courses of antibiotics in the first year of life lead to a significant increase in asthma, as they affect the microbes involved in ensuring a healthy maturation of the immune system. Similarly, in chapters 10 and 11 we discuss how antibiotics increase the rate of obesity and subsequent type 2 diabetes by altering the microbiota involved in nutrient uptake and weight gain. For these diseases, there's a direct correlation with the number of courses of antibiotics, with up to a 37 percent increased risk with multiple uses of certain antibiotics.

Unfortunately, even short-term pulses of antibiotics during early life can have an effect on the microbiota. Experiments in lab animals have shown how these pulses can cause developmental changes in mouse pups, with increased weight gain and bone growth. In studies of children, the effects of antibiotics on the microbiota are still observed six months later, and repeated use of antibiotics shift the microbiota further and further away from its initial composition. What's more, the presence of antimicrobial resistance genes can still be detected 1–3 years after the antibiotics are stopped.

As we saw in the two examples mentioned earlier, there's a significant risk of getting *C. diff* infections following antibiotics, due to the removal of microbes that normally prevent *C. diff* from taking hold in the gut. This is discussed in chapter 16 in more detail, along with the exciting (and somewhat off-putting) concept that fecal transplants can successfully treat such infections. Our lab has shown that antibiotics also make mice more susceptible to diarrheal infections caused by pathogenic *E. coli* and *Salmonella*. This is called "competitive exclusion," with the concept that the good bugs make it much harder for the bad ones to take hold in the very crowded intestinal world. Our lab also showed that antibiotics cause a thinning of the mucus layer, which serves as a protective barrier against infections and other inflammatory intestinal disorders. This is presumably because the antibiotics affect microbiota that normally eat mucus for an energy source. Another example of competitive exclusion can be seen in urinary tract infections, which often arise in women following a course of antibiotics for a different infection; the protective vaginal microbiota is disrupted, thereby allowing a pathogen to infect the urinary tract.

It turns out that antibiotics have other effects that we previously never even dreamed about. In some studies conducted in our lab, we treated mice with standard antibiotics and measured how this affected small molecules (called metabolites) in the feces and liver. Remarkably, we found that this treatment significantly changed about 60 percent of the mouse's metabolites. These are molecules involved in normal body function, including hormones, steroids, and other chemicals that allow our cells to communicate with one another. Needless to say, we were shocked when we realized that antibiotics could impact our normal physiology that much.

Immune cells are also innocent bystanders, affected when antibiotics cause changes to the microbiome. For example, in an animal model, it was shown that neutrophils, which are key microbe-hungry immune cells that clear away pathogens, were decreased in newborns treated with standard antibiotics. The researchers found that with antibiotic treatment, certain classes of microbiota that stimulate the immune system were removed, thereby resulting in a decrease of neutrophil production.

Another common side effect is diarrhea, which plagues up to

one-third of people following a course of antibiotics. Even in developing countries where diarrhea is much more common, higher levels were detected in young children (six months to three years old) following antibiotic treatment, although children who were exclusively breastfed for the first six months were protected from this side effect.

Physicians now realize the issues associated with antimicrobial resistance and are beginning to take into account the detrimental effects of antibiotics on the microbiome. Pediatric associations have developed guidelines for "antibiotic stewardship," optimizing the use of antibiotics while minimizing the unintended consequences. This certainly helps with antibiotic abuse, and it also explains why your pediatrician may be more reluctant to reach for the prescription pad than in previous years. Your doctor may ask you to wait until she gets the results back from a strep throat swab before starting antibiotic treatment—she's just being prudent, despite your screaming kid's input on the decision.

Ironically, many parents are aware of this antibiotic dilemma, and will question physicians on whether an antibiotic is necessary. One option is to get the prescription, but wait and see if the infection improves on its own before using the antibiotic.

Probiotics with Antibiotics— an Oxymoron?

Antibiotics kill microbes, so why would one intentionally take probiotics (which are live microbes) while taking antibiotics? In certain cases it's actually a good idea. As discussed previously, diarrhea and *C. diff* are common complications of antibiotic use (*C. diff* causes about a third of the types of diarrhea). A recent study showed that if certain probiotics (although not yogurt) were taken along with antibiotics, 42 percent of the participants were less likely to have diarrhea. In previous years, the counsel was to wait until the antibiotics were stopped before taking probiotics. However, *Lactobacilli* probiotics are now being marketed specifically to be taken along with antibiotics, although you're still encouraged to wait 1–2 hours after ingesting the antibiotic before taking the probiotic. An interesting concept is to take a yeast probiotic when taking antibacterials, since antibiotics do not kill yeast, as they are not bacteria. In a major meta-analysis (a summary of publications) of thirty-one probiotic studies, it was concluded that *Saccharomyces boulardii* (a yeast) worked well to decrease antimicrobial-associated diarrhea *and* to decrease *C. diff* infections. The same study found that the bacterial probiotic *Lactobacillus rhamnosus* GG helped prevent diarrhea in children, but had no effect on *C. diff* infections. They also found that mixes of two probiotics could have some efficacy against antibiotic-associated diarrhea.

The other compelling conclusion is that one needs to take large doses (more than 10 billion live microbes a day) to have any effect. Since the onset of antibiotic-induced diarrhea is usually 2–8 weeks after taking antibiotics, most patients are encouraged to continue using probiotics even after finishing the antibiotics. Collectively, this work suggests that certain probiotics can help prevent diarrhea when taken with antibiotics, given the caveats above.

Antibiotics have gone from being miracle drugs that could bring a dying person back to life to being used indiscriminately for every fever in a child. We now live in an era where their intense use has backfired in the form of antibiotic resistance. On top of that, few things shift the developing microbiome of an infant or child more than antibiotics, potentially affecting their immune system permanently. Fifty years ago no one saw any of this coming, but we're facing a reality where an antibiotic must be seen as a drug of last resort that often requires restoration of the microbiota following its use.

Dos and Don'ts

- ◆ **Don't—** assume that all infections have to be treated with antibiotics. Upper respiratory tract infections and colds are often caused by viruses, so antibacterials won't cure them. Most sore throats, especially if the child also has a runny nose and cough, are caused by viruses and don't need antibiotic therapy. If your child has a mild ear infection, it's reasonable to watch and wait for a few days to see if it gets better on its own before starting antibiotic therapy.

- ◆ **Do—** consider giving probiotics to your child if he is being given antibiotics. These could include *Saccharomyces boulardii*, a yeast-based product that decreases *C. diff*, or *Lactobacillus rhamnosus* GG, or a mixture of probiotics. This assumes the child is not immunocompromised nor has any other underlying conditions.

- ◆ **Do—** be a thoughtful steward regarding antibiotic usage. Discuss with your pediatrician why she is suggesting an antibiotic (she presumably has her reasons). Antibiotics are a remarkable treasure to medicine, but their abuse is really denting their magic, and we now realize they aren't without their own detrimental effects.

- ◆ **Do—** make your child less susceptible to ear infections by using a ventilated bottle for milk and avoiding excessive use of pacifiers in children under the age of three.

GOING BANANAS WITH AMOXICILLIN

The most commonly used children's antibiotic is amoxicillin, a type of penicillin that targets the walls that surround bacterial cells. It's often formulated to taste like bananas, with the idea that kids will like it. However, children still tend to turn up their noses at the taste, and it can be a real struggle to get them to take it! One of the most common and very appropriate uses of amoxicillin is to treat ear infections that haven't cleared up on their own after two days of watchful waiting. Recommended treatment is usually for 5–7 days.

Ear infections are commonly caused by several different bacteria, including *Streptococcus pneumoniae* (commonly called "pneumococcus"), *Haemophilus influenzae* (commonly called "H flu"), and *Moraxella catarrhalis*. However, amoxicillin kills not only these pathogens, but many other bacterial species, which has a major effect on the microbiome.

Unfortunately, amoxicillin is also commonly used, inappropriately, for viral upper respiratory infections such as the common cold and the flu. Because it targets a bacterial structure, it has no effect on viruses. And of course, resistance has become common—it used to be the first line of treatment for bacterial urinary tract infections, but now resistance is too common and other antibiotics must be used.

8: Pets:
A Microbe's Best Friend

Love at First Lick

When Nathan and Carol left the hospital with their brand-new baby, Rory, they were a bit nervous about how Milo, their three-year-old Labrador retriever, would react upon meeting the new addition to the family. On one hand, they were worried about not having enough time for Milo, who up until then would go with them on hours-long hikes and swims and on frequent camping trips. With a baby in the house, the dog was being moved to second place in the attention contest. On the other hand, Milo, like almost any other lab on the planet, liked to jump up, sniff, and lick everyone he met, and months of doggy school hadn't curbed his habits. Carol and Nathan were concerned about this seventy-five pounds of bouncing friendliness around wee little Rory. On top of that, they had conflicting opinions about how close Milo and Rory should be. Carol was happy to have him sniff and lick Rory, but Nathan, and especially Nathan's mom (a frequent visitor), were not. Nathan didn't grow up with pets in the house, so even though he was a committed owner to Milo, he would shoosh him off the couches, the nice living room carpet, and, most of all, the bed. Carol respected that Nathan was not as okay with dog slobber and hair as she was, but when Nathan wasn't home she would curl up with Milo on the couch to watch a movie, or invite Milo to bed if Nathan was spending the night away from home. Needless to say, they still needed to sort out the details on their new status as a family of three humans and a dog.

They decided to talk to a dog trainer who had helped a lot with Milo's behavior when he was a puppy. She said that although labs are usually great pets around kids, there were a few things they could do to make the transition easier and safer. First, she recommended that Milo meet Rory for the first time outside the house and have him leashed. She also said that the first encounters at home should be supervised and that they should continue giving attention to Milo when the baby was around. As for any recommendations regarding licking the baby, she quickly said, "You're on your own on that one! Everyone does it differently."

When it came time to introduce Milo to Rory, they arranged the meeting at a nearby park. Nathan's mom brought Milo on a leash, but upon seeing his owners, Milo broke free of her hold, running to greet his owners and jumping at them with excitement. Despite their efforts to calm him, when presented with his new "brother," Milo sniffed him from top to bottom, then touched Rory's face with his wet nose and very gently licked Rory's pink cheek. "No licking, Milo," said Nathan, pushing Milo's face away. "It's just a kiss, Nathan," said Carol, knowing at that moment that Rory and Milo were going to be best friends.

From the Wild to Our Couches

The partnership between humans and dogs goes way, way back. Well before humans settled down to farm, dogs would roam with packs of hunters and gatherers, possibly scavenging leftovers from their human companions' hunted game and other food. Dog fossil remains have been found in caves dating as far back as 16,000 years ago, at a time when humans would compete against saber-toothed cats to hunt mammoths across a frozen landscape. Archaeologists aren't in agreement on the exact location and timing of dog domestication, but one thing they all acknowledge is that dogs were the first species domesticated by humans, including all plants, insects, and other animals.

In the beginning they were merely tamer wolves, still feral and only seeking human interaction as a way to get food, but the two species quickly became very close. Through thousands of years of living with humans, dogs became reliable guards, hunters, herders, and carriers, and they developed an uncanny ability to communicate with humans, superior even to how chimpanzees (or any other ape) can pick up on human cues.

Cats, on the other hand, became domesticated after humans took to the farming lifestyle. They likely became useful to people as a way to control rodents in granaries, and probably acquiesced to the deal in exchange for food, shelter, and play. Compared to dogs, cats only recently split off from wild cats and some even breed with other wild feline species to this day. Their genome hasn't changed as much as the dog's genome has, and they still require a high-protein diet and can't digest human food very well. In that sense, cats remain only semidomesticated, despite having lived with humans for at least 9,000 years. Every now and then we see their wildness, whether it's when they turn up on the doormat with the offering of a dead mouse, bird, or lizard, or how they disappear for multiday escapades, only coming home to get fed. It's not surprising that cats, at least in most Western societies, are kept as full-time indoor pets, seldom allowed to roam and test out their wild sides, for fear that, one day, they simply might not come home.

Some people keep pets for their useful qualities, but most modern societies keep cats and dogs for companionship. They can require a bit of work, but their loyal friendship, silliness, and unconditional love usually make it worthwhile. There are many obvious ways that having a pet can improve your lifestyle; dogs, for example, promote exercise (daily walks, rain or shine!), encourage sociability ("Hi, what's your dog's name?"), or simply make you happy (nothing beats a wagging tail and smiling face every single time you walk through your door!). As if those aren't reasons enough, we're now beginning to learn that pets, especially dogs, also bring health into your life by bringing the outside indoors. Yes, all those dirty paws on floors, carpets, and furniture, all those stinky smells that won't go away are worth it—within all that dirt there are millions of microbes that make our clean lives that much closer to the outdoors.

The influence dogs have on our microbiota was recently documented in two studies, wherein scientists found that owning a dog (but not a cat) that roams around outside changes the composition and diversity of the human microbiota. One study showed that the microbiota among family members is more similar in families that own a dog compared to dogless families. The same study also found that the skin microbiota of dog owners had bacterial species also found in dog mouths and in the soil. The similarities between dogs and their owners were so striking that the scientists could match a dog with its owners out of a group of samples, solely based on the microbiota.

In a separate study, researchers found that the presence of a dog was associated with an increase in microbial diversity in the dust of the house that a dog calls home, and that many of the microbial species found in the household dust were also living in the dog owners' intestines. It seems that by bringing the outdoors in and by licking everyone and everything they can, dogs act as microbe delivery systems that equalize the microbiota across the household.

In both studies, cats didn't seem to influence their owners' microbiota very much, which is likely due to the behavioral differences between the two species. Dogs like to play and tumble with humans and they lick a lot. Cats? Yes, maybe, but usually only when they think that you're truly worthy of their attention. Cats also don't beg you to take them on a walk and, due to their tendency to run away for days at a time, they aren't brought outside as much as dogs. Cats and dogs are both great pets, but when it comes to the microbial gifts that pets bestow on their owners, dogs win fair and square. We'll take soil microbes over a dead mouse any day.

Bring on the Slobberfest

Like many parents and grandparents around the world, Nathan and his mom (from the anecdote above) had this notion that dogs, and dog slobber in particular, could make a baby sick. To some extent this is true. On rare occasions, dogs can pass on a disease to a child (or to anyone) because they can harbor all sorts of worms (heartworms, tapeworms, hookworms, etc.), other pathogenic bacteria, and viruses. However, these diseases are very rare among pets that are well looked after and that receive veterinary care periodically. Sure, if a dogs looks sick, has diarrhea or a skin rash or scab, it's probably a good idea to get the dog to the vet instead of letting your kid roll around with his furry friend, but there's a very low risk of catching an infectious disease from a dog that receives good care.

On the contrary, owning a dog that goes outside and allowing it to interact with children is actually beneficial for their health. Epidemiological research shows that kids that are exposed to dogs early in life have a decreased risk of developing asthma and allergies. A 2013 article published in the *Journal of Allergy and Clinical Immunology* summarized the results of twenty-one studies that aimed to figure out what factors contribute to the development of childhood allergies. What they found is that exposure to a dog during pregnancy or before the age of one decreases the risk of developing eczema (a skin disease) by 30 percent. In several other studies the presence of a dog (but, again, not a cat) is also associated with a reduced risk of asthma, decreasing the risk by about 20 percent. This recent information has surprised allergists around the world, who for years recommended removing pets from home to reduce allergies (although in certain Central and South American countries dogs have been used to cure asthma; see Chihuahuas Cure Asthma?, page 124).

Many people do develop allergies to pets, and the presence of a pet in the house can exacerbate the problem if a child is allergic to something else. In these cases it makes sense to consider finding another home for the pet. However, since studies show that the presence of a dog may prevent the development of asthma and allergies, unless Milo gets sick or someone develops an allergy to him, promoting contact between Rory and his four-legged friend is actually good parenting! Parents and grandparents everywhere please take note, though: buying a pet solely to decrease your child's risk of asthma is not a solid enough reason to own a pet. A dog is a big commitment, especially with a new baby in the house; they require a lot of attention, training, walks, and money. If you don't see yourself wanting this added responsibility, it might be a good idea to hold off on getting a pet for now, and let your baby play with a family member's or friend's dog instead.

The strong relationship between having a dog and the reduction of asthma and allergy risk certainly raises the question: What's so special about dogs? We've suggested that it's the microbes in the dirt that a dog brings into the house, but others remain skeptical, claiming that it could perhaps be something that the dog produces instead (this is a good example of the type of things scientists love to bicker about!). What settles the argument in favor of the dirt microbes theory is a study led by Dr. Susan Lynch from the University of California in San Francisco. This study collected dust samples from homes with and without dogs, and showed that upon exposing mice to the different dust samples, the mice that were given dust from homes with dogs were less likely to develop asthma. What's more, they looked at the type of bacteria in the dust samples and found a specific species, *Lactobacillus johnsonii*, associated with the improvement of asthma in mice. When they grew this bacterium in the lab and fed it to mice in the absence of any dust, they found that it lowered the risk of asthma, demonstrating that this and perhaps other species of beneficial bacteria, along with the dogs that bring them into households, are responsible for decreasing asthma risks.

These types of studies have important implications. If dogs transmit bacteria that make humans less prone to an immune disease, this implies that dogs carry around probiotic species that are beneficial for human health. What are they? Can they be grown in a lab and given to kids? We have a lot more to learn in this area, and scientists are certainly working on it. What *is* clear is that dogs and humans have a special partnership that goes beyond their loyal friendship. Dogs keep us dirtier, and as we have come to learn, kids benefit from this kind of exposure early on.

Dos and Don'ts

◆ **Do—** let your dog safely play and closely interact with your baby or small child. It's a good idea to take the dog to the vet right before the baby arrives, just to make sure your pooch is in good health. Letting your dog lick or be close to your baby is likely to decrease his risk of developing allergies and asthma, with the added benefit of providing companionship and protection, and teaching your child to be comfortable around animals.

◆ **Don't—** consider getting a dog (or any pet for that matter) simply to decrease a child's risk of asthma. Owning a pet is a lot of work and they deserve to be taken care of by committed owners that will provide food, veterinary care, and entertainment. Expose a child to a dog you know if you can't have one at home.

◆ **Don't—** shun cats just because they don't seem to offer us beneficial microbes. They make fabulous pets, too. However, cats are known to transmit parasites through their feces, so it's recommended to avoid changing the litter box during pregnancy and moving it to a place a baby can't access it.

CHIHUAHUAS CURE ASTHMA?

In many Central and South American countries there's a surprisingly widespread belief that if a baby or young toddler suffers from wheezing or allergies, buying a Chihuahua will ease the symptoms. This rumor has been around awhile (noted in medical journals in the 1950s) and has spread to the southwestern United States as well. As a result, many families of asthmatic children buy little Chihuahuas.

There are actually two versions of this story: one claims that the dog cures the disease altogether, while the other posits that the poor dog soaks up the disease from the human, becoming asthmatic in the process! What's even more unbelievable is that certain family doctors and pediatricians (especially old-school ones) have recommended this practice to patients, based on their own experience.

It would be an understatement to say that there's serious skepticism about this theory. It's true that Chihuahuas, as a short-haired breed, are considered hypoallergenic since they shed very little, but it's highly unlikely that living with a Chihuahua is going to cure someone's allergies. But hey, if you feel like snuggling in bed with an adorable little dog, go for it. It will certainly make you happier.

9: Lifestyle: Microbe Deficit Disorder

Starved for Nature

Societal changes in the past century have dramatically shifted how we live. The type of work we do, the places and types of buildings we live in, what we do for entertainment, and our family dynamics—just to name a few—are very different than only three generations ago. Because of this, being a kid now is very different than thirty, sixty, and especially a hundred years ago. Without a doubt, many of these changes have been positive. For example, 96 percent of US children go to school now, compared to only 60 percent in 1913, and the infant mortality rate has gone from 150 deaths per 1,000 births to 5 deaths per 1,000, all in a span of one hundred years—both remarkable accomplishments in societal development. However, some of the changes brought about in the past century may not be as positive. Kids have always loved to play just like kids do today, but the activities they undertake and the places in which they play are vastly different. Inadvertently, these changes have resulted in a detachment between children and the outdoors, and this has had a major influence on microbial exposure in children.

For a real-life example of this, try a simple experiment: next time you're at a family reunion or any other event where several generations are present, ask the members of various generations what they used to do for fun when they were kids. Try to get them to think not only of the activities they used to participate in, but also where they played and the amount of time they spent outside. We presume the answers you get will be along these lines: "We got home from school, ate something, and were not back in the house until it was time for dinner," or "We would roam the neighborhood in packs of ten to twenty kids, climbing trees, building forts, chasing one another until it got dark," and our favorites, "Being in the house meant we were sick or grounded" and "I remember my knees were permanently dirty and scratched." Now compare that to, "I love to watch Netflix," "Videogames!" and "IM'ing with my best friend." It seems almost unnatural to think that the childhood memories that modern children are creating mainly happen indoors, away from nature, but this is the reality.

The statistics are truly disheartening: children spend half as much time outside as they did only twenty years ago; kids ages 8–18 spend a daily average of 7 hours and 38 minutes using entertainment media or screen time; and only 6 percent of 9–13-year-olds go outside by themselves. As unrealistic as it sounds, in England more children are now taken to emergency rooms for injuries incurred by falling out of bed than falling out of trees. It's no wonder that kids are so sedentary these days, when their idea of good entertainment comes from a screen and not from running outside and physically playing with other kids. The situation is so dire that health agencies in many countries have issued minimum physical activity requirements for children, something that only thirty years ago would have seemed like a ridiculous policy. What's worse is that out of fear for their safety, or fear of them getting dirty or injured, adults are constantly supervising children or even keeping them from going outside altogether!

As we have said in the previous chapters, our resident microbes are a result of what we physically interact with and the food we eat. It's truly worrying to think that millions of children are growing up mainly exposed to indoor microbes, like the ones growing on their Wii remotes or computer keyboards. For millions of years, children have grown up exposed to a substantial number of outdoor microbes and this connection has been broken in the past couple of generations—which coincides with the time it has taken for Western lifestyle diseases to skyrocket.

Such a Thing as Too Clean

One of the main reasons parents and caretakers today have an aversion to letting kids freely play outside is the notion that they can get sick from putting dirt or dirty objects in their mouths, or from being dirty for an extended period of time. This is an ingrained perception cultivated over decades—the idea that "dirty" inevitably means the potential for infectious disease. We spent generations avoiding harmful infectious agents in the environment and cleaning up our world. The World Health Organization defines hygiene as "the conditions and practices that help to maintain health and prevent the spread of diseases," and there are indeed many proven advantages to following hygienic practices—namely, the spectacular drop in childhood mortality. However, Western societies have taken hygienic practices to the extreme. The concept of cleanliness (often cited as next to Godliness!) is not necessarily associated with health benefits but with physical appearance, and our modern societies have never been so clean. Never have there been so many brands of soap, deodorant, toothpaste, razor blades, disinfectant, shampoo, lotion, and perfume. Being clean is our standard of living; clean feels good (don't shower, wash, or shave for a week, and you will no doubt agree).

It's important to keep in mind, however, that cleanliness does not have an advantage over hygiene in preventing disease. Cleanliness is a relatively new concept; it's been part of our culture for only a hundred years or so. Before the mid-nineteenth century in the US, regular baths and teeth brushing were not common practices, and neither was using soap. The first hygienic measures took place through an organization known as the Sanitary Commission, which originated during the American Civil War. It was very successful in reducing infectious diseases and deaths by promoting washing the sick, along with their bed linens and their rooms. Back then doctors and scholars were just beginning to accept the concept that germs transmitted diseases.

Our current cleanliness practices have become more of a cultural construction based on the idea that the cleaner you are, the better it is for you. As an example, people might be disgusted by a picture of an infected wound or someone covered in dirt. In reality, the wound is an actual threat to your health because it's infected, whereas dirt is not a threat at all, it only looks unclean.

Most people become even stricter about cleanliness when they're taking care of a baby. This makes sense, as it's a natural way to protect a baby and prevent infections, but our modern sense of how clean we should be is causing babies to be brought up too sterile. When Brett was a kid, every morning after school prayer (at a public school!) he had to stand in line for hand inspection. If there was dirt under your nails, you were wrapped hard on the knuckles with a ruler, then sent to the bathroom with a brush and you couldn't come out until they were clean. This story was pretty typical at that time, and although hand inspections don't happen like this anymore, the level of cleanliness in a child is still a reflection on how well he is taken care of, so keeping them squeaky-clean is considered good parenting by lots of people.

As a result, and with the advances in cleaning technologies, gel hand sanitizers hang from almost every diaper bag; toys and pacifiers are wiped clean with antibacterial wipes if they hit the ground; bottles and utensils are sterilized before every use; babies and toddlers are often not allowed to play in the dirt or sand, and when they are, they are wiped clean immediately. Phrases like "Ewwww!" "Yuck! Don't play in the mud!" or "Don't touch that bug, it's dirty!" have become second nature. Babies and children are prevented from following their innate nature to get dirty, and in doing so they're being shielded from the microbial exposure that's essential for their development.

Recognizing that we live too cleanly may not be hard, but learning how to differentiate a potential health threat from something that only looks filthy is not always easy, and in some cases it's not a black-and-white decision. As scientists, we've studied microbes that cause diseases for many years, but as parents, knowing all that we know, it *still* hasn't been easy to make decisions regarding microbial exposure. So we polled parents to find out what they most wanted to know—their most pressing questions and concerns—and then applied current scientific knowledge to provide answers (for an extended list, visit www.letthemeatdirt.com).

Cleanliness Q&A

1. When should children wash their hands? What kind of soap should we use?

Handwashing is, without a doubt, the best hygienic practice that we can follow to prevent contracting and spreading infectious diseases. It's been shown time and again that communities with good handwashing practices stay healthier, and no one should stop washing their hands just to promote more exposure to microbes. With that said, children do not need to wash their hands all day long. Handwashing should occur before eating; after using the toilet; after being in contact with someone sick, or, if the child is sick, before she touches other people; after touching garbage or food that is suspected to be decomposing; after touching animal waste or farm animals; or after being in places frequented by many people (public transportation, malls, etc.). Children do not need to wash their hands after playing outside, unless they are about to eat; immediately after they walk into the house; or after playing with other children, unless they are sick with an infection. Children should be outside often and should be allowed to be barefoot and to get dirty, and handwashing does not necessarily need to immediately follow these activities. The above list is certainly not exhaustive, but it is aimed to differentiate the types of exposure associated with the risk of infection from the ones that are not.

Regarding the soap question, the kind you use depends on personal preference, but it's best to avoid antibacterial soap. A Food and Drug Administration (FDA) committee found that antibacterial soaps provide no benefits over regular soap and water. Except for hospitals or places where additional medical hygiene is necessary, antibacterial soap doesn't have a place in everyday use, and the same goes for antibacterial sanitizers. Plain old soap and water is enough and an alternative sanitizer (like a gel sanitizer) should only be applied if there isn't a potable source of running water and soap.

This advice is certainly counterintuitive to common practices, and it may also be hard to follow since finding soaps without antibacterial agents in them can be harder than you think (most liquid hand soaps have them). A common antibacterial chemical used in soap, called triclosan (and its derivative triclocarban), is also found in deodorants, toothpastes, cleansers, and cosmetics. Triclosan has been used for about sixty years, but recent research questions its side effects and environmental toxicity.

Besides killing bacteria, triclosan has been shown to alter hormone regulation in animals and it might contribute to the development of antibiotic resistance in bacteria. Triclosan is classified as a toxic chemical for aquatic organisms and is known not to biodegrade easily. Big companies such as Johnson & Johnson pledged to remove triclosan from all their product lines by 2015 (at the time this book was written, they hadn't yet confirmed the removal). Triclosan has been banned in Europe since 2010. The Canadian Medical Association has suggested a ban on antibacterial consumer products such as triclosan, and the FDA states that the risks associated with the long-term use of antibacterial soaps may outweigh their benefits. Still, triclosan is currently considered a safe product for human use in many places and it's up to the public to refrain from using it.

2. Should I let people hold and touch my baby?

The answer to this question is ultimately a matter of personal choice, depending on how comfortable you feel about passing your baby over to someone else. With that said, research shows that social interactions, including physical contact, is one of the ways to maintain diverse microbial communities. In one study, biologists took samples over a long period of time from two groups of African baboons that lived near each other. These two groups had the same type of diet, yet they differed in one important behavior: one group engaged in social grooming and the other did not. Interestingly, the microbiotas of these two groups were different, and the baboons that groomed one another had more similar bacterial communities than the baboons in the group that did not. This study shows that it's not just the food we eat that determines the type of microbes that grow within us, but that social interactions like physical contact play an important role. Thus, limiting physical touch, a behavioral trait that characterizes humans as a species, likely limits the exchange of microbes between a baby and his surrounding humans.

If the fear of your child getting a disease is what prevents you from letting other people hold or touch your baby, there are ways to significantly decrease this risk. First, avoid having your baby around people that are sick with an infection; second, ask everyone to wash their hands before holding and touching a very young baby. Since letting a baby be touched is one of the ways he gets exposed to microbes, it makes sense to want to avoid the chance of infection, but physical contact with healthy people is not going to be dangerous for the child, and may even be beneficial.

3. If my child is sick with a cold, should I keep her at home to avoid spreading the illness to other children, or is it advantageous for kids to share infections to toughen their immune systems?

This is definitely one of those questions that doesn't have a black-and-white answer. Although there is robust evidence that exposure to microbes during early childhood can protect against certain immune diseases like asthma and allergies, there is no evidence that we need to be exposed to pathogenic bacteria or avoid hygienic practices to prevent immune diseases later in life. With that said, it's impossible to grow up without getting an infection. Suffering infectious diseases is part of being human, and especially part of being a kid. Preventing a child from getting sick at all costs will likely result in the type of behavior that also prevents a child from being exposed to many beneficial, non–pathogenic microbes. A child should not be bubble-wrapped out of fear of them catching the common cold or any other common pediatric infection.

While it's important not to constantly worry about a child catching a cold or other infections, it's also important not to let a child become a disease transmitter. If a child is sick, the best idea is to keep her at home, simply to limit the spread of disease. No one wants to see a child with a bad cold or worse, chicken pox, show up at a birthday party, although it won't terribly harm a child to play with a kid that has a runny nose in the playground. Plus, if a child is under the weather, why not let her rest at home, giving her the best chance at a speedy recovery?

The above answer, however, is given in the context of a Western society, where most people are vaccinated. Vaccines are an artificial way to expose children to harmful microbes without them getting terribly sick. It's only because of vaccines that children these days have a very low risk of catching serious life-threatening infectious diseases such as smallpox, polio, diphtheria, etc. Fifty years ago, allowing your child to play with a friend that had a fever could have exposed her not just to the common cold viruses, but also to meningitis, whooping cough, measles, and other serious diseases. If we were to live in a world where only a subset of the population was vaccinated, the advice here would certainly be different.

Likewise, if you have decided not to vaccinate your children, understand that your children are more likely to suffer serious life-threatening diseases, as well as to carry and spread them to others. Thus, it would be prudent to limit the contact your child has with other children when she gets sick. (See chapter 15 for more on vaccinations.)

4. What about kids touching dirty surfaces?

First, not all dirt is created equal, nor does it all pose the same risk of disease. It would be almost impossible to accurately know which dirty things have pathogenic bacteria and which don't by simply looking at them, but there are a few giveaways. If something smells bad, looks slimy, or looks inflamed (in the case of a wound), it's likely harboring nasty microbes. This is especially important with food, where disease-causing bacteria love to grow, so don't touch food that smells or looks like it's decomposing, and be vigilant about expiration dates and preparing and cooking foods properly. Kids shouldn't be allowed to touch wounds or bodily fluids in general, but especially if they come from someone who's sick.

For those of us who live in cities with lots of other people, it's a given that many of those people will have an infectious disease at any given time. Picture yourself with your children riding the New York City subway one afternoon. It's very likely that someone in the same train car as you has an infection, or that someone who rode the train immediately before you sneezed and left a lot of viruses to spread on the same window that your child eagerly touches while enjoying his ride. Does this mean that we should avoid trains or any other form of public transportation, or that we should frantically apply hand sanitizer every time we touch a heavily commuted surface? No, of course not. If we were so susceptible to disease transmission that a train ride in New York City would carry a dangerous risk, the human race would have been wiped out thousands of years ago. Our amazing immune system is strong and can deal with this type of exposure; however, it pays to follow hygienic practices in order to reduce the risk of infection in heavily populated areas. This means that it's a good idea to teach your children not to play on the floor in these places, nor to lick any surfaces, and to wash their hands (with regular soap and water) when they get home or before eating.

When children are out walking or playing in a green space, it's a different situation altogether, as the risk of getting infected with microbes that carry human diseases decreases drastically. Allow your children to touch anything they want (except animal waste), including dirt, mud, trees, plants, insects, etc. Don't act on the urge to clean them right after they get dirty, either; let them stay dirty for as long as the play session lasts or until it's time to eat. In fact, our children experience so little time outdoors compared to previous generations that it's ideal to encourage them to get dirty during the little time they have outside. Bring a bucket, some water, and a shovel the next time you're at a park or on a hike—it takes only minutes before most of them start making mud pies or decide to give themselves (and you!) a mud facial. If the dirt gets in their mouths, don't freak out; they'll soon realize that dirt doesn't taste all that good and likely won't develop a habit for it. Most kids have this innate desire to get dirty, but in their modern lives this needs to be nurtured. Do your child a favor and encourage him to play with dirt.

5. Is it necessary to sterilize milk bottles and, if so, until what age?

This may come as a shocker for many of us who grew up with the idea that babies should be given only sterile bottles, but the American Academy of Pediatrics no longer recommends sterilizing bottles used for babies of any age. It recommends only a stove top method (cleaning them in boiling water) or a dishwasher with a hot cycle when using the bottles for the first time or when the water used at home is not deemed safe to drink. If the water is safe enough to drink, it is also safe enough to use to clean bottles and nipples. The same goes for any utensil or plate used to feed babies solid foods, and for pacifiers and teethers: washing them with water and soap is enough. However, be aware that milk bottles and nipples need to be washed properly as milk residues get trapped in the nooks and crannies of bottles and their accessories, and bacteria can flourish there. A bottle brush is a good idea for a proper washing.

One recent study may provide parents with an incentive to become more lax about sterilizing bottles and food utensils. A group of scientists from the University of Gothenburg in Sweden analyzed data from over 1,000 children and found that children from homes that washed their dishes by hand were less likely to develop eczema (a skin disease) by school age. The study controlled for other factors known to decrease asthma and eczema risk, such as having a family pet or breastfeeding, strengthening the validity of their finding. This study suggests that a less-efficient dishwashing method promotes more exposure to microbes early in life, which has been shown to protect children from allergies and asthma. While we may not advise tossing out your dishwasher, perhaps doing dishes manually on occasion would be appropriate (and a good way to learn to appreciate your dishwasher).

6. How often should baby/child toys be cleaned and what should be used to clean them?

This question was a popular one among the parents we interviewed. It often came accompanied with suggested answers, such as "Every day or after every use?" or "With regular disinfectant or with bleach?" But really, it's not necessary to wash toys until they are visibly dirty or after a sick child has played with them.

As for what to use, soap and water is more than enough. Harsh chemicals such as the ones in disinfectants or bleach are not necessary for this type of cleaning or for cleaning the surfaces where children play, either. (This was one of the questions that made us realize how widespread the notion is that babies need to play and develop in a pristine environment.)

7. Are sandboxes unsanitary?

Kids love sandboxes, and it's not surprising to find a dozen children playing in one all at once, making them a popular playground spot that undoubtedly has a higher concentration of microbes than other playground features (there are a couple of studies showing this). This means that a child has a higher risk of contracting an infection in the sandbox than on the swings or slide. Does this mean that they should not go in the sandbox? Absolutely not! Sandboxes are great fun and the risk of contracting a disease from one is low.

However, parents and caretakers should follow hygienic practices, like handwashing, after using the sandbox. The other possible source of infection comes from the fact that a sandbox looks like a giant litter box to many animals (read: cats) and they will use it accordingly given the chance. For private backyards, covering the sandbox after use can easily prevent this, whereas at a public playground it would be wise to inspect the sandbox before letting a child use it. If animal waste is visible, scoop it out, along with a good amount of the sand surrounding it (most of us have changed a litter box at one point or another). If the sandbox looks like it's been used by all the cats in the neighborhood, don't let a child play there and contact the local authorities to have its sand replaced (cat feces can contain parasites which can then infect humans).

8. Should my child be allowed to put something in his mouth after it's been dropped on the ground?

In general terms, putting something that has fallen on the ground back in your mouth is just fine. However, not all ground surfaces are the same and common sense applies. If a child's toy falls on the subway or mall bathroom floor, it's a good idea to give it a rinse with soap and water first, but if it falls on the floor at someone's home or while out hiking, simply remove the visible dirt (and hair) and give it back to your kid.

In fact, a recent study by the same Swedish research group that reported the association between dishwashers and an increased risk of allergies, suggests that the best way to clean a pacifier that has been dropped is to put it in your own mouth first. In this study, 184 families were interviewed when their babies were six months old. Parents were asked the question: Does your child use a pacifier, and if so, does it get sterilized, rinsed in tap water, or cleaned by parents sucking on it? Surprisingly, they found that the sixty-five babies raised by parents that cleaned their pacifiers by mouth had a significantly lower risk of developing allergies at eighteen and thirty-six months of age. This small study remains to be replicated, but it seems that by sharing mouth microbes with their child, parents are strengthening their child's immune system and preventing the development of allergies. So instead of following the "five second rule" to pick something off the ground quickly, perhaps instead we need to follow a "five second rule" in mom or dad's mouth before returning a teether or pacifier to the child. (There may be a concern with parents passing cavity-inducing microbes to their children, but this appears to be an issue only with parents who are prone to tooth decay, which can be hereditary.)

9. Is antibiotic ointment necessary in treating scratches and cuts?

Not always. Cuts, scratches, and scrapes are part of being a kid and they happen all the time. If the wound is long or deep or the edges are far apart, or if it doesn't stop bleeding after a few minutes of applying pressure, seek medical attention. Otherwise, wounds simply need to be cleaned of dirt and debris by thoroughly washing them with soap and water (or immediately rinsing them with clean water, and washing more thoroughly with soap later, when available).

The recurrent use of a little dab of antibiotic ointment may not significantly alter a child's skin microbiota, but it adds to the unnecessary use of antibiotics, which leads to antibiotic resistance. To prevent developing an infection, keep the wound clean by gently washing it daily and avoid touching it by covering it with gauze or a bandage. If, after a day or two, the wound looks red and swollen, or is oozing yellow or green pus, consider using an ointment with antibiotics. If the redness around the wound expands or there are red streaks spreading from the wound, or if there's a fever, seek medical attention.

Luckily, most cuts and scratches heal rapidly on their own due to our immune system's ability to control infections.

10. Should I allow my child to eat unwashed fruits and veggies?

In most cases, you should wash produce. Fruits and vegetables are often consumed raw, which means that any contamination that occurred during farming or storage may come in contact with whoever eats them. The irrigation systems used to water many types of crops are known to contain dangerous pathogens, and washing fruits and vegetables is an effective way to significantly reduce the risk of foodborne diseases. Foodborne diseases, also known as food poisoning, are more likely to occur in certain groups of people, including children, the elderly, and pregnant women (they all have more vulnerable immune systems). The CDC estimates that about one in every six Americans get sick from food poisoning; each year 128,000 people are admitted to hospitals, and 3,000 people die from food poisoning. Thus, this is a serious risk that ought to be reduced by following hygienic practices.

Washing food that will be consumed raw is only one of the steps to prevent foodborne disease. Other important practices are to separate raw meat, seafood, and eggs from ready-to-eat food, to cook foods to the right temperature, and to chill or refrigerate perishable foods within 1–2 hours after purchasing them.

Another good reason to rinse fruits and vegetables is to wash out pesticide residues. Some people use a fruit wash solution for this purpose, although the European Crop Protection Association and the American National Pesticide Information Center both state that using these types of products is no more effective in removing pesticides than water alone.

The only type of situation in which we would consider it okay to allow a child to eat an unwashed piece of fruit is if it was grown in her backyard or garden, watered by rain or clean water from a hose, and free of pesticides. However, this shouldn't lead to the idea that it's okay to consume unwashed, store-bought organic produce because, contrary to popular belief, organic farming does not decrease the risk of food poisoning, although it does significantly reduce the levels of pesticides in crops.

Organic foods are often fertilized with manure, which can contain pathogens. For example, there was a major outbreak of *E. coli* O157:H7 (which causes serious diarrhea and kidney failure) in apple juice made with organic apples that had been exposed to cattle feces that contained this pathogen.

11. How can parents promote the development of a healthy microbiota through diet?

This is a good one to answer because there's no better way to influence the development of a diverse microbiota than through diet. Offering a healthy diet rich in vegetables and fiber is probably even more important than not being overly clean with babies and children. As we mentioned in chapter 6, when babies start eating solid foods they should be given a diet varied in vegetables, fiber, and fermented foods. A child can be exposed to many good sources of microbes while playing and interacting with people, but if these microbes are not fed the right foods, they won't flourish in a child's gut. If a child's diet is mainly based on refined carbohydrates (white flours and sugar) and high fats, his digestive system will digest and absorb most or all of the nutrients in the upper part of the digestive tract, leaving little nourishment for the vast numbers of microbes inhabiting the large intestine farther down. The microbiota in the large intestine feed on fibers and foods that are somewhat resistant to digestion in the upper part of the digestive tract and if none of that makes it down they will starve and diversity will decrease.

While offering babies lots of vegetables, legumes, fiber, and fermented foods is a good idea, convincing a two-, three-, or four-year-old to eat their carrots and celery is a whole different game. Once babies realize that they can make their own decisions, they will try to get only what they want, and upon tasting french fries or ice cream they will undoubtedly shun anything else, especially if its green. This is when teaching them good eating habits becomes extremely important and extremely hard at the same time. Even the most dedicated parent can easily give in after the thirtieth time their toddler asks for candy, especially if this happens late in the day when the patience levels are low.

Claire found that her daughter became a lot less resistant to eating all the healthy stuff when, one day, when Marisol was three and a half, Claire told her that she had a huge collection of little bugs in her tummy. She made the story very elaborate and whimsical; the bugs all had different colors and shapes, they sang songs, had parties, and simply loved living in Marisol's tummy. They actually called her tummy their home, and they were the happiest little creatures that ever lived. They also had superimportant jobs to do, like chopping up all the food she ate into really small pieces so it could reach the rest of her body to make her grow. Her little bugs were also her poop factory and they made sure that she recovered when she got sick, too (all only slight exaggerations from the real facts!). Claire also said that these bugs were always hungry because of how busy they were and that Marisol's job was to feed them every day. Without food they would starve, get supersad, and even die. But her bugs did not like ice cream, candy, hamburgers, or french fries; they loved lentils, broccoli, kefir, beans, carrots, and tomatoes instead. This, Claire told her daughter, is the reason why we need to eat vegetables and other foods that aren't as tasty as cupcakes. "It's not for you" Claire said, "it's for your bugs. You're their home, and it's your job to feed them." The change in her daughter's attitude towards the foods she didn't like was almost immediate. She gave her tummy bugs pet names, drew elaborate pictures of what she imagined they looked like, and agreed to feed them well.

It's been almost two years since then and Claire's daughter continues to eat her vegetables. To her, the story became engrained, something matter-of-fact, just like the fact that there are four seasons in every year, that Sunday comes after Saturday, or that Santa lives in the North Pole.

As Claire's children grow, she will likely have to modify the message to one with more realistic tones and details, but the core message is the same. The microbiota is a forest that we carry inside us and it is our lifestyle choices that determine whether this forest is stable and balanced or fragile and hungry. So it's important that children eat a diet that promotes a diverse microbiota and to establish good eating habits that will hopefully last for many years to come. Teaching them to eat well, as well as teaching them that being squeaky clean is not the right way to go about life are two crucial messages that children should get. It's those early years that matter the most when it comes to microbial exposure and the development of their immune system, so it's well worth the effort to change our preconceived notions regarding diet and cleanliness to promote a healthier future for them.

PART THREE

Collateral Damage

10: Obesity:
The World Is Getting Heavier

Body Weight and the Microbiome

Jack Sprat could eat no fat,
his wife could eat no lean.
And so between them both, you see,
they licked the platter clean.

We all know that increased body weight, especially in children, is a huge problem (pun intended). The statistics are downright scary: childhood obesity has more than doubled, and has even quadrupled in adolescents in the past thirty years. Between one-quarter and one-third of all American children are either overweight or obese—and the numbers continue to increase. There is also, not surprisingly, a direct correlation between childhood obesity and adult obesity. The average American woman now weighs the same (166 pounds) as the average male weighed in the 1960s (yes, men are equally guilty of major average weight increases). The problem is that packing on the pounds translates into serious health problems, including cardiovascular (heart) disease, strokes, diabetes, and cancer. Most health experts agree that it's the biggest epidemic facing the health of the developed world, and it has happened rather suddenly (within the past three decades). This time span suggests that it isn't due to a recent mutation in the human genome, but that something in the environment has changed to cause this sudden surge in human body weight. Ask any expert why this is happening and they will cite three factors: a change and increase in our diet, a decrease in our activities, and some unlucky genes.

Decreased activities are definitely a lifestyle change. Ask your parents or, even better, your grandparents what their daily routine was when they were your age. Answers might include hard physical labor on the farm, working outside, walking often, etc. Better yet, ask them what they did as a kid. Their daily activities probably would have consisted of a lot of running around, playing outside, bike riding to and from school and friends' houses, soccer, baseball, climbing trees, jumping rope, and so on. Now look around at kids these days. They are driven to school and they spend their spare time (and even class time) on computer screens, which all adds up, comparatively, to very little exercise.

Besides inactivity, the other factor affecting body weight is what we eat, and how much we eat. There has been a massive shift in our diet from unprocessed foods rich in vegetables and fiber to heavily processed foods with high levels of sugar. Corn syrup is a very inexpensive sweetener, and has found its way into many foods that were never before as sweet as they are now. High-calorie foods are cheap (especially fast food), and we consume a lot more of them than we used to. The average size of a bagel, cheeseburger, soda, or blueberry muffin has more than doubled in only twenty years, and children are getting used to these portions and adjusting their

appetite accordingly. In an effort to curb the mindless intake of liquid calories, whole cities have passed laws to limit the size of sugary
beverages that businesses can sell—a first for humankind.

So what actually governs how we process our foods and convert
them into energy? We have accepted that diet and exercise are the
main factors that influence weight, but is it as simple as that? This
formula doesn't explain how many people don't lose weight even
when they follow a strict diet and exercise regime. You've probably
guessed by now that the microbiome plays a major role in all this.
We know that diet changes result in microbiota changes. However,
could changes in gut microbes be responsible (at least in part) for
the obesity crisis?

We're now learning that even early life microbiota can have a
profound effect on a child's weight. In chapter 3, we saw that if a
pregnant woman gained more weight than normal, the child was
much more likely to become overweight. If a mother smokes, her
child is also more likely to become obese (this hasn't been directly
linked to the microbiota . . . yet!) Children born by C- section are at
a higher risk for obesity than vaginally delivered children. Children
fed infant formula are twice as likely to become obese than those
who are breastfed (see chapter 5). All of these events directly affect
the microbiota, and the gut microbiota play a major role in regulating energy extraction and metabolism from ingested food, affecting
both weight gain and loss.

Fat Mice

One of the most satisfying things in science is when fairly simple
experiments have obvious outcomes. A few such simple experiments

conducted by Dr. Jeff Gordon's group at Washington University in St. Louis and by other labs were fundamental in demonstrating once and for all that the microbiota have a major effect on body weight. Germ-free (GF) mice have 40 percent less body fat than normal (microbe-containing) mice, even though the GF mice consume 29 percent more calories than the control mice; that is, they eat more yet weigh less. GF animals can even be put on a high-fat diet and they're protected from obesity. However, if the GF mice are colonized with fecal microbiota from normal mice (mice are coprophagic, which means they like to eat their poop, making the experiment pretty easy), the newly colonized mice increased their body fat by 60 percent in two weeks, just by gaining gut microbes. This alone shows that microbiota really affects body fat.

To take it a step further, they colonized GF mice with feces from obese mice, and the newly colonized mice gained much more weight than GF mice colonized with microbes from a normal weight mouse. From this experiment we can see that the microbiome from obese animals is more efficient at harnessing energy from food than the microbiome from animals with a normal weight. That a simple fecal transfer can seriously affect body weight is a groundbreaking finding.

Of Mice and Men

Although we can't ethically repeat the above experiment in germ-free people, several studies indicate that similar things occur in humans. For example, using human twins in which one was obese and the other wasn't, Dr. Gordon's group found that by transferring the human feces into GF mice, the mice that were fed microbiota from the obese twin grew heavier and gained more body fat than those that received microbiota from the lean twin. They also found that the microbiota from the obese donors was less diverse in its microbial composition compared to that of the lean donors, and this was also true in the colonized animals. On an encouraging note, they discovered that if they transferred in a set of microbes from the lean donors, they decreased the weight gain in the animals with the obese microbiota; in other words, the lean microbiota won over the obesity-inducing microbes. Studies of overweight children also found differences in their microbiota compared to that of normal weight children, and showed that that these changes preceded actual weight gain. All these studies suggest that the bugs in our gut really do affect our weight. But how?

This is where things get more complex, and unfortunately we have to leave the simplicity and beauty of fecal transfer experiments behind. Several reasons have been proposed as to how and why microbes affect our weight, but at this time they're really just theories and, frankly, we don't yet know exactly how this all works. We know the microbes do the bulk of the hard work in breaking down our food (humans don't even bother having the genes that produce the enzymes needed for digestion of certain foods, as we know the microbes have it covered). As shown in the GF mice studies, the obese microbiome seems much more efficient at harnessing the energy from food. These microbes have more enzymes dedicated to food breakdown and energy harvesting than microbes from lean people. Some of these breakdown products trigger a hormone release in our body that affects whether we feel full or not (from a microbe's point of view, what a great way to get more food!).

Another potential reason involves inflammation, a condition that occurs in obesity, and which is thought to induce obesity-associated diseases like type 2 diabetes and insulin resistance. Inflammation is an immune response to pathogens or tissue injury. In the context of obesity, inflammation is thought to occur because high-fat diets trigger an increase in gut permeability, making the gut leakier and allowing microbes or microbial molecules to pass through the gut wall, which in turn triggers a general inflammatory state (this is discussed in more detail in chapter 12, where we talk a lot about gut permeability and gut diseases).

Many necessary details are still lacking in order to explain exactly how certain microbiota compositions favor weight gain and obesity, but the general concept that's emerging is that the gut microbiota modulate the body's ability to absorb energy through various methods, all as a result of the by-products of food degradation by the microbes.

A Microbiota Diet

So, knowing all this, can we tweak our microbiota to affect our weight? Although tempting, nobody has done the experiment yet in which you ask your thin spouse for his or her feces for a transplant (think Jack Sprat). However, attempts are now being made to directly alter the gut microbes, as well as the diet, that instigate changes in the microbiota. Prebiotics (dietary fibers that our body can't digest but that feed and stimulate particular microbes) have been given to healthy humans, and it turns out that they reduce hunger and make one feel full. This is probably because these prebiotics are modified by the microbiota, and they then affect the body's production of hormones, preventing a hunger signal shortly after consuming them.

The typical processed meal associated with obese individuals, on the other hand, contains only nutrients that are digested by human enzymes in the small intestine. By the time this meal reaches the colon, where most microbes live, there are very few food sources for them, which is thought to trigger hunger signals despite the body having obtained sufficient calories. Thus, it seems that in order to feel full you must not only feed yourself, but also your microbes.

Studies regarding infants and prebiotics are just getting under way, and it appears that they do cause an increase in beneficial microbes in young children. Some infant formulas now contain prebiotics in addition to probiotics.

Likewise, there have only been four randomized control studies done on probiotics in humans regarding body weight changes, and the data were inconclusive due to the small sample size. However, as we learn more about how microbes break down food, and the various roles they play in lean and obese individuals, it's quite likely that in the future we'll have probiotic-like mixtures of microbes that can be taken to decrease weight gain.

For now, we do know that when obese humans are put on a fat-restricted or carbohydrate-restricted low-calorie diet, a beneficial shift in their microbial population occurs, moving it towards a microbiota composition that promotes less weight gain. These changes occur rapidly, usually starting within twenty-four hours. It's also been shown that the physical responses of overweight adolescents to diet and exercise weight loss programs depends on their microbiota composition prior to treatment. This probably explains why some individuals are more successful on diets than others. It also again emphasizes how important the microbiome is in affecting body weight.

Antibiotics and Childhood Weight

As we saw in chapter 7, antibiotics are very good at killing microbes, both good and bad. And while these drugs remain wonderful at controlling serious bacterial infections, we must discuss a major dark side to them: they seem to promote weight gain, and their use may be directly contributing to the obesity epidemic.

Approximately seventy years ago, veterinarians made the observation that using antibiotics in subclinical doses (amounts that are less than would be used to treat an infection, but that will still affect some microbes) caused animals to gain weight by 10–15 percent. This effect is seen in pigs, sheep, cows, poultry, and even fish. It's made a huge difference in massive farming operations, which dose their livestock with antibiotics in order to get more meat off their chattel. This practice has become a cornerstone of agriculture in North America, and it now accounts for more than 80 percent of antibiotic consumption. However, this practice has dramatically increased the rate of antibiotic resistance, which is a major issue. It also leads to large amounts of antibiotics entering the environment. Europe has wisely banned the use of antibiotics as growth supplements in animals, but the United States and Canada stubbornly refuse to follow suit.

Initially it was thought that the antibiotics might control infectious microbes, thereby decreasing infections and allowing the animals to grow more quickly. However, the reason now appears to be more complex, and it's related to microbiota changes. Studies of animals showed that subclinical doses of antibiotics, no matter what kind, do indeed alter the microbiota to a population more conducive to weight gain, including an increase in energy-harvesting microbial genes. Several experiments also suggest that the weight gain is more pronounced if the antibiotics are given early in life, rather than later, hinting yet again at the critical role of early-life microbiota.

At the risk of offending proud parents, what do the results regarding pigs, sheep, cows, and chickens have in common with antibiotics and children? Recent evidence suggests quite a bit (plus common sense dictates that if it happens in so many diverse animals, it would logically affect humans, too—we are just animals, biologically speaking). Some compelling data show that the states in the US with the highest antibiotic usage also have the highest obesity rates. In a large Danish study involving more than 28,000 mother-child pairs, antibiotic exposure during the first six months of life was associated with an increased risk in the child being overweight at age seven, especially if the mothers were not overweight. In a Canadian study, antibiotics administered in the first year of life increased the likelihood of a child being overweight at nine and twelve years of age. The list of studies goes on and on, and they all overwhelmingly point to the fact that antibiotics, especially given early on, affect the microbiota, which in turn increases weight gain and risk of obesity.

We don't know as much about the effects of low-dose antibiotics in humans. The animal data are extremely convincing though, and presumably the high levels of antibiotic use in both society and agriculture suggest that even children who don't directly receive antibiotics may be inadvertently exposed to smaller doses that could still affect their weight. These exposures could come from environmental sources such as water—remember, tons and tons of antibiotics are used every day in agriculture, and consequently end up in our groundwater—or even from eating meat from animals raised with antibiotics.

Malnutrition

In direct opposition to the obesity epidemic that burdens wealthier nations, malnutrition continues to be a major problem in poorer areas of developed countries and worldwide (although obesity is surging in some of these areas as well). Malnutrition has major detrimental effects on a child's physical and mental development, including stunted growth and even impairment in brain development. Historically it was assumed that malnutrition was the result of a lack of calories, and the solution was to simply provide more food. However, this solution often does not work (it has been tried many times by feeding children in impoverished areas without success). A study done a few years ago showed that if the children were treated with antibiotics first, then many more of them gained weight, hinting at the role of the microbiota. Experiments have been conducted in which feces were taken from Malawian twins, one of whom was extremely malnourished and the other not, and transferred into germ-free mice. Similar to the results of the obesity studies discussed earlier, it was found that this fecal transfer also transferred the malnourished characteristics to the mice, which strongly supports the idea that the microbiota has a large role in malnourishment.

Work done in our laboratory has also confirmed the role of microbiota in moderate malnourishment. In an effort to develop a realistic animal model with which to study this major worldwide childhood problem, we fed mice two different diets that contained an equal numbers of calories, but were either rich in protein and fat (typical of a Western diet) or high in carbohydrates (typical of a developing country diet). As expected, changing the diet alone was not sufficient to mimic the features of malnourishment. However, we know that children in developing countries frequently live in a less sanitary environment, so are often exposed to feces. We also know that these children have more microbes in their small intestine (just below the stomach) that resemble the microbiota normally found lower down in the large intestine. These children presumably acquire these microbes orally via fecal contamination in their water or other sources.

Based on this, we fed mice feces from other mice that were on the two diets, and found that, remarkably, all the features of a malnourished child were found in the mice with a high carbohydrate diet, as long as they also were fed feces. We were also able to identify select microbes in the feces that caused this effect (we tried many, including probiotics, but only specific ones had the effect we were looking for). This suggests that certain microbes play a major role in malnourishment and that we finally have a good animal model with which to study malnourishment. Hopefully we can use this knowledge to develop therapies for this major global problem in the future.

Anorexia Nervosa

To be the thinnest. That is the tragic and tormenting goal of an increasing number of young girls and boys, mainly in affluent cities around the world (although it's starting to become more widespread). Anorexia nervosa has been called a silent epidemic because there's little awareness about this disorder, despite the fact that it has seen an increase of more than 50 percent in the past five years in North America and the UK. Anorexia nervosa (also called just anorexia) is a neurological condition that is characterized by self-starvation. It often occurs with other neurological issues such as depression (up to 80 percent of people with this disease also suffer major depression) and anxiety (75 percent have anxiety disorders). It's most common in adolescents, affecting 3 million Americans.

Unfortunately, anorexia has serious side effects on the heart as well as the entire body, and has a tragic 5 percent fatality rate, the highest death rate of any psychological disorder. Treatments for anorexia always include dietary interventions, but they're not always effective and relapses often occur. Recently, it has been suggested that the microbiota may be involved, based on two lines of reasoning: 1) the neurological involvement in the disease (which microbes affect; see chapter 14), and 2) the major weight-loss issues. A few small studies have indicated that the microbiota in patients with anorexia is different than that of control subjects.

A recent study out of North Carolina looked at fecal samples from sixteen anorexic women when they were admitted for treatment, and again when they were discharged (having reached 85 percent of normal body weight). The study found that the microbes were quite different between sample times, and that the microbe population in the women when they were admitted was not nearly as diverse as when they were released (although it still didn't reach the diversity seen in healthy people). As the patients were treated and gained weight, the study also found that their moods improved, and it discovered a correlation between microbiota diversity and the presence of certain microbes and a decrease in depression and anxiety. Again, this study doesn't prove that microbes cause anorexia and its associated depression and anxiety, but a strong correlation can be made. These types of studies will pave the way to more extensive analysis and ultimately determine whether microbes can affect the outcome of this tragic disease.

Given the role microbes play in food metabolism and weight gain/loss, and their impact on depression and anxiety (chapter 14), it's most likely that microbes will play a central role in managing this disease in the future.

Dos and Don'ts

◆ **Do—** avoid unnecessary exposure to antibiotics during pregnancy and early childhood. Antibiotic usage is increasingly associated with obesity, and, just like farm animals, our children are gaining weight at a much faster rate, setting them on a course for obesity and all its problems. On the other hand, antibiotics are a wonder drug for serious bacterial infections, and need to be used in certain cases. If your child is treated with antibiotics, you should consider various measures to promote the health of the gut microbes after the treatment is stopped. This could include breastfeeding, prebiotics, probiotics, and a varied diet rich in plant fibers.

◆ **Don't—** let your children spend their days in front of electronics. Get them out of the house, and promote physical exercise through walking, a trip to the playground, or an organized sport or other activity. And don't let the seasons be an excuse to keep your family indoors: swimming is a great way to keep cool during the summer and ice-skating can warm up your toes during winter months.

◆ **Do—** purge your kitchen cupboards of junk food and stock your shelves with healthy foods. Throw out those sweetened beverages, too, opting instead for plain water. By eating healthier you're not only treating obesity, you're preventing it by giving the microbes in your lower intestines something to eat so they don't send you signals to keep eating!

◆ **Do—** start reading the labels on the meat, eggs, and dairy you buy, and opt for products from animals not treated with antibiotics. Sometimes these products cost a little more, but it's worth it in the long run, especially since so much is still unknown about the effects of the antibiotics we consume through our food.

◆ **Do—** engage your children in the story of their gut. Even at a young age, your child can understand that there are good bugs in his tummy and they work hard to keep him healthy. Making our children responsible for their bodies and health at a young age is a great step towards a long life of good habits. Plus, they might be inclined to eat their vegetables because they want to, not just because you told them to.

THE 5210 DIET

Unfortunately, despite what you read on the Web, there's no magic bullet for dieting. We can't control our genetics (blame your parents), but we can control diet and exercise, and these are the two factors we have to focus on, especially as parents. Increased physical activity and decreased screen time are critical for both burning calories and helping the body develop (we know that exercise promotes a favorable microbiota). Eating healthy foods with plenty of plant fibers and avoiding sugar, such as that found in sweetened beverages, is also key to a healthy diet.

A dietary and lifestyle program called 5210 promotes just that. It's easy to remember the guidelines: 5210 suggests that every day your child should eat at least five fruits or vegetables, spend two hours or less on screen time, have at least one hour of physical activity, and consume zero sweetened beverages. It's also nice because you can hold up the number of fingers on one hand for each category to help your child count. The program is equally applicable to encouraging a healthy lifestyle for adults (although many of us would have to quit our jobs to avoid all that screen time).

The fact is, if you can maintain your child's weight in an optimal range for their age and height, you're doing them a huge favor for later in life, both in terms of maintaining a healthy weight, but also in terms of preventing major diseases such as diabetes, cancer, cardiovascular disease, and stroke.

11: Diabetes:
Microbes Have a Sweet Tooth

A Disease on the Rise

Glucose is the most widely used sugar in living organisms and is our body's main energy source. It's taken from our blood by our cells in order to energize them. However, glucose cannot get into the cells by itself. The pancreas releases a hormone called insulin right after we eat, and it attaches to the cells just like a key attaches to a door. Insulin signals the cells to start absorbing glucose and in this way it regulates sugar levels, keeping them from becoming too high or too low. An excess of glucose in the blood is not good, and if these levels remain high for a prolonged time, it can turn into a disease called diabetes mellitus (*mellitus* means "honey-sweet" in Latin), but we usually just refer to it as diabetes.

There are three main types of diabetes: gestational diabetes, which occurs during pregnancy with no prior history of the disease; type 1 diabetes, in which the body destroys specific cells in the pancreas that produce insulin, so cells are not stimulated to take up glucose; and type 2 diabetes, which causes the cells to become resistant to the effects of insulin—this particular form is closely linked to obesity. All three types of diabetes result from high blood glucose levels because cells are not able to absorb the blood glucose for various reasons. The hallmark symptoms of diabetes are frequent urination (because the high sugar levels pulls more water out of the body and into the urine) and a marked increase in thirst (since the body is trying to replace the fluids that are being lost). High blood glucose levels can cause serious long term complications that include cardiovascular disease, kidney failure, foot ulcers and amputation, strokes, and blindness.

It's estimated that more than 380 million people worldwide have diabetes (there are an estimated 100 million cases in China alone), and this is expected to grow to 600 million by 2035, mainly due to type 2 diabetes, associated with increased obesity rates. The disease is thought to have killed over 5 million people in 2013 (to put this in perspective, HIV kills about 1 million people yearly, and a total of about 54 million people die worldwide each year), and up to one-third of the population in some areas of the world suffer from diabetes. Given that diabetes involves sugar uptake, which comes from food being digested in the gut, and also involves the host immune system, it's no surprise that the microbiota is increasingly thought to play a role in this disease.

A Sugarcoated Pregnancy

It's estimated that between 2–10 percent of pregnant women temporarily develop gestational diabetes. It resolves almost immediately after birth, but about 10 percent of these women will go on to have type 2 diabetes later in life. Once it's detected, gestational diabetes can be managed with diet, and sometimes insulin. The trick to detecting it lies in identifying an elevated glucose level, hence the urine and blood tests for sugar that are administered during pregnancy. For the glucose tolerance test, a mother-to-be must drink a horrible-tasting bottle of glucose solution, and precisely two hours later, her blood is sampled to see how the body handled this sugar load. When pregnant, there's a significant increase in the body's energy production (and consumption) in order to feed the developing fetus. Although it's not known exactly why some women are unable to control their glucose levels during pregnancy, it's speculated that the placenta somehow affects the body's sensitivity to insulin.

There's limited information about how gut microbes might affect gestational diabetes. Nutritional counseling and proper diets play a role in controlling this disease, which would of course also affect the microbiota. In one study of 256 women in their first trimester of pregnancy, women who received both nutritional counseling and standard probiotics (*Lactobacillus rhamnosus* and *Bifidobacterium lactis*) had the lowest rates of gestational diabetes, with decreased blood glucose levels both during pregnancy and for a year after.

We're still in the early days of defining the role of the microbiota in this disease, but the microbial contribution to energy production is well established, so in the future there may be ways to optimize this relationship in order to decrease the risk of diabetes in pregnant women.

Finger Pricks and Insulin Pumps

The hormone insulin is produced by specialized cells in the pancreas (called beta islet cells) and is then secreted into the blood, where it promotes glucose uptake by cells in the body. In some people, the body's immune system attacks the beta cells (a type of autoimmune reaction), destroying them and thereby stopping normal insulin production. When this happens, you have type 1 diabetes (T1D).

T1D is usually diagnosed in people younger than thirty, so historically it's been referred to as "juvenile-onset diabetes." It's one of the most common metabolic disorders in children and young adults. The prevalence of this disease has doubled in the past twenty years, and is set to double again by 2020. In Europe, it's increasing 3–4 percent per year, with the fastest rate of increase in children less than five years old.

Fortunately, patients with T1D can have a nearly normal lifestyle by regularly checking and monitoring their blood glucose levels with finger pricks and injecting insulin every day, or by having an insulin pump surgically implanted. A promising new therapy involves transplanting in new islet cells, which can then produce the needed insulin.

Like many diseases, there's a genetic component to T1D, and several genetic markers have been linked to increased susceptibility. However, less than 10 percent of those who have these susceptible markers will develop the disease, which hints at—you guessed it—environmental factors such as microbes playing a role. We also know that the rapidly rising rates cannot be explained by genetic changes alone, as humans just can't genetically change that quickly.

Other clues that suggest microbiota involvement include: the increased risk of having the disease if one is delivered by C-section; dietary changes early in life; and perhaps antibiotic use (animal data convincingly show that antibiotics increase risk, but it hasn't been proven in humans yet). The theory that breastfeeding decreases the risk of T1D remains controversial.

The microbiota of children with T1D is different than the microbiota of those who don't have it—it's less diverse and less stable, and lacks the microbes that produce butyrate (an anti-inflammatory molecule that improves gut health). These differences are also seen in prediabetic children, indicating that microbial changes precede the disease. Although the number of studies is still small, using animal models of this disease shows very strong evidence that gut microbes play a role, as we've seen the microbiota undergo changes in diabetic animals. Interestingly, antibiotic treatment of mice that are genetically prone to this disease actually protects against T1D.

As we will see in the next chapter, gut permeability (or gut leakiness) is an issue of many intestinal diseases. However, it also seems to contribute to T1D. People (and animals) who have the genes that make them more susceptible to T1D also have increased gut permeability. How this might contribute to diabetes is not yet known, but a leaky gut may allow microbial molecules from the gut to get through and somehow affect the body's immune response to insulin.

Early-life diet also seems to play a large role in T1D by modulating the body's immune system and encouraging its attack on its own beta cells. If infants who have the genetic risk factors for T1D are weaned on extensively broken down or hydrolyzed casein formulas, they have a decreased risk of developing T1D by age ten. On the other hand, if infants have a short breastfeeding period and are then fed cow's milk in nonhydrolyzed formulas, they have an increased risk of T1D later in life. Furthermore, in mice that are at risk for T1D, a gluten-free diet had a dramatic reduction in the disease, as well as an accompanying microbiota shift. How all this ties in to affecting the body's attack on its islet cells still isn't exactly understood. But we know that the body's immune system is responsible for this autoimmune attack, and that early-age microbiota play a key role in the immune's system development, which may contribute to T1D. However, microbiota also affect gut permeability, which could also affect disease. Unfortunately, we also don't yet have reliable data on whether probiotics could influence T1D later in life.

The Western Diet: A Life Too Sweet

As we saw in the previous chapter, due to high-calorie diets and lack of exercise, people these days are gaining too much weight. One of the most direct consequences of obesity is type 2 diabetes (T2D). It's actually quite difficult to uncouple the two diseases as they generally go hand in hand. T2D begins with the body's cells becoming resistant to insulin, which makes it not as effective at triggering glucose uptake. As time passes, the body may also decrease insulin production, and the liver may increase glucose production. Together, these cause increased blood glucose, leading to T2D and all its awful side effects, such as foot amputation and blindness.

T2D now accounts for more than 90 percent of diabetes cases in adults. Originally thought to be a disease that only affected adults, with the surge in childhood obesity, T2D is now appearing in children as young as three years old. We've also discovered that children born from mothers with gestational diabetes have an increased risk of T2D later in life.

Not surprisingly, just as the microbiota is linked to obesity, gut microbes play a central role in T2D. If feces taken from obese mice are transplanted into GF mice, they develop higher insulin resistance. Several microbiome studies have been done on humans with T2D, and the results show that the changes to their gut bacteria are very similar to those seen in obesity, which is certainly not surprising. Again, as we saw with T1D microbiota, there's a decline in butyrate producers; presumably there's a lack of butyrate as well, and a corresponding increase in microbes that could cause diseases.

A major role of butyrate is to dampen inflammation, and continual low-grade inflammation is a hallmark of both obesity and T2D, so lacking the organisms that make butyrate may contribute to these diseases. Feeding butyrate directly to mice was shown to improve their insulin sensitivity. Butyrate also decreases gut permeability, which may help prevent inflammation by keeping pieces of bacteria from passing across the gut wall and triggering inflammation.

There's increasingly good data to suggest that the microbiota associated with glucose tolerance play a role in the early stages of T2D. Some scientists have gone so far as to say that microbiota analysis could be used as an early diagnostic for T2D, since some of the changes occur before the full-blown disease. However, there are significant differences in gut microbe composition in different areas of the world due to diet, genetic background, age, cultural differences, etc. For example, when the gut microbes of European women were compared to those of Chinese women, the major differences made finding common marker microbes difficult. However, in both populations changes were noted in the microbiota of those with decreased insulin sensitivity, and in both groups there was a decrease in butyrate producers and an increase in potentially pathogenic microbes.

Can we manipulate the microbiota to affect the rates of T2D? There are several studies that suggest this could be possible. In one series of experiments, feces from lean male donors were transferred into obese males with poor insulin sensitivity, and six weeks later they found a significant improvement in insulin sensitivity in the recipients. They also saw an increase in microbial diversity, as well as an increase in butyrate producers. However, the effect seemed to depend a lot on the particular fecal donor, as not all lean donors had the same effect. These studies suggest that once we figure out the right bugs to transplant, this might be a viable therapy to decrease T2D.

Diet is another obvious, promising way to change the microbiota. In one study, six obese volunteers with T2D were put on a strict vegetarian diet, which improved their insulin sensitivity, and their microbiome shifted to a more regular composition. Likewise, consuming probiotic yogurt for six weeks led to a marked improvement in T2D patients, causing their circulating glucose levels to decrease. Probiotics could possibly be used at treatment, since they're known to tighten up gut permeability and decrease inflammation, which could help resolve the disease.

Results using diet, pre- and probiotics, and metformin treatment of T2D (see Drugging the Bugs—Metformin, page 170) are certainly causing pharmaceutical companies to sit up and take notice. In the future, we'll likely see more microbiota-altering therapies developed, not only for T2D, but for several diseases of the Western world.

Dos and Don'ts

- ◆ **Don't—** let your diet—and that of your children—get out of control. The complications of obesity and T2D are terrible, and they last a lifetime. Eating healthily and including foods that encourage diverse and healthy microbiota are important, not only for diabetes, but for many health-related problems throughout life.

- ◆ **Do—** have your blood sugars checked regularly during pregnancy if you're deemed high-risk, which includes excessive maternal weight gain, a large baby, excessive amniotic fluid, or being older than thirty-five. If undetected, gestational diabetes can lead to a very large baby, which can then lead to a C-section and/or obstetrical trauma (damage) to the mother. After delivery, these babies' sugar levels plummet, requiring serious medical attention, such as intravenous sugar intake. Gestational diabetes can be controlled if detected, but it has to be detected first.

DRUGGING THE BUGS – METFORMIN

Metformin is a drug commonly used to treat T2D by lowering blood glucose levels. Despite being approved for human use and being used extensively, exactly how this drugs works is not known, but there are strong hints that it may act via gut microbes. If the drug is delivered directly into the blood (intravenously), therefore bypassing the gut, it doesn't work. Also, the drug isn't effective in mice treated with antibiotics. Metformin causes a profound shift in the gut microbe composition to a healthier profile, and at least one of these healthier microbes can be given directly to mice in order to decrease T2D.

A recent large international study showed that T2D patients taking metformin had a different microbiome profile than those who were not taking the drug. This included an increase in microbes that produce short-chain fatty acids, which are known to decrease blood glucose levels.

We often don't take the gut microbes into consideration when we're thinking about a drug that works well on humans, but the concept of drugging the bugs is a new and exciting angle to potentially treat diseases.

12: Intestinal Diseases: Fire in the Gut!

The Gut: A Thirty-Foot Tube, but Mind the Gap

Nearly every discussion about the microbiota includes some aspect of the gut, including gut health and gut diseases. As we've seen, this is where incredible numbers of microbes happily live while we feed and water them every day and, for the most part, both we and the microbes seem happy with the arrangement. However, the intestines also provide an important barrier between the microbiota and your body, and sometimes there are problems with this barrier, resulting in nasty diseases. We've all heard of inflammatory bowel diseases (IBDs for short) such as Crohn's and ulcerative colitis, which are characterized by severe inflammation of the gut.

However, there are several other less obvious gut problems that the microbiota also seem to have a hand in. Some of these include colic (yes, think screaming, wailing infants and ultrastressed parents), celiac disease (gluten intolerance), and irritable bowel syndrome

(IBS). We're beginning to realize that the gut microbiota has an important role in all of these diseases.

You've probably never given it much thought, but the intestinal tract is an amazing organ that plays a critical role in our body. At a gross level (pun intended), we have a thirty-foot (nine-meter) tube running though our body that starts at the mouth, hooking up to the stomach, then the small and large intestines, and finally ending at the anus. Anything within that tube is actually not considered "inside" us, but transiting through us. The gut, including the intestine, has two main functions. The first is to form a barrier to keep everything we ingest inside the tube (food as well as all the microbes). However, its other function, which is contrary to its barrier function, is to digest and absorb nutrients and fluid from the intestine. Thus it has a tricky balancing act of being both permeable to things we want to take up, but impermeable to things that we don't. Luckily, nature has sorted this problem out, and a normal gut performs both of these functions very well.

Proper intestinal barrier function is critical for health. If spaces between cells widen, the permeability increases, and the general contents of the gut (including microbes and their molecules) can directly enter the body. This triggers a strong inflammatory reaction from the body, which is a common feature in IBD. Ironically, inflammation also seems to increase gut permeability, which can worsen the symptoms.

Similarly, as we will see when we discuss celiac disease, increased gut permeability presumably lets through more food particles, such as gluten, which causes the body to react to it. However, it seems that the presence of certain microbes tightens up the gaps between intestinal cells and decreases gut permeability. Studies show that the colonization of germ-free (GF) mice with bacteria early in life seals up their guts, while those that remain microbe-free have leaky guts—indicating that this is yet another function we rely on our microbes for.

To enhance its ability to absorb and transport fluids and nutrients, the gut has a very large surface area. Several reports suggest that if the intestine were flattened, it would cover an area nearly the size of a tennis court (850–1,000 sq. ft. or 260–300 m²)! A more recent study, using sophisticated microscopy and measuring techniques, showed that the human gut has a surface area of "only" 100 sq. ft. (30 m²), about the size of a studio apartment. Still, imagine cutting out a very thin cloth the size of an apartment and stuffing that into a tube the width of a sock. Anyone who has tried to roll up a tent and put it into its stuff sack will know exactly what we mean! The way the body achieves this remarkable feat is to have folds in the cloth (called *villi*), and then have many tiny fingerlike projections (called *microvilli*) on each fold. Think of a shag rug—each little projection on the rug is like a microvillus finger. You can see how this increases the surface area remarkably. However, remember that this entire area is exposed to microbes, and it cannot have any holes in it at all. Oh, and now also make each projection move in unison—that is how gut motility works. Finally, to move things along, the body coats the entire large intestine in mucus, a slimy substance made of proteins and sugars. Mucus also keeps many of the microbes at bay, at a distance from the microvilli. A thick mucus layer is associated with a healthy gut, and in intestinal diseases such as IBD or diarrhea it becomes thin. Many types of bacteria feed on mucus, and this actually helps maintain a thick mucus layer, as the body produces more in response to microbial mucus munching (mmm . . .).

As we've mentioned before, babies are not born with a fully functioning gut. In fact, quite the opposite: at birth babies are pretty much sterile, their gut is leaky, and their microvilli are not fully formed yet. Microbes kick-start many physiological processes in the gut, but the initial weeks of life are a period of adjustment while the gut settles into a more controlled environment. Recent studies point to the assembly of the initial microbial communities in a baby's gut as a factor that influences how a baby's intestines adjust to the first weeks of life.

For Crying Out Loud

We all know babies can cry quite a bit, and this noise certainly gets our attention (plus that of everyone else on the plane). It's been suggested that the reason a baby's cry is so noticeable is just that—she ensures she isn't ignored, since she can't speak up when her diaper needs changing or when she's hungry. However, some babies seem to cry nearly all the time, and this is called colic, something that can turn the early days of parenting into a complete nightmare for even the most doting parents.

Anamaria and Pedro were at their wits end. Their beautiful daughter, Sofia, was one month old and she cried day and night. Sofia would often cry so hard that she would turn blue. Their poor little baby would cry to the point of losing her voice until she eventually fell asleep, completely exhausted. Even while she was sleeping, Anamaria and Pedro would notice that Sofia was uncomfortable and fussy, and she would often wake up crying. During her whole second month of life, Anamaria recalls Sofia as either crying or sleeping; there were no cooing sounds, no baby grins or relaxed playtime.

To make matters worse, Sofia would flare up in rashes around her mouth and her chest. *This just can't be normal*, they thought.

As first-time parents, Anamaria and Pedro knew that taking care of an infant was supposed to be hard and that babies cry a lot, but Sofia acted like she was in pain all the time. Anamaria asked her mother for help and advice, but her mother just kept saying that Anamaria's brother was like that when he was a baby and that colicky babies simply need to be constantly held. Despite her mother's well-intended advice, they felt they needed medical help.

Throughout the following two months, they went to a total of five pediatricians to seek advice. The first three doctors dismissed their worries by saying that some babies just cry, and that Anamaria and Pedro were stressed because they were new parents and tired. Another doctor finally diagnosed Sofia with severe colic—a condition defined as excessive and inconsolable crying lasting three or more hours per day. He suggested that Anamaria's milk was causing Sofia's stomach pain, making Anamaria feel awful and guilty. He also said that she should stop breastfeeding and start using an extensively hydrolyzed (broken down) hypoallergenic formula, which they immediately did. The first time they tried the formula they could see an improvement in Sofia's colic, which gave them hope that the worst was over.

They followed up with a visit to a physician who specialized in pediatric gastroenterology. She tested a sample of Sofia's feces and when the results came back she determined that Sofia was allergic to a protein in cow's milk. The doctor was emphatic that breastfeeding should continue, but Anamaria had to follow a strict diet, avoiding dairy and soy products. Seriously committed to resume breastfeeding, Anamaria followed the diet and began to pump milk religiously. However, it had been a few weeks since she had stopped breastfeeding and despite all her efforts, she wasn't able to keep up with Sofia's feeding needs. As an alternative, the doctor suggested a different type of formula, to which Sofia did not react well at all. Her crying resumed and her rashes came back. After trying a couple more formulas, they found the right one for Sofia.

Anamaria and Pedro are not alone; one in five families deals with babies with severe colic, a condition that continues to increase in incidence around the world. Infant colic peaks at around six weeks and wanes after the infant is 3–4 months old. Although it lasts only a couple of months, it can be a devastating ordeal for the whole family and it has serious consequences, as severe colic can lead to parental emotional distress, anxiety, depression, and, on occasions, even child abuse. These families often require psychological therapy to deal with the stress related to colic.

In a sense, Anamaria and Pedro were lucky because they found the cause, but only about 5–10 percent of colic is a result of cow's milk allergies—the other 90–95 percent of cases have no known cause. There's no correlation with the sex of the infant, mode of delivery, breastfeeding, or birth weight. However, the microbiome still comes into the picture.

Studies of very young infants (in their first one hundred days) showed that infants with colic have a decreased diversity in their microbiota (remember, microbiota diversity is a good thing), plus a decrease in certain infant microbes acquired from the mother (*Lactobacilli* and *Bifidobacteria*) and an increase in bacteria that potentially produce gas and intestinal problems (*Proteobacteria*). They found that these differences start at one to two weeks of age, and that they were evident in all kids by one month of age, which is before colic sets in.

The types of bacteria that infants with colic lack are responsible for making substances that have anti-inflammatory effects and decrease pain (butyrate). In addition, colicky babies have more microbes that cause intestinal inflammation and gas production. The change in intestinal microbiota may also affect gut motor function, which could contribute to abdominal pain (and more crying). Thus, it's now being suggested that colic (at least those cases not caused by cow's milk allergies) could be thought of as a microbiota problem.

So, the million-dollar question to the suffering parents of children with colic is: If the problem is with the gut microbes, can we fix it? Maybe. Two recent small studies using a probiotic (*Lactobacillus reuteri*) found that it decreased crying by twofold (heck, one and a half hours of crying versus three hours? Sign us up!). However, another recent study used the same probiotic and found it had no effect. Clearly the jury is still out regarding the treatment of colic with probiotics.

There are two main messages we can take home here: 1) the microbiota is different in kids with colic, and this change is detectable right after birth, even before colic sets in; and 2) altering the microbiota, if done right (and early in life) might just improve it. As with so many of the topics in this book, we're just realizing that the microbiota play a pivotal role, and we still don't know the exact good and bad bugs involved, or the perfect combination of bugs to fix it. The fact that studies are under way to test the administration of probiotics to prevent colic is of little consolation to those parents who have an infant with colic right now, but hopefully we'll know more soon—or the kid will grow out of it.

Chewing on Gluten: Microbes and Celiac Disease

Gluten is a natural complex protein (actually two proteins together) that is found in wheat, rye, barley, and other grain products. It gives bread its chewiness and helps bind baked goods together. The next time you're at a pizza house watching the staff toss a pizza crust high in the air and stretch it, remember it's the gluten that's holding it all together. However, wheat gluten has increasingly been associated with immune reactions against it, including celiac disease (barley and rye have similar proteins that can also cause this disease). Between 1–3 percent of the population suffers from celiac, and this disease has increased more than fourfold in the last fifty years, lumping it in with the ever-growing incidence of "western lifestyle" diseases. It requires a human genetic component (which means you have to have the right genes to potentially get it), as well as environmental factors (which include the microbiome).

About one-third of the human population carries the genes that are a prerequisite for celiac (called HLA-DQ2 and HLA-DQ8). Having these genes means you're at risk for the disease, but it certainly doesn't mean you'll get it. It's a disease of the small intestine, where the presence of gluten triggers a strong immune response, causing gut inflammation and damage to the surface of the small intestine. The symptoms are miserable, and include diarrhea, abdominal pain, and weight loss. They can also be subtle, causing stunted growth in kids and iron deficiency. The most obvious way to handle this disease is to avoid foods containing gluten, also known as a gluten-free diet. Such diets are usually quite effective at treating the symptoms, but not always. Because wheat is a moderately recent addition to the human diet (we figured out how to cultivate it about ten thousand years ago), gluten-free diets are sometimes considered to be more like the diets we evolved on (scavenging nuts, seeds, etc.), and are thus becoming popular in some circles, even for those that don't have celiac disease.

Several smoking guns implicate the microbiota in this disease, but unfortunately, we don't have the full picture yet. A major risk factor for celiac disease in children (assuming they have the right genes) is an infection or being treated with antibiotics, especially penicillins and cephalosporins, in the first year of life. However, taking antibiotics during pregnancy did not seem to affect the risk of the child developing the disease.

Having an elective C-section also increases the risk, but an emergency C-section does not, which at first seems weird. However, many emergency C-sections are done in the second half of labor, once the baby is on its way down the birth canal and after the membranes ("water") are broken. It's thought that during such C-sections, the infant would have already been in the birth canal and thus was exposed to the maternal microbiota. Breastfeeding is also recognized to decrease celiac disease.

When the intestinal microbiota is studied in people with celiac disease, differences are found when compared to people without the disease, but the results vary widely between studies and there's no consensus regarding the composition of a celiac microbiota other than that it's different. Then again, celiac sufferers have an inflamed gut that would contain different microbiota anyway. Also, celiac disease is a disease of the small intestine, and fecal samples (from which these studies are done) are more indicative of large intestinal microbiota (the small intestine is much harder to sample). One trend that does come through is that the microbiota changes are generally similar to those described above for colic, with decreases in the good bacteria and increases in the inflammatory ones.

Putting people on gluten-free diets often, but not always, re-stored their microbiota to a more normal type, but again this could simply be due to the decrease in gut inflammation. Similarly, people on gluten-free diets who still show symptoms have altered micro-biota, as well as gut inflammation.

What might the microbiota be doing? Again, there's little hard data on this, and much speculation. The microbiota may be gener-ating products that affect the immune system's response to gluten, especially early in life. They may be making toxic products that increase intestinal permeability, which could lead to an increase in gluten penetrating into the body, triggering an immune response to it, or influencing the immune system in some other way that affects its tolerance to gluten.

The good news is that there's excellent data coming out of studies defining when gluten should be introduced into a child's diet. Stud-ies done in the 1970s showed that introducing gluten and solid foods into a diet before four months of age increased the occurrence of celiac disease, possibly by exposing the child to gluten too early (into a leaky gut that is still developing) and triggering an intolerance. Other studies showed that introducing gluten at seven months of age or older also increased the risk of disease, possibly because these kids can eat more and may be getting a large dose of gluten the first time it's introduced. It appears that the sweet spot for gluten introduc-tion is between four and seven months of age. It's important to add gluten to the diet in small amounts *and* to continue breastfeeding. Breastfeeding may introduce small amounts of maternal gluten or gluten antibodies to the child, decrease early infection rates, or just affect the microbiota composition (nobody really knows).

What about probiotics? In animal models of celiac disease, probi-otics worked by decreasing inflammation and the disease. However, data for humans are very scarce, with only one study; it showed that probiotics had a slightly beneficial effect, but it wasn't statistically significant. As the role of the microbiota is further understood for this disease, it's likely that more targeted probiotics will be developed, which will include microbes that specifically digest and break down gluten.

Irritable Bowel Syndrome

By far the most commonly diagnosed gut problem is irritable bowel syndrome (IBS), affecting up to one in five people, often teenagers. It's not a single disease, but rather a set of symptoms that, as the name aptly implies, irritate one's bowels, or gut. The symptoms vary from diarrhea to constipation (or both), bloating, excess gas, and abdominal pain —all of which makes a person very uncomfortable. Unlike the other diseases in this chapter, there don't appear to be structural changes to the intestine with IBS. Stress, anxiety, and depression are often associated with it, hinting at links between the gut and the brain (see chapter 14).

Stress is the most commonly acknowledged risk factor of IBS (we know that stress affects the microbiota based on a study of university students writing final exams). Antidepressants are commonly used in moderate to severe cases of IBS, helping with pain perception, mood, and gut motility.

There are strongly established correlations between the microbiota and IBS. About one-quarter of new IBS cases happen after an intestinal infection, which we know impacts the gut microbiota. Antibiotics and diet changes, which alter the microbiota, can also trigger IBS, and patients often claim a particular food seems to be a trigger. Probably the most compelling evidence comes from experiments in GF animals. When human feces from IBS patients were transplanted into microbe-free mice, they developed IBS symptoms. However, when feces from healthy individuals were transplanted, no IBS symptoms were observed. There are certainly differences in the microbiota composition in IBS patients, although no defined "microbial signature" has been identified yet. IBS patients have a decreased diversity in their microbiota, and bacteria normally found in the large intestine are often found in the small intestine in IBS patients, reflecting a microbiota imbalance.

Certain clues suggest that if one alters the microbiota, symptoms of IBS may decrease. For example, treatment with an antibiotic that remains in the gut and is poorly absorbed (rifaximin) decreased intestinal symptoms, although they did redevelop later. Similarly, there are reports of using prebiotics and probiotics with some level of success in treating symptoms, but because of the small trial size and varying probiotics, currently no general recommendations can be made for probiotics and IBS. There has been one small fecal transfer trial of thirteen people, and it resolved symptoms in 70 percent of the patients—this was after dietary modifications, antibiotics, probiotics, and/or antidepressants had all failed! This is particularly good news for a miserable and tough-to-treat disease, and begs further trials in this area.

Another treatment that shows much promise is a particular diet, low in FODMAPs (fermentable oligosaccharides, disaccharides, monosaccharides, and polyols—jargon for foods that contain certain sugars and sugar alcohols). These small compounds are poorly absorbed in the small intestine and accumulate in the gut, causing water to increase in the intestine, which leads to diarrhea. However, they are readily digested by gut bacteria, producing gases that can also contribute to IBS symptoms, such as bloating and pain. Studies in Australia have shown that this diet reduced IBS symptoms, and it is now being recommended to treat IBS. A low-FODMAP diet is not a DIY diet, and needs to be undertaken with a dietician specially trained in this area. In the beginning, all high-FODMAP foods are omitted from a person's diet. It takes 6–8 weeks before certain foods are reintroduced, and eventually only a small number of foods are usually excluded. It should also be noted that some FODMAPs are needed for our body (and our microbes) to function properly, and it is *not* a zero-FODMAP diet, just one that has fewer of these components.

Inflammatory Bowel Diseases

The last of the major gut diseases we'll discuss are the inflammatory bowel diseases (IBDs), which include two major, related intestinal diseases: Crohn's and ulcerative colitis. As the name IBD implies, they feature inflamed intestines that don't resolve, making patients' lives miserable with persistent diarrhea, rectal bleeding, abdominal cramps, pain, and weight loss. About 1 in 150 individuals suffer from these diseases in Canada and 1 in 300 in the US. Attempts have been made to control the gut inflammation with anti-inflammatory drugs, but these have limited success, and about 75 percent of Crohn's patients ultimately need bowel surgery to remove damaged and destroyed portions of their intestine.

The onset of these diseases is usually in young adults (ages 15–30), although children can also have them. The rate of IBD in the US and Canada has plateaued, but it is increasing rapidly worldwide as other countries become more developed and adopt a Western lifestyle.

Like most gut diseases, both host genetics and the environment (including the microbiome) play a major role. Scientists have identified 163 human gene mutations that increase the risk of disease, but no single gene has been identified as causing it. Instead, there's a collection of risk factors, both human and environmental. The genetic mutations often occur in biochemical processes associated with inflammation. It's thought that these genes play a role in controlling the gut microbiota and, if mutated, they're less able to contain the microbiota within the intestine. Any loss in gut barrier allows microbes to penetrate through, which then triggers extensive inflammation as the body tries to repel the microbial invaders.

A major problem with IBD is the need to go to the bathroom frequently (up to twenty times per day!). This is tough if you're travelling, or out of your regular neighborhood. Crohn's and Colitis Canada has started a program called Go Here, placing signs in windows of establishments where you can use a washroom immediately, no questions asked. The best part is that there's an app for it, so a smartphone can rapidly identify the location of the nearest washroom.

The microbiota is implicated in IBD, but once again no causative microbiota have been definitively established. We know that GF mice do not get IBD, presumably because there are no microbes to breach the barrier and trigger inflammation, even though GF mice have a leaky gut. However, when particular microbes are introduced into GF mice, they do cause varying degrees of IBD.

As expected, the gut microbiota is altered in IBD patients at the time of diagnosis, before treatment has been started. It's known that antibiotic usage can trigger IBD, presumably by altering the microbiota. As we've seen with the other diseases in this chapter, as microbiota diversity is threatened, the number of inflammatory microbes increases, and the number of beneficial, anti-inflammatory microbes decreases.

There has been a major effort to control and treat IBD diseases by implementing fecal transfers and recommending specific diets. Although fecal transfers have been used since 1983 to treat *C. diff* disease, with great success (at a >90 percent cure rate; see chapter 16 for more on this), only recently have they become extensively used for IBD, and the results have been mixed. However, this makes sense because, with *C. diff*, it doesn't really matter which microbes are used or who the donor is—as long as some microbes are added, they will displace *C. diff*. However, in IBD, the person already has an inflamed gut, so these transfers are like putting microbes into a fire and asking them to survive and put it out. It appears to matter which donor is used for a successful outcome. When we look collectively at the several small trials that have been done for IBD, they indicate that fecal transfers work better for Crohn's disease than colitis, and better in pediatric populations than with adults. Together, they have about a 45 percent clinical remission rate, which is promising, but still needs more work (see The Scoop on Poop Transfers, page 187).

Finally, similar to IBS, the low-FODMAP diet is being tried for those with IBD, and in one small trial, 50 percent of patients saw a marked reduction in their symptoms (although this doesn't treat their disease, just their symptoms). There are several more low-FODMAP diet trials under way, so we should know fairly soon how effective it really is at improving symptoms.

Dos and Don'ts

◆ **Do—** ask your doctor if a cow's milk allergy might be responsible for your child's colic. This is the one cause of colic that we know can be treated (by removing cow's milk from the mom's diet). Also discuss a course of probiotics for your child with your doctor, as these are safe, and they may significantly improve the condition. And remember: the silver lining of colic is that it spontaneously goes away after a few months of wailing—the trick is to survive those few months.

◆ **Do—** introduce small amounts of gluten into a child's diet between 4–7 months of age, while continuing to breastfeed; avoid introducing gluten before four months or in large quantities for the first time after seven months. By continuing to breastfeed while adding solid foods, you are decreasing your child's chances of developing celiac, as well as continuing to improve your child's microbiota.

◆ **Don't—** try to tackle a low-FODMAP diet yourself, as it's quite complex and requires the supervision of a trained dietician. However, talk to your gastroenterologist about whether a low-FODMAP diet might give you results, and get a referral for the right professional. Such diets are known to improve symptoms in a significant number of patients.

◆ **Don't—** attempt a fecal transfer yourself (do we really have to say this?), but consult with your physician regarding whether this might be an option for you, if you suffer from IBD or IBS.

THE SCOOP ON POOP TRANSFERS

Fecal transfers for IBD have a nearly 50 percent success rate over-all, which is great . . . but can it be improved? Just recently two major clinical trials of fecal transfers for ulcerative colitis were published. Although one study didn't see any beneficial effect, the other one was quite interesting: it wasn't working until Donor B, who apparently had the right fecal microbial mix, showed up and then the trial was a success. Hopefully, they're assessing Donor B's microbiota (as well as those from donors that didn't work) in order to define the characteristics of a good donor. Five more major clinical trials are under way, so we'll soon have a much better idea about the use of fecal transfers to treat IBD.

There are several ways one could potentially improve the odds of a successful fecal transfer, and charitable groups that are dedicated to improving treatments for IBD, such as the Kenneth Rainin Founda-tion, are looking closely at the various options. These include using donors like Donor B (we would love to know what they eat and drink!), who are known to "have the right stuff" (although we don't know exactly what that is yet).

Another possibility is to decrease gut inflammation with anti-inflammatory medication before the transfer is given. Because IBD, by definition, means inflammation in the gut, this might provide the incoming microbes a fighting chance to successfully colonize, and perhaps displace the previous microbes. Similarly, putting patients on a favorable diet could alter the microbiota and decrease inflamma-tion prior to a fecal transfer. One might consider antibiotic treatment to get rid of the "bad" bugs and clear out the intestine prior to trans-plantation to allow the incoming bugs the opportunity to colonize. Or one could use repeated fecal transfers to try to increase the odds of seeing a beneficial effect.

Unfortunately, like most things in medicine, to try all these permutations and combinations in a major clinical setting will take many hospitals, numerous patients, a lot of money, and multiple years to complete before the ideal fecal transfer setting is established. However, given the major impact of this disease, we're optimistic that such efforts will succeed.

13: Asthma and Allergies: Microbes Keep Us Breathing Easy

The Burden of Asthma

When she was a child, Claire would wake up in the middle of the night to the sound of her sister's wheeze far too often. The forced, fast-paced whistling raised her sister's upper body with every exhalation, seemingly taking all her energy with every breath. This sister, Stephanie, was only ten years old (Claire was eight), but had dealt with sleepless nights like this all her life. Their mother would usually give her medicine before going to bed, propping Stephanie up with two or three pillows and singing to her until the child fell asleep.

Even at that young age, Stephanie would practice different techniques before going to bed to prevent her asthma attack from getting out of hand; she would try to remain calm (hard to do when you feel like you're asphyxiating), avoid gasping, and, above all, control the need to cry or cough, as these only made things worse. Yet sometimes her breathing would become unmanageable and she would ask for help.

Claire shared a room with her sister for as long as she could remember, and she could tell when Stephanie couldn't deal with the asthma attack anymore and it was time to wake their parents. They had bought Stephanie a personal nebulizer, which probably cost an arm and a leg, but was probably the best purchase they ever made and was a game changer for their family (home nebulizers had just become available in the mid-1980s). Before owning a nebulizer, a night like this one would invariably result in a trip to the ER that would either last until the morning or result in a multiday hospitalization.

Claire remembers the night her mother asked her to prepare Stephanie's nebulizer by herself for the first time. It involved attaching a syringe to a needle, drawing up a small amount of Ventolin (a common asthma medication), another amount of saline solution, and combining them in a small cup. This little cup would then get plugged into an air compressor at one end and into a mask on the other end, which Stephanie would strap to her face.

To this day, Claire remembers how the noise from the compressor sounded and how the sweet-smelling steam from the medication would fill up the room within minutes, allowing her sister's breathing to become a bit deeper and slower. Little by little, her breathing would relax until everyone fell asleep until the morning.

For Stephanie, asthma season lasted from about August to December, when the rain was the heaviest and the humidity in their native Costa Rica caused all sorts of allergens to trigger her asthma. Her attacks could be extremely severe and land her in the hospital for days or weeks, making their mom refer to the local children's hospital as their second home. On two occasions her asthma got so bad that she went into respiratory and cardiac arrest and was brought back to life by paramedics. It was a frightening time for Claire's entire family.

Fortunately, asthma treatment has improved over the years, and now Stephanie, almost forty, keeps her asthma in check with a medicine cabinet full of inhalers and pills. But about once every year she can't escape a bad cold that spirals downward into another asthma crisis. Her lungs have accumulated too much damage over the years, leaving her with less than half of normal lung capacity. Undeniably, asthma is a terrible disease that takes a toll on a person's physical, emotional, social, and academic life, and on their family as well. Imagine if there was a way to prevent asthma from developing in the first place.

Too many families have had to deal with the consequences of having an asthmatic child. This chronic disease of the lungs is characterized by inflammation of the airways and a sudden narrowing of the bronchioles, the smallest branches of the airways, which then makes breathing difficult. It's thought that asthma develops due to a combination of genetic and environmental factors, an explanation that scientists commonly use when there are just too many things involved in a disease and we don't yet know how it all works.

What is known is that the rates of asthma have skyrocketed in certain parts of the world. A case as severe as Claire's sister's was considered rare back in the 1980s, but asthma has been increasing in incidence for the past three decades, and so has its severity. In just one generation, rates of asthma have tripled, with 10–20 percent of children currently suffering from asthma in North America, Australia, and most Western European countries, affecting an estimated 300 million children worldwide. It's the most prevalent chronic pediatric disease in the world, the number one cause for hospitalizations of children and missed days of school. In contrast, the rates of asthma in underdeveloped countries did not change much during that period.

Today asthma cannot be cured. It's also difficult to treat, and once treated is even more difficult to control and prevent from flaring up again, making it one of the most expensive diseases for public health systems around the world. Even more worrying, asthma rates in highly populous, less developed countries are now beginning to rise. All this makes us wonder: What is going to happen two generations from now, if asthma continues to increase at these rates? Are asthma inhalers in the future for every single one of our grandchildren or great-grandchildren? Is this disease going to become part of the human condition just like dental cavities?

Searching for the Culprits

The fact that this disease, along with the other Western lifestyle diseases discussed in this book, has increased so sharply in such little time is a true enigma. There are known genetic predispositions for asthma: for example, in Claire's family asthma can be traced back four generations. Genome-wide association studies (also called GWAS, a name given to large studies that look for gene and DNA variation in people with and without certain diseases) found several genes associated with asthma, but they don't explain why the majority of asthma cases are inherited. More importantly, these genetic alterations fail to explain why asthma has become such an endemic disease. Our genetic makeup simply has not changed that much in just one generation—it has to be due to something else that has changed in the environment.

Research groups have looked at things such as diet, socioeconomic status, sun exposure, pollution, pollen, ethnicity, contact with animals, urban vs. rural environments, exposure to specific insects, and more. Many of these studies have yielded strong associations between a particular environmental factor and the risk of developing asthma, but the strongest and most consistent one is the farm environment. People who grow up on farms have a much lower risk of developing asthma than anyone else in Western societies. Something in their lifestyles has protected them from the dramatic increase in asthma, but what is it?

In trying to figure this out, one of the most studied groups are the North American Amish, who have the lowest reported prevalence of asthma and allergies of any Western population. Dr. Mark Holbreich, an allergist from Indianapolis, noticed the low rates in allergies in Amish communities in Northern Indiana after treating them for over twenty years. He is the lead author of a study published in 2012 in which children from 157 Amish families were compared to 3,000 Swiss farming families and 11,000 Swiss nonfarming families. They found that only 5 percent of the Amish children had asthma, compared to 6.8 percent of Swiss farm children and 11 percent of Swiss nonfarm children. The incidence of asthma in Amish children is comparable to what it was for everyone else a few generations ago, making them an ideal population to examine further.

The Amish and their traditional lives, without cars, electricity, or modern appliances, feels like going back to the mid-1800s. Entire families work the land and tend to farm animals, and live a technology-free life. Children as young as five years old milk cows, three-month-old babies visit the cowsheds, and many of them even learn to walk there. Contact with farm animals and dirt clearly occurs very early in life. Pregnant mothers work in the cowshed throughout their pregnancy, potentially exposing their babies to microbes prenatally as well.

However, not all farms are created equal in terms of asthma protection. An interesting epidemiological study that looked at the rate of asthma and allergies in Hutterite communities found that they do not appear to be as protected from the increasing asthma incidence as the Amish people are. Hutterites share certain similarities with the Amish; they both live communal, self-sufficient lives based on strict religious beliefs, and they're both of German descent. However, Hutterites have an important distinction: unlike the Amish, they welcome technological advances, use state-of-the-art farming equipment, and usually run quite large farms. Hutterites also use antibiotics in their farm animals in a way similar to most other modern farmers.

Studies of European farms, many of them led by Dr. Erika von Mutius at Munich University in Germany, have shown that farms with a higher diversity of microbes and in which the microbes from the cowshed reached the home offered the strongest protection from asthma. In one study, the amount of microbes in the mother's mattress was inversely correlated with the risk of her children developing eczema, an allergic skin condition often associated with asthma. All these studies clearly suggest that, when the farm's microbial residents come in contact with the human residents early in life, children are somehow protected from developing this disease.

A separate study, also led by Dr. von Mutius's group, found that newborn babies from farms are born with an immune advantage over those not from farms. Researchers isolated immune cells from the umbilical cord blood and found that newborns whose mothers lived on farms during their pregnancy had an increased number of regulatory T cells, which are important immune cells that serve as the peacekeepers and modulators of the immune system (see chapter 2), making sure that the immune system does not overreact towards a particular intruder. They're also known to be crucial in preventing asthma and allergies. Since the researchers compared groups that share the same ethnicity, the differences they found are not due to genetic variations, but likely due to differences in environment. The thought is that farm exposure in pregnant mothers reflects a more natural way of fetal immune development, one that involves the stimulation of a particular group of cells necessary for immune control and for prevention of allergies. By being in contact with cows, chickens, pigs, and dirt, mothers are essentially immunizing their children to tolerate future allergens (substances that cause allergies) by more closely mimicking how humans evolved as a species, in close contact with animals and the natural environment. Removing this exposure during pregnancy, as happens in most modern Western environments, is likely one of the drivers of the sharp increase in asthma risk in everyone but farmers and their families.

From the Gut to the Lung

All these studies about the patterns of asthma incidence made it clear to us that, in order to figure out what the protective factor is and how it works, scientists needed to focus on microbial exposure and how it interacts with the immune system. An idea on how to do this came to Brett while he was having dinner with his wife, Jane. She has worked as a pediatrician for thirty years and she told him that there was some evidence that children who receive antibiotics early in life have a higher risk of developing asthma. Brett thought this concept was fascinating, but was skeptical. But as always, Jane was right—the literature backed her up.

Brett became very interested in this and brought the discussion to his lab. At that time, they were using antibiotics as a tool to shift microbiota and were then looking at the effect of these shifts on diarrheal infections in the gut. Brett talked to one of his students, Shannon, and she decided to tackle a very important question during her PhD research: Do changes in intestinal microbiota affect susceptibility to asthma? This was only five years ago, but back then it was considered a far-fetched idea and she even had some side projects on the back burner, in case this didn't work.

Shannon decided to do a simple experiment. She set up two groups of mice: one group would receive antibiotics starting from birth until they became adults and the other group would receive antibiotics starting at around three weeks of age, when mice are weaned and not considered babies anymore (there was also a control group that didn't receive any antibiotics). When Shannon tested the mice for asthma, the results surprised everyone. The mice that received antibiotics when they were babies had worse asthma than the mice that received them only in adulthood or those that didn't receive them at all. Furthermore, she analyzed the type of immune cells in these mice and found that, similar to what had been seen in humans living on farms, the mice that received antibiotics during infancy had lower amounts of regulatory T cells (the peacekeepers) in the intestine. The antibiotic that Shannon used was given orally and does not get absorbed into the bloodstream and the rest of the body, which means that somehow the shift in microbes in the intestine changed the outcome of an immune disease that occurs in the lungs. Her finding that regulatory T cells are involved also suggests that what she observed in mice probably has a connection with what occurs in humans. Shannon went on to do many more experiments, and she showed that the effect of antibiotics on asthma was limited to very early in life, from right after the mouse pup was born until it was weaned, somewhat equivalent to the first few months of a human infant.

Claire joined the lab at this point, and she was eager to carry on with this project using human samples. Our lab was fortunate to form a partnership with a national study called the Canadian Healthy Infant Longitudinal Development (CHILD) Study, which had been looking at factors that might affect asthma in children. The CHILD Study had been collecting samples across Canada since 2009 and its goal was to collect and characterize samples from 3,500 children from birth until five years of age (a massive effort). Claire and her collaborators requested fecal samples from 350 children, which were collected both at three months and one year of age.

Our question was simple: Are there intestinal microbial differences in babies that go on to develop asthma earlier in life compared to babies that do not develop asthma? Using a new method to survey bacteria (16S analysis; see chapter 16), we sequenced the intestinal microbiota in these samples and found something that really surprised us. Three-month-old babies at a high risk of developing asthma later in life were missing four types of bacteria, but when we looked at the microbiota at one year of age, the differences were essentially gone. All of this—plus the other studies about antibiotics, farms, etc.— points to a critical window of time in which microbial changes in the intestine have long-term immune consequences in the lung. In addition, we found that the differences were not limited to the type of bacteria found in feces but were also observed in some of the compounds they produce. Interestingly, only one of these compounds was a bacterial product made in the gut (acetate; see chapter 2), and many of them were bacterial compounds detected in the urine of these babies— another proof that bacterial metabolites go everywhere in our body.

Our lab is still trying to figure out how these four bacteria (which are present in low numbers and which we nicknamed FLVR) lead to asthma, but one additional set of experiments in mice suggests that FLVR is directly involved with mediating this, as opposed to just showing a correlation. Claire gave one group of GF mice a fecal transplant from a baby who had no FLVR in his feces (and also developed asthma later in life), whereas a second group of mice received the same fecal transplant plus added FLVR. When the mice became adults, the group that received FLVR had much less lung inflammation and other markers of asthma.

We're still far from considering this as a preventative therapy for human asthma, but it opens a few doors that might give us a major advantage in combatting the asthma epidemic. The first is that, assuming our findings hold true in other populations, we should be able to identify infants that are at very high risk for developing asthma. Even more exciting, we may be able to give those high-risk infants certain microbes or microbial products as a way to prevent asthma. It really is a bizarre and almost unorthodox concept to think that changes in intestinal microbiota in a three-month-old could affect an allergic lung disease in a school-age child several years later. Stay tuned!

Allergies and Eczema, Too?

When Claire's daughter, Marisol, was two months old, her skin would break out in red glossy patches and she would cry while constantly trying to rub her skin. By the time she was six months old the patches were in almost every one of her chubby body folds. Her scratching got so bad that she would break her fragile skin and Claire would sometimes find blood on her crib sheets. Claire tried different creams and lotions recommended by their pediatrician, but when her daughter's rash was severe, the only treatments that would control it were topical corticosteroids (which decrease inflammation) and fewer baths.

During the dry winter months, Claire would bathe her daughter only once a week, moisturize daily, and apply anti-inflammatory creams if her skin flared up.

When Marisol turned one she experienced a reaction the second time she was offered eggs. Claire and her husband didn't introduce allergenic foods until then, as they knew that Marisol was an allergy-prone child, given her early start with eczema (those red patches described above). They realized that she probably had a higher chance of developing food allergies and/or asthma later on, so they needed to be careful. The statistics were not in Marisol's favor; she had a family history of allergy and asthma on both sides—she's what is called an atopic child. Atopy is a predisposition towards developing allergic diseases, driven by an overly reactive immune system. Luckily, Marisol's eczema is kept under control fairly easily these days (she's five now); she completely outgrew her egg allergy; and, most importantly, she hasn't developed asthma.

At the heart of all this allergic disease is an unbalanced immune system. The immune system can affect different organ systems, causing eczema, asthma, and hay fever. This is why understanding the microbiome is so important. The microbiome has the potential to train the immune system and prevent all forms of allergic disease.

Atopic children can manifest skin, respiratory, or food allergies, separately or in any combination. More often than not, the appearance of these conditions follows an order known as the atopic march. This "march" often starts with eczema (also known as atopic dermatitis), followed by food allergies, allergic rhinitis (hay fever), and finally asthma. All of these diseases have been increasing in incidence, with eczema affecting approximately 20 percent of children in wealthier countries of the world; many of these children go on to develop asthma. It's been shown that the likelihood of a child with eczema to develop asthma is associated with how severe the skin condition is, making efforts to control or treat eczema an important step in preventing worse allergic conditions from developing.

Unfortunately, eczema doesn't have a cure, but it can be effectively treated using a combination of bathing techniques, avoidance of skin irritants, and anti-inflammatory treatment (such as corticosteroid cream). However, recent data show that the early gut microbiota of children that later develop eczema is different and less diverse than that of control children. In addition, the microbiota of affected skin areas is different than that of healthy skin. Researchers still don't know whether the difference in skin microbiota is a cause or an effect of eczema (the inflamed skin may alter the microbes), but it opens the possibility of treating the microbiota as a way to control eczema. Experimental approaches to treat eczema by treating the skin microbiota look promising and a few products are already in the market, but additional research is needed to demonstrate that it is indeed effective.

Some studies have used oral probiotics to prevent or treat eczema, but the results have been mixed. The strongest evidence shows that probiotic treatment during pregnancy and early infancy is effective at decreasing eczema risk (even in cases with strong family histories), but the evidence for probiotics as an actual treatment is weaker. Still, this area of research is extremely new, and the reason many of these treatments appear to be ineffective may be because current probiotics contain bacterial or yeast species that are not involved in the development of eczema or related allergic diseases. Our research on asthmatic children shows that the bacteria involved in protecting against asthma are not found in any probiotic formulation (yet). Needless to say, our laboratory and others are working hard to determine which microbial species should be targeted.

Another promising finding comes from a large study of 5,000

children, in which it was discovered that babies who were deficient in vitamin D were significantly more likely to develop food and respiratory allergies. Vitamin D is a known anti-inflammatory that suppresses the action of immune cells. Emerging evidence shows that vitamin D is also necessary for some of the cross talk that occurs between our gut bacteria and our immune cells.

Dos and Don'ts

♦ **Do—** take measures early on to prevent the development of asthma and allergies in your child, even before she's born! Maintain a healthy diet, avoiding tobacco smoke and antibiotics if possible. Given the chance, opt for a vaginal birth, and prolong breastfeeding.

♦ **Do—** give your baby and toddler a vitamin D drops supplement daily. Deficiency of vitamin D has been shown to increase the risk of food and environmental allergies and is also a crucial component in regulating the microbiota.

♦ **Don't—** rush out to the nearest farm if you're a city dweller and are pregnant or have a young child and bed down with the cows. While farm families should be considered fortunate to live in such an environment, some studies show that occasional visits to farms may actually exacerbate any pre-existing allergy tendencies. It seems that prolonged continual exposure to this environment early in life is what provides protection from asthma and allergies. But if you have a dog, snuggle up with it frequently—and let your child do the same, to bring some of the outdoors into your family (see chapter 8).

◆ **Do—** stay on top of literature on this topic. It's clear that the gut (and possibly the skin) microbiota is directly involved in asthma and allergy development, but it's still not known how microbial alterations can be used as a way to prevent these diseases.

ALLERGENIC FOODS: AVOID OR FIGHT BACK?

The common advice to parents of children with food allergies is: avoid that food! However, a relatively new treatment option is to face a food allergy head-on, under careful medical surveillance, until a child's body no longer reacts as fiercely to the food.

This therapy is known as allergy desensitization and is being offered experimentally in a handful of hospitals and doctor's offices. The treatment consists of first finding out the amount of allergen (e.g., peanuts, milk, etc.) that will provoke a reaction, and then beginning to feed the child increasing amounts of the food every week over the course of a few months. The goal of this therapy is not necessarily to cure the allergy, but to reduce the risk of a severe anaphylactic reaction if the child eats a small amount of the allergic food.

Allergy desensitization is quite effective, but upon "graduation" it needs to be maintained. Children that have been successfully desensitized must eat a small amount of the allergenic food every day (e.g., a few peanuts or a few sips of milk) to keep the allergy from returning.

This is a promising but potentially dangerous approach to treat a severe allergy, so it should be attempted only under the care of an allergist trained in this therapy.

14: Gut Feelings: Microbiota and the Brain

Bottom-Up Thinking

So far we've seen that changes in a developing microbiota can alter our lungs, skin, intestines, pancreas, liver, and fat tissue. What about the organ that controls our senses, intelligence, and behavior? Can microbes in our gut affect brain development and be involved in neurological disease? Well, sure.

We've known for a long time that there's a link between the gut and the brain. The gut has the second highest number of neurons (millions of them)—next to the brain, of course. These neurons combine to connect the entire digestive system, from the stomach all the way down to the anus, reporting back to the brain on how our digestive system is doing, and affecting gut movement and other digestive functions. An important nerve, called the vagus nerve, serves as a direct neural connection between the brain and the nerves in the gut; the vagus nerve links gut sensations to the brain. Remember the last time you peered over a steep cliff and felt that funny sensation in your stomach? Or when you had butterflies because you had to speak in front of people? What does your gut have to do with your physical responses to any of these situations? A tingling tummy won't help you avoid falling off a cliff or make your speech any better, but it does prove that your gut and your brain are in sync.

Until recently, this connection (called the gut–brain axis) was thought of in terms of a "top-down" concept, with the brain doing all the talking and the gut following orders. For example, in irritable bowel syndrome (IBS), it's thought that the brain contributes to the pain and other bowel symptoms, as there's often no known direct cause within the gut, and because things like stress affect it. But that's all changing, and we now realize that the gut, and especially the microbes in the gut, have a lot to say; a "bottom-up" approach is beginning to influence how we think about the brain and its functions.

Nature is full of amazing examples of how microbial pathogens can affect their host's behavior, with the microbe driving its host to do things that actually help the microbe (but not necessarily the host). We've all heard of the frothing, snapping rabid dog that tries to bite anything that moves. Rabies is caused by a virus that's transmitted through bites that inject the virus into a new host, where it enters the nerves and climbs up into the brain. Unless treated early, rabies is fatal. So how does a rabies virus find a new brain before its current host dies and is buried deep in the dirt? By affecting the brain, convincing the animal to bite as many potential hosts as possible. In humans, rabies has also been linked to aggressive behavior and, strangely, to hydrophobia, which is an intense fear of drinking water—sufferers cannot even bring a cup of water to their mouth (watch the YouTube videos for a scary demonstration).

Another fascinating example of behavioral change is carried

out by a parasite called *Toxoplasma gondii* that lives in the brains of mammals. *Toxoplasma* likes to live in cats and is fairly common around the world. It undergoes sexual reproduction in the guts of cats, which is key to its life cycle. *Toxoplasma* also lives in the brains of rodents such as mice and rats. So how does it get from a mouse brain to a cat gut? When this parasite is in a rodent brain, it somehow reprograms that brain to do two things: it decreases fear so that the rodent leaves its hiding places and wanders out in broad daylight, and also it attracts the rodent to cat odors! Humans are not the preferred host for *Toxoplasma*, but infection does occur (this is the parasite that pregnant women aim to avoid when instructed by their doctor not to change the cat litter box). Humans who have been infected with this parasite have a higher risk of suffering behavioral changes and schizophrenia.

There are many more examples of behavioral modifications by microbes, including convincing fish to seek out warmer water (to let the parasite grow faster), persuading grasshoppers to leap more and jump into water (for the parasite to mate and lay its eggs), and there's even a fungus that turns ants into zombies so they wander around aimlessly and then chomp on to a leaf and die, holding the leaf in a death grip (which the fungus then uses as a food source). These are pathogenic microbes commandeering their host. So what about our resident microbes—are they chatting with the brain, and could this affect our development, our behavior, and even have an impact on diseases of the nervous system? To a certain extent, yes. The microbiota can affect anxiety, depression, social recognition, and stress responses; it may even play a role in common mental disorders such as autism.

The Microbes Made Me Do It!

So how can a microbe that lives in the gut talk to the brain? As mentioned above, our nerves are hooked up to all parts of the digestive tract, which then feed into the vagus nerve that not only brings signals down from the brain to the gut, but also feeds gut information back up to the brain (again, the gut–brain axis). Any of these nerves can be used to send a message from the gut to the brain, using compounds known as neurotransmitters. Alternatively, microbes can either make molecules, or alter existing molecules, and send them as messages directly to the brain using the bloodstream. The third way is for microbes to tweak the immune system (something they're very good at), which is directly hooked up to the nervous system, and send messages this way.

Neurotransmitters are chemicals that send signals between two nerves, passing along a message in an incredibly fast relay-race fashion. It's thanks to them that when you stub your toe, your brain quickly gets the signal ("ouch!"). It's also why you start hopping on one foot and saying words you wish your kids didn't hear (the brain promoting uninhibited behavior, again by neurotransmitters). Microbes seem to have figured out this concept, and whether they make or modify our neurotransmitters, they can rapidly join in the conversations. Although there are a few examples of microbes specifically producing neurotransmitters, this has been observed only when they were grown in a test tube, and there are no known examples of them actually doing this in our guts (yet). However, there are many examples of the microbiota affecting our body's neurotransmitter levels, which can in turn affect how our brain works. This is beginning to provide molecular clues as to how the microbiota may affect people's behavior, depression, and stress.

This area could be an entire book on its own (actually, there is at least one), and many neurotransmitters with complicated names and abbreviations such as BDNF, dopamine, GABA, G-CSF, serotonin, acetylcholine, and norepinephrine are all affected by the microbiota. All play major roles in our everyday brain and nervous system functions, and alterations in these neurotransmitters can lead to neurological problems and mental illnesses. Instead of going through all these mechanisms in detail, we'll use just one as an example to illustrate how things might work.

Let's pretend you and your spouse hire a babysitter and go out to a nice restaurant for a quiet dinner (finally!). Picture that satisfied feeling of being able to finish your meal without interruptions, while it's still warm! The reason you feel that way is due to a neurotransmitter called serotonin (plus the fact that you're not witnessing your two kids fighting with each other—that's the babysitter's problem). Nearly all of your body's serotonin is made by cells in the gut, where it regulates intestinal movements (needed for proper gut functions). In the brain, serotonin regulates mood, appetite, and even sleep. Serotonin is made from tryptophan, a building block of proteins. So if tryptophan levels are affected, so are serotonin levels. How do the microbes dial in? They affect tryptophan levels, as well as the chemicals that control serotonin synthesis. In germ-free (GF) animals, serotonin levels are decreased in both the blood and the colon, and this can be corrected by adding back certain microbes to the animals. These microbes make small molecules (metabolites) that affect serotonin production, and if you add these metabolites back to GF animals, even this can restore serotonin levels.

Experiments with GF mice have been quite helpful in demonstrating the large impact the microbiota has on brain function. For example, GF mice are more exploratory, less anxious, and significantly less fearful than mice born and raised with regular microbe-rich conditions (not good from a mouse's point of view, but great news for a cat!). It's believed that certain microbes regulate the secretion of an important neurotransmitter molecule, known as GABA (gamma-aminobutyric acid), demonstrating yet another critical human function delegated to microbes.

GABA is necessary to inhibit feelings of fear and anxiety, two essential emotions that all animals need to control as a primal survival strategy. Amazingly, colonizing GF mice with bacteria, or even doing fecal transfers from a normal mouse to a bold GF one restores their behavior back to normal. However, this is time-limited, as only GF animals recolonized early in their lives can restore normal behavior, emphasizing once again the importance of microbes early in life.

Other neurological aspects that are altered in GF mice are cognitive functions, learning, memory, and decision-making, which are all pretty important things when thinking about childhood development.

The above-mentioned experiments were done in the absence of microbes, which, as we have mentioned before, is an unrealistic scenario. However, there have been other behavior-changing experiments done in the presence of *different* microbes. For example, feeding lab flies a specific bacterium (a probiotic) makes them prefer to mate with other flies colonized with this particular microbe, while they avoid potential mates that lack this bug. It's thought that the bacterium provides building blocks to molecules called pheromones that are scent attractants.

Similarly, hyenas hang out in tribes or social groups, much like teens hang out in cliques. These animals have scent glands (which stink, unless you're a hyena), and specific tribes have their own set of microbiota living in these scent glands that affect which pheromones are made. While this has implications in hyena tribal structure, it has even bigger implications in that the microbes are affecting mate selection, which could actually drive evolution.

Microbes and Moods

At present little is known about how microbes directly affect human behavior, but given the tantalizing animal data, studies are now under way. With that said, other clues emphasize that the microbiota is needed for a healthy mental state in humans. Many psychiatric illnesses and mood disorders are associated with gastrointestinal disorders. For example, depression is common in people with irritable bowel syndrome and they are sometimes treated with antidepressants.

In any given supermarket, there always seems to be a two-year-old having a temper tantrum (perhaps even your own child). Although not a psychiatric condition (many parents might disagree), the unruly and defiant behavior of most two-year-olds is a developmental stage to be reckoned with. For most parents, the so-called terrible twos is something one doesn't easily forget. Going from a sweet, smiling child to a wailing, flailing irrational toddler is a remarkable change in behavior. Luckily, it's a time-limited event, and kids outgrow it.

Could the microbiota play a role in this? There's very little data at this point, but one recent study of the microbiota in children between 18–27 months showed that the presence of certain bacteria is associated with a reduced ability to control emotions, especially in boys. We also know this is a time in a child's life when remarkable changes are occurring in their microbiota as they shift from being an infant to a toddler. Thus, it's entirely possible that bacteria may be involved in this dramatic shift in temperament. This doesn't mean that the microbiota is causing those changes; it may very well be that the microbiota is responding to changes in stress-related hormones. Either way, the next time your toddler is rolling around in the super-market aisle desperately demanding a toy, you can practice patience by thinking perhaps you can blame it on your toddler's microbes and not on your parenting skills.

If microbes are so good at interacting with our brain, could they actually affect the way it develops and functions? Recent studies suggest that early-life changes in the microbiota may impair memory. Experiments that tried to mimic the Western diet of developed countries showed that if you put mice on a high-sugar diet early in their development, it affects both long- and short-term memory later in life. It also causes changes in the composition of the microbiota, which led the authors of the experiment to state: "These results suggest that changes in the microbiome may contribute to cognitive changes associated with eating a Western diet." Scary stuff, when considering so many children in our society follow a high-sugar diet.

In other experiments, in which the microbiota was modified by an infection instead of diet, mice infected with a harmful microbe and then subjected to stress showed memory dysfunction, which could then be prevented by probiotic treatment. Additional experiments with GF mice showed that their memory was impaired compared to normal mice, whether they were stressed or not, indicating that our resident bacteria somehow impact the capacity to memorize and remember. Still, without human data it's far too early to definitively say whether the microbiota directly impacts human memory or other aspects of intelligence, but there are some tantalizing clues that could have profound societal implications if they prove true.

Stress, Depression, and Anxiety

We live in a stressful world, and this has all sorts of implications for our microbiota and brain. It starts in the womb. For example, maternal stress and infections during pregnancy are linked to neurological disorders such as schizophrenia and autism spectrum disorder in children. It's thought that the ability to cope with stress is set up early in one's life. This system is immature at birth, and goes on to develop at a time when the microbiota is changing rapidly. Stress during this formative time is also linked to later brain disorders.

Moreover, early-life stress changes the microbiota. We know that the separation of rats from their mothers early in life leads to long-term changes in their ability to deal with stress, as well as long-term changes to their microbiota. Adult GF mice show exaggerated stress responses that can be fixed if microbes are encountered early in life, but this cannot be reversed if microbes are added later. However, when mice were fed a particular probiotic, these animals were less stressed. In fact, the effect of the probiotic was as strong as giving them antianxiety medication. They also made more cortisol, a hormone that helps cope with stress. If the vagus nerve connection was severed, the effect was lost, again implying a strong gut-brain connection. Similarly, GF mice have decreased cortisol, which helps explain why they have increased stress reactivity. This is obviously a new and experimental field, with nearly all the results coming from animal studies. However, it does have significant implications for raising children in a stressful world. It's interesting to contemplate a world in which an infant's microbiota could be modified to improve their ability to deal with stress and anxiety later in life.

Depression and anxiety may also have links with the microbiota. One school of thought is that depression is an inflammatory disorder (which would affect the microbiota). Damage to the gut, which causes inflammation, has been linked to depression (as well as schizophrenia and autism spectrum disorder). Incredibly, anxiety can be transmitted in mice simply by doing a fecal transfer between animals. In a small human study of fifty-five people (thirty-seven had depressive disorders and eighteen were controls), it was found that there were several correlations between emotions and changes in the microbiota. Larger studies addressing this are warranted, and fortunately these are now under way.

Some probiotics have shown beneficial effects by decreasing depression-like behavior to the same extent as antidepressant medication, but again, this has been studied only in animal models thus far.

Autism Spectrum Disorders

Autism, a neurological disorder affecting more and more children every year, has deservedly received a lot of public attention. Autism spectrum disorders (ASD) are a collection of neurodevelopmental disorders characterized by social and communication difficulties, repetitive behaviors, and sometimes cognitive delays. They include autism, Asperger's syndrome, and other related disorders. Unfortunately, the rates of these disorders are increasing; the World Health Organization estimates that 1 in 160 children have ASD and this number continues to increase. Because these rates have risen so rapidly, it's unlikely that it's caused by genetic changes alone. A common reason cited for the rise in autism is the increased diagnosis of ASD—the argument that we simply weren't identifying the disorder before—but again this doesn't fully account for such a steep rise (and where are all the autistic sixty-year-olds now that weren't diagnosed long ago?). In any case, the debate over whether it's an increase in diagnoses or in the actual number of cases is irrelevant when there's an epidemic of ASD affecting such a large proportion of children.

Though theories abound, there's no known cause of ASD. However, links to the microbiota have recently gained much attention. It's been well established that maternal stress and infections during pregnancy are linked to neurological disorders, including ASD and, as we previously discussed, this is accompanied by microbiota changes. Children with ASD often have serious gastrointestinal problems, such as diarrhea, constipation, bloating, abdominal pain, and increased intestinal permeability, all of which could affect the microbiota–gut–brain axis. These symptoms are especially present in children diagnosed with regressive-onset ASD (children who were developing normally, but started showing symptoms as toddlers, around 15–30 months of age) with abnormal behavior and loss of previously attained skills. Furthermore, several reports suggest that the microbiota of ASD children is different from the normal control groups. One hypothesis being put forward is that ASD may be either caused or enhanced by a general imbalance of the microbiota.

Based on this hypothesis, there was a small study in which the antibiotic vancomycin was given to ten children with regressive-onset autism who had also manifested gastrointestinal symptoms. Eight of these children showed clear improvements after treatment, but these were short-lived. In a similar study in France, 80 percent of regressive-onset autistic children reported dramatic improvement in their symptoms but, as with the previous study, it didn't last long. Could it be that once the antibiotic treatment stopped, the microbiota reverted back to its previous altered state? The following compelling real-life story suggests that this just might be the case.

Leah was a normal little girl who at five years of age developed an infection, as many five-year-olds do. A week later, her parents started noticing very odd and serious symptoms and took her back to her pediatrician, who became worried after seeing her autistic-like behaviors. Leah had begun walking on her toes, flapping her hands constantly, screaming, avoiding eye contact, showing minimal verbal skills, etc. As parents ourselves, we can only imagine the severe anguish her parents must have felt after seeing their healthy daughter's personality transform this way in just a few days.

Her father, a university professor, sat in front of the computer for days and gathered all the information he could, then went back to Leah's doctor. He managed to convince him to prescribe Leah with a short dose of the antibiotic vancomycin to change her microbiota. Antibiotic treatment produced a night-and-day effect in this little girl. While on vancomycin her symptoms almost disappeared, yet off vancomycin the autistic symptoms returned! At this point they were convinced they could treat Leah's symptoms by modifying her gut bacteria, but they couldn't keep Leah on antibiotics all her life, so they needed to find a better solution.

Leah's father contacted the original scientist who had published about this and he suggested they find a doctor who would be willing to do a fecal microbiota transplant, or FMT. As the name suggests, this procedure involves the transplantation of fecal matter from a healthy individual into a recipient. Unfortunately, it was exceedingly difficult to find someone in the US who would do this procedure at that time, so the family traveled to Canada, where they had managed to convince a gastroenterologist to do the procedure (try telling that story to the border guards when asked if you have anything to declare). Leah responded incredibly well to the transplant, not showing any symptoms for months. She did have a mild relapse after another infection, but an additional FMT made those disappear again.

The scientist and doctor involved in Leah's case analyzed her fecal samples before and after the FMT and found that her microbiota had definitely changed. Interestingly, her sample before treatment showed an unusually high number of a bacterium called *Clostridium bolteae*, a microbe that had been previously associated with autism. The story of the discovery of this bacterium is very unusual, but hopeful.

Let's also look at the case of Andrew. He's now twenty-three, but when he was about eighteen months old, he started showing signs of autism and severe gastrointestinal issues. He had received numerous bouts of antibiotics to treat fluid in his ear over the span of three months. During the sixth course of antibiotics his mother, Ellen, started noticing significant behavioral changes. Andrew became withdrawn, not wanting to be held or touched, even by her. Ellen soon suspected there might be a connection between the antibiotics and Andrew's altered behavior. Just like so many parents whose children have been diagnosed with regressive-onset autism, Ellen bounced from doctor to doctor, and was met with a lot of disbelief and skepticism. So, just like many parents in this same situation, she devoted herself to investigating if there was a link between antibiotic treatments and autism. However, unlike most parents, Ellen became so dedicated to her research that in 1999, and without any formal scientific training, she published a paper proposing that autism could be a disease of microbial origins. When Andrew was five, she decided to take all her research findings and her own publication to a pediatric gastroenterologist—the thirty-seventh doctor she had seen since Andrew was diagnosed. It was this doctor who finally agreed to prescribe Andrew an eight-week treatment course of vancomycin.

Ellen and her doctor started documenting the effects this antibiotic had on Andrew through videos and additional doctor visits.

The changes immediately after starting the treatment were overwhelmingly positive. Within a couple of weeks Andrew was using speech again, he could wear clothes, he was hugging his mom, he was even able to potty train in the span of a week. As Ellen eloquently puts it, "Andrew was aware of his surroundings again—he was starting to come back." Andrew's doctor could not believe that the child who had previously been so irritated and out-of-control in former visits was the same child now patiently sitting down waiting for the doctor to see him.

After seeing how successful Andrew's treatment was, a clinical trial was immediately started. However, as the weeks passed, Ellen's enormous relief and happiness turned into heartbreak, since at the end of the antibiotic treatment Andrew's behavior started spiraling downward again. Nevertheless, they had discovered that his autistic symptoms could be modified with an antibiotic, something completely unrecognized before.

Her efforts of course didn't stop there; she became involved in several clinical studies led by Dr. Sydney Finegold, a medical researcher from the University of California, Los Angeles. Together, they discovered that children with regressive autism often had an unusually high abundance of a previously unknown bacterium, which they named *Clostridium bolteae*, in honor of Ellen's heroic efforts for this cause (Ellen's last name is Bolte). These bacteria are extremely resilient to antibiotic treatment, and once the treatment stops they always come back.

Since then, Ellen has tried to modify Andrew's microbiota through other methods such as diet and strong doses of probiotics. Recently, they subjected Andrew to an FMT, which unfortunately did not improve his autism. They believe it didn't work because it was done when he was twenty years old, a very long a time after his symptoms started. They speculate that over time the processes affecting the nervous system that responded to a microbiota treatment might have become unchangeable. Still, the severity of Andrew's autism has decreased dramatically from the time they started treating his intestinal microbiota. He now has a much less severe form of autism, something unusual in ASD.

These stories are truly compelling, but it is important to emphasize that it has not been proven yet that this or any other bacteria *cause* autism. It could be that the microbiota of children diagnosed with ASD change due to the elevated anxiety and dietary changes that these children undergo. More importantly, not every case of autism improves after antibiotic treatment, likely because ASD includes a large group of disorders with similar symptoms but different causes. Parents of children diagnosed with regressive autism should definitely consider bringing up this information with their health practitioner, yet at the same time refrain from becoming overly optimistic that this type of treatment will work for their particular child.

A lot more human research is necessary to determine the true role of certain gut bacteria in ASD development. However, there are very persuasive arguments about microbiota involvement in ASD coming from, yet again, germ-free (GF) animal studies. In a well-known study from Sarkis Mazmanian's lab at the California Institute of Technology, researchers used a mouse model of ASD to probe the role of the microbiota. As in humans, pregnant mice exposed to stress produce offspring with some of the behavioral and physiological changes seen in ASD, including repetitive movements like obsessively burying marbles, and showing less communication and fewer interactions with other mice. They also found that the microbiota in mice with ASD-like symptoms was altered and they had increased gut permeability, just as in children with ASD. Digging deeper, the scientists identified a small molecule that was more prevalent in the ASD mice and that is similar to one found in human ASD. This molecule, which is produced by microbes in the intestine, triggered ASD symptoms when given alone to mice, suggesting for the first time that a bacterial compound can cause or at least trigger ASD in mice. They went on to suggest that because of the increase in intestinal permeability in ASD mice, this molecule could also get to the brain more easily. In support of this idea, when they gave a probiotic known to decrease intestinal permeability, it restored the microbiota, decreased the production of this metabolite, and remarkably, also decreased the ASD symptoms in these animals.

So where do we stand regarding microbiota and ASD in children today? There are an awful lot of smoking guns hinting that the microbiota might be involved, as we have seen in the above cases. However, nothing has been proven in humans. Neurologists are now aware of these many correlations, and several studies are being undertaken to test this hypothesis in humans, but it's not yet common medical practice to consider the gut microbiota when treating ASD. It seems strange indeed to consider that something as simple as a probiotic or a fecal transfer could influence such complex neurological disorders such as ASD, but given the animal studies and many other human correlations, it may not be such a far-fetched proposition if we get the probiotic strain(s) right.

Attention Deficit Hyperactivity Disorder

Another childhood behavioral disorder that's becoming very common is Attention Deficit Hyperactivity Disorder (ADHD), with up to 12 percent of boys and 5 percent of girls being diagnosed with it in the US. The average age of diagnosis is seven years, although these children are often hyperactive in the womb. This disorder is characterized by lack of attention, impulsiveness, and hyperactivity. Like ASD, this disorder has a big spectrum and wide degrees of severity. There are children who show mild symptoms and children with very severe manifestations that prevent them from attending school or carrying on a normal life. Such kids may tend towards risky behavior as teens, including drunk driving, substance abuse, unprotected sex, etc.

Many factors are thought to contribute to ADHD, including a genetic component (it can run in families) and factors encountered during pregnancy, such as low birth weight, prematurity, and prenatal exposure to alcohol and smoking. Is there a microbiota link? We don't know for certain, but again there are several hints. Children with food allergies, eczema, or asthma (all associated with microbiota) have increased rates of ADHD. We also know that diet changes can sometimes reduce the symptoms of ADHD, which would also affect the microbiota.

In a remarkable (but very small) study, forty children were given a probiotic for the first six months of their life, while a placebo group of thirty-five children were not given the probiotic. Thirteen years later (it was a long experiment!), they found that 17 percent of the control children had ADHD and/or Asperger's syndrome (three had ADHD, one had Asperger's, and two children had both), while remarkably not one child in the probiotic group had either of these disorders (0 percent)! They did look at the microbiota, but could not define a single microbe that correlated with this, although thirteen years ago the tools available for microbiota analysis were not very good. We want to stress again these are very small numbers and only one trial, but they are extraordinary if indeed they hold true.

The Road to a Better Brain

Can we make our brains even healthier, or fix neurological problems by changing the microbiota? Maybe. There are currently three main methods that seem to show promise in improving brain health. The first is diet (which of course also alters the microbiota). We've all heard about the beneficial effects of omega-3 fatty acids, and eating lots of healthy foods such as fruits and increased fiber (see Food for Thought on page 223). The second way is exercise. Studies have shown that exercise, even in small amounts, has anti-inflammatory effects. It also leads to changes in the microbiota that appear to be beneficial. Perhaps there's a reason kids roar around so much. The final way, which is being studied extensively, is to more directly modify the microbiota using antibiotics, prebiotics, and fecal transfers. However, most of the focus is on probiotics. These living microbes seem to have beneficial effects, and if they're used for affecting the brain, they now have their own new name called "psychobiotics."

Probiotics are a huge market worldwide, estimated at over $20 billion in sales. Despite their large consumption, there have been only a few human studies done on their effects on the brain. However, there is some very compelling data coming from animal experiments that we should look at before we briefly discuss what's known in humans. For example, if mice are given a probiotic for twenty-eight days, it decreases their anxiety-like behavior. Similar experiments were also performed on rats, and one study found that two probiotics given together had the equivalent effect to antianxiety medication. In other rat and mouse models, scientists found that probiotics decreased depression to the same extent as antidepressants. Studies have also shown probiotics to improve learning and memory in mice.

Unfortunately, in humans there just aren't sufficient studies yet to conclusively recommend a particular probiotic for certain conditions or brain health. There is preliminary research that suggests probiotics may help cognition and stress-related conditions such as anxiety, autism, depression, and schizophrenia, and even multiple sclerosis. Great! We should all take probiotics, including our kids, right? Well you certainly can, given that probiotics are safe and will not cause an adverse reaction in children. The problem is we don't really know which ones do what in people, and whether they really work for neurological disorders. The bottom line is that there are some very promising results done in very small trials, but what we need are larger well-designed clinical studies to figure out whether they really do work and which probiotic(s) should be used and for what.

Dos and Don'ts

◆ **Do—** get your kids to eat healthily and exercise regularly, since studies show how this benefits the gut microbiota, as well as improves brain development (they're probably linked). Remember that a healthier gut means a healthier brain, as the microbiota have much more to do with our brain than we previously thought.

◆ **Don't—** worry, be happy! Same for your kids. Easier said than done, but we do know that stress has a detrimental effect on the gut microbiota, which in turn can affect the brain.

◆ **Do—** follow the press and literature about probiotics, and look for large-scale controlled clinical studies that suggest they work. The field is currently very active, and could change quickly in the next few years. It often takes a long time to move from a successful trial into regular medical practice. This is especially true for more "unconventional" treatments such a probiotics, given their history and lack of regulation. It's also going to take a while to convince a neurologist or psychologist to treat a neurological disease with intestinal microbes.

◆ **Do—** consult with your child's pediatrician or psychiatrist, if she suffers from ASD or ADHD, about their opinion on the impact of the microbiota, but please don't expect microbiota alterations to be a certain cure. Consider consulting a different health practitioner, such as a gastroenterologist or a naturopath, who has been involved in treating ASD or ADHD patients. They may agree to try certain minimal risk options, such as antibiotics, probiotic treatment, or a fecal microbiota transplant (for ASD).

FOOD FOR THOUGHT

Can we improve brain function in children by feeding them certain foods? We know that malnourished children have decreased cognitive function, presumably because of nutrient deprivation, hampering full brain development.

Diet plays a major role in maintaining brain health in older people, as well. Eating plenty of plant fiber and other antioxidants decreases the risk of dementia and other neurological diseases such as Alzheimer's and Parkinson's later in life (if this interests you, you might read *Brainmaker* by David Perlmutter).

We also know that the brains of germ-free mice don't develop normally. So obviously good nutrition and microbes are key for proper brain function. But can you actually improve the brain of your child through diet? There are no definitive answers yet, but given all that we're learning about microbes, diet, and brain function, a fascinating (but very controversial) study would be comparing microbiota composition, diet, and IQ scores. Until someone does the experiment and finds out if there are "smart bugs," make sure your child eats as healthfully as possible, including lots of fruits and vegetables—just tell them it's "brain candy!"

15: Vaccines Work!

The Not-So-Magical Kingdom

Five-year-old Ethan was so excited he barely slept that night, anxiously tossing and turning, and waiting for the first rays of morning sunshine—he was going to Disneyland! He could finally see the big castle where Mickey Mouse lived, and check out the Pirates of the Caribbean (he had been obsessed with this movie lately and wore his pirate hat everywhere). At last, his mom and dad pushed his bedroom door open and he sat up like a spring. However, his giant smile was met by his parents' concerned faces. His father sat on his bed and gently said, "Ethan, we're very sorry, but we can't go to Disneyland right now."

On a tantrum scale of one to ten, Ethan threw an eleven. Finally, when Ethan had calmed down, he was able to listen to the reasons his parents were trying to give him: "There's a very bad disease called measles that is infecting children at Disneyland right now. You're protected from it, but your sister Olivia isn't old enough yet and she could get very, very sick." Ethan's parents were right; Olivia was only nine months, not old enough to be fully vaccinated and protected against measles. "We have to stay home," they said. To which Ethan promptly answered: "It's not faaaaaaiiiiiiiiir!"

Sure enough, during the spring of 2015 a measles outbreak that started at Disneyland infected 131 children. Disneyland was, for the first time ever, recommending families not to come unless every visitor was fully vaccinated. It was simply too risky. That winter, the Happiest Place on Earth was not quite so.

Measles is an extremely contagious childhood disease. One infected child can infect another eighteen kids, and approximately 90 percent of unvaccinated people will contract the disease if exposed to the virus. Measles is miserable! Fever, cough, runny red-rimmed eyes, and an extensive rash are the main symptoms. In some cases, measles cause lung infections (pneumonia), brain infections (encephalitis), and even death (1 in 2000 cases). The CDC calls it "the deadliest of all childhood rash/fever illnesses."

Fortunately, since 1970 there has been a childhood vaccine that has nearly eradicated this nasty disease, which affected 900,000 people per year prior to vaccination in the US. The vaccine is one part of the MMR vaccine (measles, mumps, and rubella) that's given to children at 12–15 months of age, and then a booster shot is given again at 4–6 years of age. Those of us born prior to 1970 will never forget how miserable this childhood disease is, as nearly all children suffered through it. The vaccine is produced by crippling a live virus to the point that it can't infect very well, yet it can still "tickle" the immune system and cause it to "remember" the virus. The immune system is very good at remembering these things, and if a vaccinated child is exposed to measles, he is protected 98 percent of the time.

Vaccines, along with antibiotics and sanitation, have been fantastic tools to combat infectious diseases. Vaccination has been remarkably successful, and has rid the world completely of smallpox, and very nearly of polio. Ironically, their success has also been their downfall, as living without these diseases have led some people to believe that vaccines aren't that necessary.

A year before the Disneyland outbreak, in the spring of 2014, an area just east of Vancouver, Canada, called the Fraser Valley, started to see cases of measles. Within four weeks there were over four hundred cases of measles, which is more than the province of British Columbia had seen in the last fifteen years combined and the worst outbreak in thirty years. When investigators started looking closely at the cause, they found that the cases were clustered at a Christian school (which had to be closed for a while) that's populated by the Reformed Congregation of North America. This group does not believe in vaccines, saying they're not safe and citing other religious reasons. Consequently, this area has a high rate of unvaccinated children. Fortunately, most of the surrounding area (and most of Canada) has over 90 percent vaccine compliance rates, which prevented further spread of the disease.

Despite this outbreak, according to one survey, 80 percent of anti-vaccine parents remain "not at all likely" to vaccinate against these nasty childhood diseases, citing health and religious concerns. You could say, "Fine, that's their choice." The problem is, it could also affect your kids, because to eliminate a disease from a population and break the infection cycle, the large majority has to be vaccinated. Children younger than twelve months, like Olivia, are not vaccinated and are at risk. In addition, the vaccine doesn't "take" in some kids, leaving a portion of people unprotected from the disease.

A Parent's Nightmare – What Do I Do?

A situation like this hit too close to home a few years ago, with a different preventable disease. Claire's son was only six weeks old when her husband took him to a doctor's appointment. It was her husband who needed to see the doctor, but he was in charge of their son that morning, so he came along for the ride. Strapped in his car seat, he slept through the whole thing like only a six-week-old can. There was nothing out of the ordinary until three days later, when the doctor's receptionist phoned them at home. She said that they had to bring their son in immediately to see the doctor because he had been exposed to whooping cough. It turned out that a different doctor from the same clinic had seen two school-age siblings with whooping cough an hour before Claire's husband's appointment. Claire assured the receptionist that her son was doing great and that he hadn't even been examined that day. Still, "Your son was here, I remember seeing him, this is why I'm calling you," she insisted. She said that because he was so young, he must be treated with antibiotics to prevent a possible infection.

Claire got dressed, asked her neighbor to look after her two-year-old daughter, and took her son to see the doctor. While in the waiting room, Claire read recent studies on whooping cough. She was familiar with the disease, but wanted to be more informed before she saw the doctor. Whooping cough, also known as pertussis, is transmitted by the bacterium *Bordetella pertussis*, a very contagious pathogen that causes severe coughing and difficulty breathing. In older children and adults it can last a long time (often called the hundred-day cough). In adults it's usually treated at home, often requiring antibiotics. However, infants often require hospitalization, and in this age group complications include pneumonia, seizures, brain damage, and even death.

Once Claire saw the doctor, he explained that the risk of contagion was extremely low, as her son never left the car seat, and he was never in the same room where the infected children were examined. His potential exposure came from being in the same large waiting room where these children had been an hour before. Although the doctor assured her that it was incredibly unlikely that her son would be infected, provincial health policy required him to recommend and prescribe her son a preventative dose of antibiotics. Claire wasn't very sure she agreed with what she was being told. So, right before the end of the consult, she asked him if he would give the antibiotics to one of his own children, to which he answered that he probably wouldn't. Once she got back to the car she opened the doctor's prescription note and read: *azithromycin for ten days.*

Now, Claire is not the type of patient to refuse medical treatment, but in this case she needed a second opinion. After all, her child seemed perfectly healthy and the risk of infection was minimal. Not only did she consider it unnecessary, she was obviously very aware of what ten days of azithromycin could do to her son's microbiota. She was already immersed in the microbiota field, reading study after study showing how antibiotics early in life increase the risk for all sorts of diseases. Plus, Claire's son was already genetically predisposed to asthma (with very severe cases of the illness in the immediate family), so the last thing he needed was a big shift in his microbiota at such a young age.

But after consulting with no less than six other physicians and researchers (mostly friends and friends of the family), the recommendations were completely mixed regarding whether or not Claire should give her son the antibiotics. So Claire came up with plan B:

for the next five days she was going to monitor her son's temperature every four hours around the clock. If at any point there was a small increase in temperature, she would start him on antibiotics. She followed the plan meticulously (just like doing an experiment), which made for five very long days and nights that she will likely always remember.

Fortunately, her son had not been exposed and he never developed whooping cough. Yet to this day and after imagining all possible scenarios, she still doesn't know if she made the right decision or not. As a mother of young children Claire has become familiar with the idea that parenting comes hand-in-hand with making decisions that you might regret later. However, in this particular case she knew that she shouldn't have been put in this position.

This happened in December 2012, a winter that saw the worst whooping cough epidemic in the Pacific Northwest since 1942. It was that day, waiting for Claire's doctor to see them, that she really understood how dangerously successful the anti–vaccination movement was. Those two kids in the waiting room had not been vaccinated, and as a result were a real threat to her son. More and more people are becoming convinced that growing up without vaccines is a safer option than getting vaccinated. How can this be? How can a parent believe that the risk of an extremely rare vaccine reaction is bigger than the risk of contracting a very dangerous infection?

Unlike popular belief from pro-vaccination activists, we don't think "anti-vaxxers" are acting out of a lack of information or mere stupidity. On the contrary, a quick online search on why some people choose not to vaccinate their children yields not hundreds, but thousands of articles with scary stories of vaccine reactions and reasons to avoid vaccines. In fact, some of the information we saw was somewhat compelling, and if we didn't have a strong scientific background it wouldn't be that hard to succumb to these theories. After all, Western lifestyle diseases are increasing dramatically and they are affecting more and younger children, and something must be causing this. As one parent said, "How can we believe vaccines don't cause autism, if science hasn't told us what does?" A very valid question from someone who is dealing with the overwhelming reality of having an autistic child, and who still has no answers to why this terrible disease is on the rise (see chapter 14 for a discussion on autism spectrum disorders).

Unfortunately, not knowing an answer does not make a wrong statement right, and the science behind anti-vaccination theories is not solid. There just isn't a single validated study that, as scientists, we can believe shows that vaccines cause diseases. Yes, vaccines are associated with certain reactions, and in very exceptional cases they can be severe, but in the end, it comes down to assessing risk. According to the World Health Organization, the risk associated with severe neurological reactions to the DPT vaccine (against diphtheria, polio, and pertussis) is extremely low—it occurs once in every 5 million cases. Compare that to the risk of your child developing whooping cough in the state of Washington, where there are 62 cases per 100,000 residents, and the number continues to rise every year. That is one thousand times more likely. Moreover, add the risk of not only having your child suffer a severe infection, but also passing this infection to other kids, who are either too young or too sick to be vaccinated, like a six-week-old visiting a health clinic.

We're currently parenting during the digital age, where any information we want (and don't want) is at the tip of our fingers. It's not easy to make decisions with this amount of information around, but please, don't fall for the quick-access blogs and "health-oriented" articles promoting an all-natural approach to protecting your children from vaccines. Not unless you decide to move to the middle of the woods, away from society. Infectious diseases are a reality of living in large groups of people; they have been around as long as we have. The only reason our children don't suffer from them now is because of vaccines, and without vaccination there's no other alternative than having these diseases come back. So, despite extremely rare cases in which vaccines cause severe reactions, vaccines work and are one of the safest medications in the world.

Vaccines and Microbiota— Is there a Connection?

Vaccines work, yes, but they are not perfect. Why don't vaccines protect 100 percent of the people that take them? Most vaccines work about 85–98 percent of the time, which leaves a significant number of people semi- or unprotected. A good example is the seasonal flu vaccine. We all know people who get the vaccine, but then still get the flu. We also know that vaccines that work really well in developed countries often work poorly in less developed places where these diseases have a major toll on children. This is true for polio, rotavirus, and cholera vaccines. Remember, these children have a very different microbiota—could this be influencing their vaccine responses? The microbiota has an important role to play in the way our immune system functions.

As an example, when animals are given antibiotics their antibody response changes. Antibodies are an essential aspect of a vaccine response and antibiotics are a great method to shift the microbiota. Furthermore, animals that are raised germ-free have poor responses to vaccines, and feeding probiotics or prebiotics to mice affects their subsequent responses to vaccines. All these results suggest that the microbiota may affect vaccine responses.

This makes sense, as the microbiota is critical for your child's immune system to develop normally. In fact, specific microbes have been identified that shift the immune response in different ways. Throughout this book we've seen that our microbiota plays an important role in many diseases because of their ability to influence our immune system. From this perspective, the microbiota can also influence a vaccine response just as it does with other aspects of our immune function.

Scientifically understanding the role of the microbiota in vaccine responses is in its very early days. There are only a few experiments thus far, although they are increasing in number. We do know that in people given a typhoid fever vaccine, the vaccine did not affect the microbiota. Although we expected it, this is good news. The microbe that causes typhoid fever (a type of *Salmonella*) is not part of the normal microbiota, and is usually only present when it causes disease, so the vaccine targets that microbe only. However, this does raise a bigger question—what happens when we take a microbial species out of the normal population? Does this change the overall microbiota structure? Could something worse crawl in?

The experiment has been done once, with the elimination of the smallpox virus from the world, and luckily this scenario has not happened—no new nasty virus has appeared in its place. With the recent introduction of pneumococcal (pneumonia) vaccines, which target fairly common microbes of the respiratory tract that cause ear infections, we will be watching closely to see what, if anything, replaces them. In experiments done on macaque monkeys, it was found that monkeys with the most diverse microbiota responded the best to certain diarrheal vaccines. We presume a similar concept holds true for children, again arguing for letting our kids eat dirt, and whatever else they can safely put in their mouths. It seems that having a diverse microbiota is beneficial all-around and we should provide our children with opportunities to diversify their microbiota.

What about the less fortunate parts of the world? There are major groups, such as the Gates Foundation and Gavi, The Vaccine Alliance, that are trying to get as many children as possible vaccinated against common childhood diseases. However, as mentioned above, children in developing countries do not respond as well to vaccines as those in developed countries (where the vaccines are often developed and tested).

With the aim to study how to improve this, our laboratory has developed a mouse model that mimics some of the conditions observed in developing countries. By shifting their diets to have more carbohydrates and less protein and fats (similar to certain childhood diets in developing countries), and then feeding specific intestinal microbes to these mice (kids in developing countries live in less hygienic conditions), these mice have remarkably similar symptoms to malnourished children—they have stunted growth, intestinal inflammation, chronic diarrhea, poor development, and all the things one normally sees in these children. Intriguingly, these animals have very different immune responses, too. This puts us in a position to experimentally examine how the microbiota affects the immune response, hopefully enabling us to work on making even better vaccines and delivery systems for impoverished children of this world.

As mentioned before, besides not being 100 percent effective, vaccines, like any medicine, are not 100 percent safe. Sometimes there can be side effects, although rarely are they serious. Most of the side effects are minor—pain (hey, needles hurt), swelling, and redness. As anyone who has taken a first aid course knows, these are the cardinal signs of inflammation. Ironically, this is probably good, as it means you're tweaking the immune system, and it needs to activate and remember it (much like one certainly remembers better if you are kicked in the shins, rather than gently tapped on the shoulder). Fever and irritability can accompany vaccination, although these can be dealt with fairly easily. It's the serious complications such as seizures and even a risk of dying that one worries about. Using the MMR vaccine as an example, there has been just one death in more than fourteen years, although it wasn't directly attributed to the vaccine. Serious events for most vaccines average about one in a million. However, before the MMR vaccine was introduced, one in a thousand kids died of measles, which means there was a thousand-fold higher chance of dying before the measles vaccine was developed. The problem is that we don't see children dying from these diseases because of vaccines, so even one in a million odds seem high when dealing with your own child.

How could the microbiota play into adverse vaccine events? Again, we know the microbiota impact on our immune system development, which is tweaked by vaccines. There's currently no information either way about how microbiota might affect such outcomes, but given everything we're learning about how much these microbes do in our body, there's certainly a possibility that they might be part of the equation. Time will tell. Until then, it's best to act wisely and vaccinate our children and try our best to keep their microbiota healthy.

Dos and Don'ts

- **Do–** get your child vaccinated according to the standard guidelines (these vary by state, province, and country—the standard vaccine schedules for different areas are all available online). If you remain skeptical about vaccines, have a serious discussion with your pediatrician about the risks associated with childhood vaccines.

- **Don't–** believe everything you read on the web about vaccines causing autism and other diseases. Some of this is based on fraudulent science that has been retracted, and other hype that is not based on science at all. There is a very small risk of adverse reactions, but these are minimal compared to the problems caused by the actual disease. Talk to your doctor, or your grandparents that lived before vaccines—they were terrified of polio and other diseases we no longer even hear about.

- **Do–** follow our advice throughout this book on how to maintain your children's microbiota in good health. A diverse microbiota promotes healthy immune responses, and increases resistance to infectious diseases.

A MYTH THAT HAS LASTED
FAR TOO LONG

The MMR vaccine is a highly effective vaccine that is routinely given to all children as part of the normal vaccine schedule to prevent measles, mumps, and rubella. In 1998, Andrew Wakefield and colleagues published a paper in the prestigious medical journal the *Lancet*, suggesting that the MMR vaccine causes autism spectrum disorder (autism), based on a very small study of twelve children. The media picked up the story, and very rapidly the rates of vaccinations in the UK and Ireland dropped, resulting in a jump in measles and mumps cases, along with the deaths and long-term damage associated with these diseases.

Unfortunately, the scientific study was a fraud. Wakefield did not declare the numerous conflicts of interest regarding his sources of funding, plus he manipulated evidence and breached several ethical codes. He was stripped of his medical license in the UK. As study after study came out showing no correlation between MMR vaccines and autism, the *Lancet* partially retracted the paper in 2004, and fully retracted it in 2010.

At least a dozen studies since have shown no correlation between autism and the MMR vaccine. The latest, published in 2015, looked at 95,000 US children, including younger siblings of autistic kids (who have a higher chance of getting autism), and again concluded there was no correlation. One medical journal article called the original study "perhaps the most damaging medical hoax in the last 100 years."

The question is: Why is this false myth still propagated, in the face of overwhelming scientific evidence to the contrary? It's an interesting commentary on science and society, the media, and parents desperate to find the source of a heart-wrenching disease. The media has been blamed for significantly fanning the story, and continues to mention it. For example, there were five times the number of evening news stories on it in 2010 than in 2001, long after the science had thoroughly debunked it. Even celebrities got on the bandwagon—like Jenny McCarthy, who publicly blamed her son's autism on the MMR vaccine—which then spurred even more misinformation and Internet coverage. Many sites on the Internet also continue to promote this link (there are typically no referees or other quality controls on websites), and those opposed to vaccines continue to cite the original study.

A cornerstone to any scientific finding is the ability to be repeated by others and stand the test of time. The link between the MMR vaccine and autism does neither, and MMR continues to be a highly effective vaccine preventing diseases that, prior to the vaccine, maimed and killed countless children.

16: Bugs As Drugs

The Future

Sometime in the near future:

Doctor: Congratulations, the tests confirm that you're pregnant! Your fecal microbiota tests also suggest that there may be ways to improve the development of your baby during your pregnancy. Our nutritionist will suggest a modified diet that will alter your intestinal microbiota to enhance the health of your baby.

Doctor: Congratulations, it's going to be a girl! Everything looks great on the ultrasound, but your baby is still in a breech position and we might have to opt for a C-section if she doesn't turn around in time. One thing we may consider during a C-section is to take a vaginal swab before delivery and wipe your baby's mouth with it in order to give your baby the microbes you would have given her if she had been born vaginally. This will make her healthier in the long run.

Doctor: Your baby turned around just in time, and your vaginal delivery went great! As you know, you've been given antibiotics for Group B strep as a preventative measure, which may have altered your newborn's microbiota. Since you're breastfeeding, we'd like you to put a few drops of this solution on your breasts just prior to feeding. It contains a few probiotics that will replenish your baby's microbiota with microbes that are beneficial at this stage for your child's early mental and immune development.

Doctor: Don't worry, urinary tract infections like the one your daughter has are fairly common at six months of age. The antibiotic treatment worked nicely and the infection is now cleared. However, we noticed that because of the antibiotic we gave her, your child's gut is now missing some good bugs, putting her at risk for allergies and asthma. Here is a solution of four microbes that can be given orally that will replenish these organisms.

Doctor: Your toddler is developing beautifully! We noticed, however, that she has a certain microbe that might put her at increased risk for autism—nothing to worry too much about, especially because we can give her this pill that gets rid of this bad bug and leaves the good ones alone.

Doctor: Thanks for coming in for a three-year-old checkup! We noticed that when we tested your daughter's urine, there were certain molecules called metabolites that put her at a higher risk for inflammatory bowel disease. Luckily, we're going to give you this medicine that will help push her microbiota to a healthier population that will decrease her risk.

Doctor: I can't believe your daughter is starting kindergarten! It's great you're here for her vaccinations. We noticed she didn't have a really strong response to the measles vaccine we gave her when she was one, so we're going to give her this specially designed probiotic, which will help the booster shots work better.

Doctor: Thanks for bringing your daughter for her checkup before she starts first grade. She's doing fantastic, but we did notice that she has been gaining more weight than is ideal for her age. By having her wear the glucose monitor and doing an extensive analysis of the food she ate, we were able to determine which foods triggered a glucose spike based on her microbiota composition. The good news is that we're able to recommend a specific diet for her—this is her very own personalized diet, and yes, she can eat ice cream and pizza. However, there are some other foods that she should avoid, as they're the ones that trigger the weight gain.

Could these fictional conversations become reality in the future, indeed in a few short years? Definitely! All of the above examples are based on concepts that are already well under way in labs around the world, and are rapidly being developed and commercialized as new ways to promote health and prevent or treat diseases.

Understanding the Microbiome

The underlying concepts of the above conversations are that we can: a) rapidly determine a person's microbial composition, and b) actually do something to correct it if it needs altering. The first is relatively easy, and there are several companies that can analyze a person's microbiota for as little as $100 per sample. The trick is to know how to make sense of that analysis. The second concept is much more profound. Every person is born with the same DNA they die with—it is impossible to change our *Homo sapiens* DNA in our lifetime, other than perhaps through some skin cell mutations picked up from a bad sunburn in our youth (and even these changes won't alter the DNA that we pass on to our offspring). We do not evolve in our lifetime; it takes many generations to select and pass along genetic changes that then become part of the population.

However, we already know that we can rapidly shift the microbiota in and on our bodies—diet changes, antibiotics, probiotics, and fecal transfers all result in microbial changes. Specific gene therapy to correct a defective human gene is still an experimental technique, and unfortunately an early trial had a fatal outcome, which has slowed progress. But given that microbes have at least 150 times more genes than we do, and that we can readily change the microbes (and their genes), there is terrific excitement and promise in manipulating the microbiome in certain ways in children, and even in adults. You already do it every time you eat yogurt with probiotics in it, or when you travel to another country with a very different cuisine.

Research is getting closer to finding more effective ways of targeting and modifying our microbiome. This chapter explores some of these promising methods and their implications, and what is being done to make those conversations with the doctor a reality.

As DNA sequencing began to come on line in the early 1990s, we started to talk about sequencing the entire human genome. In the early days it was pie-in-the-sky discussion, much like the fictional idea from Jurassic Park (sequence a dinosaur's genome from a fossilized insect). However, the sequencing technology improved rapidly and, with a combined massive international effort, by 2003 we knew the entire genome sequence of humans. This was truly a remarkable scientific milestone (although to this day we still don't know exactly how many genes there really are). However, with the completion of the Human Genome Project, there were an awful lot of scientists with DNA sequencing machines wondering what they could sequence next. The microbiome represented a wonderfully appealing target because of its size—much bigger than the human genome and with lots more to sequence. This was only a few years ago, but at that time we knew very little about the composition of the microbiome; what we knew was solely based on our ability to grow a handful of microbes in the lab.

As a result the Human Microbiome Project was launched, with one of its major goals being to establish the composition and sequence of the human microbiome, much like the hugely successful human genome sequence. Samples from five different body sites (airways, intestines, mouth, skin, and vagina) were taken from more than a hundred "normal" people. What is "normal"? Young, healthy, no antibiotics—but of course there is still debate about what a normal human is. From this large and ambitious project we learned much of what we know about the human microbiota today, and hundreds of new bacterial genomes were sequenced. The results from this project came out in 2012, which is considered old news in this incredibly fast-paced field.

By the time the project began, scientists were starting to realize that the microbiota was important, and we generally assumed we would identify a core human microbiome that we all shared. Wrong! What we learned is that each person has his or her own set of microbes. This was spectacularly confusing (and frustrating), but it did keep the sequencers happy, as they had lots more to sequence. The finding that each individual has their own microbiota has held true, and for the vast majority of one's life it stays fairly constant. So how do we deal with the complexity of each person having a different microbiota? How can we come up with general microbiome therapies that work on most people if everyone is so different? Surprisingly, it isn't as impossible as it sounds.

A golden rule in biology is that if the function of something is important, it is used widely again and again (i.e., it is conserved). Based on this concept, if the microbiota carries out important functions, there must be something these microbes do that is in common. When scientists analyzed the microbiota not by the identity of its members but by what they do, the picture became much less confusing. Given that there's probably a common core of microbial genes that need to be turned on while living off a human, different microbes should have similar genes that do the same job. In other words, it doesn't really matter which microbe the gene is in, as long as it's there and its product is being made. When one looks at the microbiome from this functional point of view, there really does seem to be a set of core genes that are needed to make us the normal functioning human beings we are.

Analysis of *Your* Microbiome

There are two main methods for determining the composition of a person's microbiome. The first is to take a sample (i.e., feces) and sequence all the DNA in it; subtract out the human sequences and what's left are the microbial sequences. This is a labor-intensive (and expensive) way of doing it, and may be realistic for only a handful of people.

The alternative, and by far the more common way, is to only sequence a gene found in all bacteria called the 16S rRNA gene. Some parts of this gene are the same in all bacteria (this gives us a common handle to grab on to and is needed for sequencing), and other parts of the gene are different in different bacterial species, which gives a unique fingerprint for each microbe. The major advantage of this is that we don't have to grow the microbe in the lab (we still can't grow many of the microbes in the human body), and the amount of sequencing data obtained is a lot more manageable (by manageable, we mean being about half a million sequences per sample!). The companies offering to sequence your microbiota for a small fee are regularly doing this kind of sequencing—all you have to do is mail them a small fecal sample (lucky postman!).

The problem with both methods is that you need to figure out what this massive dump of data really means. This is where the science is these days: Many microbiologists are culturing fewer bacteria and behaving more like computer scientists, sitting in front of computer screens for a good part of the day. Bioinformatics, the science that uses computers to handle large biological data, plays a major part in this, as the data are extremely complex. Bioinformatic platforms need to be built, combining many different programs, with the output ideally telling you a) the composition of the microbiota in that sample, and more importantly b) what it means—is it good, bad, or do we just not know yet (yet is the operative word here). As discussed previously, because of the large differences between each person, this is actually quite difficult to do.

Beyond Genes: Microbial Metabolites

There's a third method of analysis that's developing rapidly, and will complement and possibly even replace the two DNA analyses mentioned above. Every microbe goes about its business, making small molecules (metabolites) as part of its normal life of breaking down food, generating energy, and just generally living. In the past decade science has made tremendous advances in analyzing small molecules using sophisticated machines called mass spectrometers ("mass spec"). These powerful machines can take a complex mixture of molecules and figure out the weight of each molecule in the mixture. Nearly every molecule has a unique weight, so this gives us a fingerprint of what's in the mix. The problem is that we need to know what molecule each mass represents, which is a problem if the mixture contains molecules that no one has seen before. So far, we can confidently detect about 20 percent of the human metabolites, and less than 1 percent of the microbial metabolites. However, this is where the action is, because these small molecules produced by microbes, by humans, or by both, are telling us how we interact with our microbes. Knowing the names of the microbes or even the names of their genes only tells us what these microbes *may* be doing. In contrast, information on all our metabolites, also known as metabolomics, tells us what the microbes are actually doing. This, in our view, is where the future is.

In one of the futuristic examples given above, the doctor mentioned analyzing a urine sample for metabolites, and based on that, figuring out what microbes might be in the gut—these are what we call disease biomarkers. In our work, we've found that we can identify metabolites in the urine from three-month-old children that are indicative of increased asthma risk years before a child actually gets the disease. Strikingly, some of the metabolites found in urine are produced by intestinal microbes. How did they get there? They travel. Although gut bacteria live in the gut, the molecules they make or modify can enter the human body and end up in the urine (or brain, or the placenta, or anywhere else in the body). This is how microbes talk to us, and how we listen to what they have to say. Luckily scientists are slowly getting better at listening to their signals—an area that could have a huge impact on diagnostics.

Although metabolomics is still in an early phase, as we figure out which metabolites are important and which bugs they come from, it will make a very powerful technique indeed to analyze our microbiota and how they relate to health or disease. In the not-so-distant future we'll be able to predict a child's risk for certain diseases before they occur, based on "the talk" from his or her microbiota.

Second Generation Probiotics

Okay, now that we can figure out your child's microbiome, and decide that perhaps it needs tweaking for optimal health, how do we go about it? As we said before, the good thing is that the microbiota is much easier to change than human genes. There are several methods already under way, with more sophisticated ones coming.

By far the oldest method is to use probiotics. These are live bacterial strains that you put in food or drink, that won't harm you, and that may or may not have any effect on you. We've been doing this ever since humans started eating fermented food such as sauerkraut centuries ago (microbes cause the "fermented" part of these foods). However, probiotic bacteria don't stick to an already full house in the gut, and can't easily become part of your microbiota. The solution? You have to take them daily in great numbers (10 billion plus), which is of course of great corporate benefit but not necessarily as effective as it could be.

There's an entire field dedicated to probiotics, with claims that they improve nearly everything you can imagine. The problem with the current field is twofold. First, it's usually only one strain that a company champions (e.g., certain *Lactobacillus* or *E. coli* strains), and it's hard to believe that a single, noncompetitive microbe can have so many profound health benefits when we know that the microbiota is such a complex ecological community. Second, probiotics are not currently regulated by the FDA, and do not have to go through the incredibly rigorous clinical trials that drugs do. This also means that beneficial claims made for a probiotic are not necessarily backed up by extensive clinical trials. Probiotics are also often designed for longer shelf life stability and ease of manufacturing, rather than for medicinal properties (since they aren't regulated). Of all of the dozens of probiotic options you see in a store, very few of them follow rigorous microbiological methods to ensure that the microbes will stay alive and active by the time you take them. However, as we have seen throughout this book, many hints and small studies indicate that probiotics have some beneficial effects. Ask your health practitioner to recommend probiotics that have been tested in clinical trials.

With the increasing knowledge of microbiota and how they work, second generation probiotics are going to play a major role in our health and disease fairly soon. Work is already under way to create even better probiotics, by using mixtures of several microbes together, instead of just one or two species, and by using microbes that are normal residents of the microbiota and have a documented beneficial role. This makes sense, as the aim is to deliver an entire functioning microbial community, which should probably colonize and work better than the current single strains.

Probiotics of the future will include microbes that are quite happy to colonize and remain within you. Current probiotics are also being altered to express anti-inflammatory products, or encode adhesins that promote their colonization in the gut. This has significant implications, as you wouldn't have to take them daily, but it will increase safety concerns and affect marketing. Whether these will be FDA regulated (to ensure safety and prove claims of efficacy) is a heated discussion these days.

Prebiotics

Another area that's seeing increased attention is the use of prebiotics. These are usually complex carbohydrates or sugars, such as fiber, that serve as a food source for certain types of microbes. The concept is that if you eat this specific microbe food, you will enhance those particular microbes.

Again, this concept has been around for a while, and has seen varying degrees of success. It's difficult to find a carbohydrate that enriches a single organism, so their effects tend to be broader. Like probiotics, prebiotics are not regulated, so verification of the claims in controlled clinical trials is usually lacking. However, with our increased knowledge of microbiota and the effects of different diets, it's not hard to imagine getting human volunteers to eat various sugars and follow specific diets, and then analyze their microbiota to define exactly how these prebiotics work. Diet changes alter microbiota, so if we can figure out exactly what changes different prebiotics cause, they could show much future promise. In the context of the microbiota, you truly are what you eat.

Back to the Future: Fecal Transfers

Throughout this book, we have discussed the concept that a child may have a microbiota composition that puts them at risk for IBD, obesity, asthma, or other disease, or one that has been altered and unbalanced by antibiotics, gut inflammation, etc. Significant inroads are currently being made to correct this. Fecal microbiota transfers (FMTs) have completely changed how we think about manipulating the microbiota. These involve the transfer of feces (and all the microbes that they contain) from one person to another, either orally or by enema. Although they have been used in China since the fourth century to treat diarrhea and other diseases, they have recently gained huge notoriety because a) in certain cases they work very well, and b) it is a very gross concept. However, for people who have long-suffered a debilitating disease, opting for an FMT is an easy choice.

Most people undergoing surgery are given antibiotics to prevent secondary infections. However, as we have discussed, antibiotics carpet bomb the microbiota, which can allow potentially harmful microbes to gain a foothold, especially if the person is old, sick, or otherwise more susceptible. One such bacterial pathogen seen under these conditions is *Clostridium difficile* or *C. diff.* This has caused major problems in hospitals all over Canada, the US, and elsewhere (see Poop vs. *C. Diff* on page 257 for a recent study on *C. diff* cases in children). The real problem is that antibiotics given to treat the *C. diff* infection are less than 20 percent effective (they actually caused the problem in the first place, so why would they work now?), and this is a deadly disease. Ironically, *C. diff* is a relatively wimpy pathogen, and is easily outcompeted by pretty much any normal microbe, which in a way explains why it doesn't usually cause infections in healthy people. Several studies have now shown that simple fecal transfers are over 90 percent effective to treat *C. diff* infections. The simple act of delivering fecal microbes, either through a nasal tube to the gut or by enema, cures a potentially fatal disease.

This is the true proof that microbiota manipulations have a real place in modern medicine. However, because fecal transfers are really body fluid transfers, concerns have been raised, and rightfully so. Remember the issues with blood donors and hepatitis C several years ago? The medical community didn't know about the hepatitis C virus then, so blood was not tested for it, resulting in many hepatitis cases after blood transfusions. The same happened with HIV in the 1980s. As a result, the FDA has now put very tight restrictions on fecal transfers, so that a physician or company has to do all the paperwork that is normally required for an investigational new drug, which amounts to roomfuls of documents. In the US, this has severely dampened the clinical use of fecal transfers. Because of its simplicity, there are now even YouTube videos showing how to do it yourself—don't! This is causing significant medical concern, as there are still risks associated with the process, especially in a non-clinical setting.

In addition, many experts on fecal transfers believe it's necessary to give a heavy dose of antibiotics before the transfer, in order to remove the unwanted microbiota and increase the chances of the donor microbiota sticking. Logically, this cannot be achieved if someone opts for the DIY method.

The spectacular success of fecal transfers for *C. diff* has unleashed a flurry of fecal transfer clinical trials for other microbiota-associated diseases, such as IBD and autism. Thus far the results have been mixed, and not nearly as successful as with *C. diff* infections. However, this makes sense. In *C. diff* cases, there is one known cause of the disease, so it really doesn't matter much which organisms you put in (i.e., whose feces you use), as long as *C. diff* gets booted out of your system. With IBD, it's a big ask of the incoming microbes—they have to be able to colonize an already inflamed gut (inflammation kills microbes), displace the current resident population, and then dampen the inflammation in a human genetically predisposed to this disease. Although we're starting to identify microbes that seem to be beneficial for IBD, it's important whose feces you use as a donor, and we just don't know enough about this yet.

RePOOPulating Our Gut

OK, feces are gross—enough already! There should be nicer ways to alter our microbes, right? There are certainly people working on this. Among them, Dr. Emma Allen-Vercoe, a collaborator of ours at the University of Guelph in Canada, has become very skilled at growing fecal microbes in fermenters (containers without air). Although a bit smelly (she has an entire floor dedicated to this for obvious reasons), she can grow defined cultures of twenty to thirty human microbiota species together in a fermenter, or as she calls them, a "robogut." Her team of scientists has even put this defined population into two people with *C. diff*, with the great advantage that it doesn't contain bodily fluids, nasty viruses, etc. In both of these cases the transplant worked well. Emma calls this concept rePOOPulating people, which is a wonderful term as far as we are concerned. Her team is now working at producing these microbes under special clean conditions so that they're pharmaceutical grade.

In addition, others are working at packing feces or microbial cultures into gelatin-coated capsules. You have to take a fair number of

these "pills," but they can be sugarcoated, and are not nearly as ob-
noxious as the alternative. There's no doubt that fecal transfers will
soon become a thing of the past as these more sophisticated methods
are further developed, but they have already served their purpose
by demonstrating how powerful a change in the microbiota can be.

Crystal Ball Time

Where are we going? We can now analyze our microbiota, but we
need intelligent ways of manipulating them. Commercial interest
in this area is exploding, with multimillion dollar deals being announced between pharmaceutical companies and smaller biotech
companies that are developing potential therapies. Some are working
on delivering specific microbial populations for defined benefits. We
know the microbiota has a profound influence early in life in determining how the immune system and the brain develop. This is an
area of huge potential therapy, and several companies are defining
groups of microbes that have direct effects on immune development.
They plan to use these as potential therapies for IBD, asthma, vaccine responses, and a multitude of other diseases.

More sophisticated methods are being developed to specifically
alter microbiota populations. In a recent example, a compound was
developed that specifically binds to a cavity-causing bacterium that
lives in the mouth. In addition to targeting that bacterium, the compound also had an antibacterial activity coupled to it so it would kill
only the microbes it bound. Using this technique one could specifically target a single microbe and remove it from the population,
another first in the microbiology field. This breakthrough suggests

that in diseases in which just one or a small number of microbes are associated with the disease, these microbes could be targeted without altering the rest of the microbial community, which is much different than the concept of antibiotics that cause huge collateral damage.

Another area that is being touted is "phage therapy." Bacteria, like us, have viruses that attack specific species. These are called bacteriophages. Because each phage targets a particular bacterial species, there's optimism that if viruses specific to a bad bug can be identified, produced, and delivered, they can be used to target that microbe in the body. It's a very appealing concept (let the phage do the hard work), but, like most things, has its problems. Phage therapy has been around for a long time (it was explored extensively in Russia to try to cure infections), but has not been integrated into Western medicine. The problem is that the bacteria don't really like to be blown apart by a phage (who does?), and quickly mutate to become resistant to the phage. Thus, if phage therapy is used extensively, we will see rapid resistance, just as we have seen with antibiotic resistance, rendering the phage therapy useless.

Recently there's been a major discovery of a system that can be used to target specific genes in most organisms, called CRISPR/Cas9. This system can be used to target a gene and cut it, killing the organism. Bacteria have used systems such as this as a type of immune system to defend against the viruses that infect them. Data are emerging to indicate we may be able to use this system to specifically target a unique microbe within the microbiota. This has obvious applications, but is still in its very early stages of development.

Personalized Diets

Let's finish discussing the future with what the microbiota does best—breaking down our food and making energy for us (and themselves). As we discussed in the chapter on obesity, microbiota play a role in this worldwide problem. As we all know, perhaps from personal experience, diets just don't work very well—we lose a bit of weight, but then we gain it back. Also, given what we now know, it's hard to believe that one diet would work for everyone, given the differences in our microbiota and what they do. We also know that some people seem to have all the luck—they can eat anything yet stay skinny as a rail, while most of us can only look on enviously. All this suggests that we should think about personalized diets tailored to our microbiota.

There's work already under way in Israel by Dr. Eran Elinav and his colleagues at the Weizmann Institute of Science, where they're working at personalizing an individual's diet based on their microbiome. They're using massive data analysis to correlate the microbiota and glucose (sugar) spikes in different people, and are finding, not surprisingly, that different individuals (depending on their microbiota) respond differently in their glucose spikes with different foods. They really did find people who could eat ice cream and pizza without causing glucose spikes (now that is one feces every kid should want!). Our presumption is that work such as this will completely change the entire dieting world, as we become much more sophisticated at designing personalized diets, with hopefully much better outcomes in controlling weight and other aspects of our health.

Given all this, we hope it's apparent that the fictional physician conversations at the beginning of this chapter are based on what's already being tested in labs around the world. This area is changing very fast, and the results will be in the clinic or on the market much faster than standard drugs for the reasons previously discussed. It's truly an exciting time, and we believe this will result in a major revolution in medicine, empowering us to deal with many of the most common health problems plaguing modern society in ways we couldn't even dream of a few short years ago. Hang on, it's going to be a fun ride!

Dos and Don'ts

+ **Don't—** believe everything you hear about the microbiome; trust your physician to stay on top of what has been proven medically. It takes a long time from the eureka moment in the lab until something becomes common medical practice. There's an awful lot of information available that has no scientific backing, which makes things much more confusing. However, if a treatment passes full randomized clinical trials, and is FDA-approved, you can be certain it has been extensively tested, and the claims are real.

+ **Do—** look into a fecal transfer, if you or anyone you know gets *C. diff* (but do *not* do it yourself!). Unfortunately, this is a common hospital infection, usually following surgery and antibiotics. The use of antibiotics to treat it has had a very poor success rate. There are now good clinical trials that prove that fecal transfers are much superior to antibiotics, even in kids, although there are still regulatory hurdles to go through as this becomes more mainstream.

+ **Do—** pay attention: this is a rapidly changing field, and new treatments could come on line quickly, and might be of use for your child, perhaps even as part of a clinical trial. Because of the extensive testing needed for full clinical approval, one often has a pretty good idea if something works long before it is officially approved. By getting involved in clinical trials at an earlier stage, you can often benefit from these treatments before they're fully approved. If the treatment works spectacularly well, they will stop the trial before it's done, and even treat the controls. This happened for fecal transfers and *C. diff*. It was unethical to keep the controls untreated, as it was obvious it worked so well.

◆ **Don't—** believe that microbiota will cure everything—they can have effects, but these are part of complex interactions between complex populations of microbes with the environment and our genes. This is complex science, with many factors involved. Because of the complexity of the microbiota, and its overlapping functions, it's going to take a lot of science to figure out exactly how things work. However, there are more than enough examples to indicate that the microbiota plays a major role in both our health and disease. Right now it's an incredibly popular area of science, and there's a major bandwagon effect. As science plods on, it will tease out which effects are real and which ones might not be.

And finally, as we've said all along:

◆ **Do—** let your kid be a kid and interact with their world, and develop as kids have for the past million years. Let them eat dirt!

POOP VS. *C. DIFF*

C. diff infections in children are becoming more recurrent and severe. What's more alarming is that this infection is not just occurring in hospitals, but also in day cares and schools.

To address this, the Mayo Clinic in Rochester, Minnesota, began a fecal transplant program for children in 2013, something that had been avoided in the past due to safety concerns in dealing with pediatric patients. The results have been outstanding. Every one of the twenty-seven children treated so far dramatically improved almost immediately after the transplant (a 100 percent cure rate!). Many of the parents simply could not believe that a cure could occur so fast and so simply. Some parents came to this clinical study after their pediatricians told them that FMTs were dangerous, so it's crucial that more doctors are aware of these results and that more clinics start performing these transplants safely across North America.

Acknowledgments

First and foremost, we would like to thank our spouses and lifelong partners in parenting, Jane Finlay and Esteban Acuña. Their support, input, critical reading and editing, and help in so many other ways ("Can you get the kids . . . again?") has made this possible. We especially would like to thank Jane, who carefully read and edited all the chapters from the perspective of a practicing pediatrician and a certified infectious diseases expert to ensure the medical accuracy of the book's contexts and bring a real-life pediatrician point of view to it all, and to keep us PhDs on track.

So many people have been incredibly helpful in so many ways. Janis Sarra, Pieter Cullis, and Joel Bakan demystified the book agent process, and Joel introduced us to our wonderful literary agents, John Pearce and Chris Casuccio from Westwood Creative Artists. From the moment we sent our first pages to John and Chris, they were as committed to this project as we were. They have been a treat to work with and we are very grateful for all their intelligent, creative, and on-point suggestions; this book became a reality in great part thanks to them. More importantly, it was through them that we were able to work with Andra Miller from Algonquin Books and Nancy Flight from Greystone Books, our extremely talented editors who did so much to simplify this process for us, and provided such insightful, helpful guidance and editing of the entire book.

A wide range of people from numerous different areas of expertise helped us improve the manuscript, as this was our first attempt at writing a public book, rather than a scientific paper. We figured if they could understand it, we were on the right track. These include Nancy Gallini (an economist), Janis Sarra (a legal scholar), Toph Marshall (a Greek papyrus scholar!), Lara O'Donnell (a scientific program manager), Shaylih Muehlmann (an anthropologist and new mother who studies the Mexican drug trade while breastfeeding her son), Stefanie Vogt (a fellow microbiologist), Rocio Pazos (a high school teacher), Wendy Colling (a high school teacher), Mariela Podolski (a psychiatrist), Melania Acuña (a dentist), Shelly Blessin (an inventory analyst), Adriana Arrieta (an economist), Edmond Chan (a pediatrician), Stuart Turvey (a pediatrician-scientist), and Tobias Rees (an anthropologist).

So many other wonderfully generous people supplied personal and moving anecdotes, questions, and experiences, which are by far the most interesting part of the book. These include Pat Patrick, Ellen Bolte, Carley Akehurst, Thomas Louie, Emma Allen-Vercoe, Eran Elinav, Marjorie Harris, Julia Ewaschuk, Malcolm and Jennie Kendall, Veronica Niehaus, Margo Nelson, Joey Dubuc, Ivonne Montealegre, Anamaria Castillo, Astrid Antillón, Kristie Keeney, Jennifer Sweeten, Amanda Webster, Agnes Wong, Navkiran Gill, Amanda Roe, Erin Sawyer, Marilyn Robertson, Roxana Ramírez, and Fiorella Chinchilla.

As scientists, we have had the good fortune to have grant funding from the Canadian Institutes of Health Research (CIHR), Canadian Institute for Advanced Research (CIFAR), and other agencies that have allowed us to mentally and experimentally explore several of the topics discussed in this book. Brett would also like to thank the Peter Wall Institute for Advanced Studies for his endowed Chair and the Carnegie Foundation for funding his sabbatical stay in Scotland where much of this book was written, and much good whiskey was experimentally sampled (all in the name of science, of course).

After finishing this book we continue to be very interested in the topics at hand and welcome any questions or comments about what is covered in these pages. We have created www.letthemeatdirt.com, which provides a selection of practical information, as well as our contact information.

Selected References

Chapter 1

1. Cox, L. M., and M. J. Blaser. "Antibiotics in early life and obesity." *Nature Reviews Endocrinology* 11, no. 3 (2014): 182–90.

2. Strachan, D. P. "Hay fever, hygiene, and household size." *BMJ* 299, no. 6710 (1989): 1259–60.

3. Mutius, E. von. "Allergies, infections and the hygiene hypothesis—the epidemiological evidence." *Immunobiology* 212, no. 6 (2007): 433–9.

Chapter 2

4. Arrieta, M. C., and B. B. Finlay. "The commensal microbiota drives immune homeostasis." *Frontiers in Immunology* 3 (2012): 33.

5. Arrieta, M. C., L. T. Stiemsma, N. Amenyogbe, E. M. Brown, and B. B. Finlay. "The intestinal microbiome in early life: health and disease." *Frontiers in Immunolology* 5 (2014): 427.

6. Wrangham, R., and R. Carmody. "Human adaptation to the control of fire." *Evolutionary Anthropology* 19, no. 5 (2010): 187–199.

7. Clemente, J. C., L. K. Ursell, L. W. Parfrey, and R. Knight. "The impact of the gut microbiota on human health: an integrative view." *Cell* 148, no. 6 (2012): 1258–70.

8. Dominguez-Bello, M. G., M. J. Blaser, R. E. Ley, and R. Knight. "Development of the human gastrointestinal microbiota and insights from high-throughput sequencing." *Gastroenterology* 140, no. 6 (2011): 1713–9.

9. Wlodarska, M., A. D. Kostic, and R. J. Xavier. "An integrative view of microbiome-host interactions in inflammatory bowel diseases." *Cell Host & Microbe* 17, no. 5 (2015): 577–91.

10. Sekirov, I., S. L. Russell, L. C. M. Antunes, and B. B. Finlay. "Gut microbiota in health and disease." *Physiological Reviews* 90, no. 3 (2010): 859–904.

Chapter 3

11. Aagaard, K., J. Ma, K. M. Antony, R. Ganu, J. Petrosino, and J. Versalovic. "The placenta harbors a unique microbiome." *Science Translational Medicine* 6, no. 237 (2014): 237ra65.

12. Aagaard, K., K. Riehle, J. Ma, N. Segata, T. A. Mistretta, C. Coarfa, S. Raza, S. Rosenbaum, I. van den Veyver, A. Milosavljevic, D. Gevers, C. Huttenhower, J. Petrosino, and J. Versalovic. "A metagenomic approach to characterization of the vaginal microbiome signature in pregnancy." *PLOS ONE* 7, no. 6 (2012): e36466.

13. Jašarević, E., C. L. Howerton, C. D. Howard, and T. L. Bale. "Alterations in the Vaginal Microbiome by Maternal Stress Are Associated With Metabolic Reprogramming of the Offspring Gut and Brain." *Endocrinology* 156, no. 9 (2015): 3265–76.

14. Koren, O., J. K. Goodrich, T. C. Cullender, A. Spor, K. Laitinen, H. Kling Bäckhed, A. Gonzalez, J. Werner, L. Angenent, R. Knight, F. Bäckhed, E. Isolauri, S. Salminen, and R. Ley. "Host remodeling of the gut microbiome and metabolic changes during pregnancy." *Cell* 150, no. 3 (2012): 470–80.

15. Zijlmans, M. A. C., K. Korpela, J. M. Riksen-Walraven, W. M. de Vos, and C. de Weerth. "Maternal prenatal stress is associated with the infant intestinal microbiota." *Psychoneuroendocrinology* 53 (2015): 233–45.

Chapter 4

16. Dominguez-Bello, M. G., E. K. Costello, M. Contreras, M. Magris, G. Hidalgo, N. Fierer, and R. Knight. "Delivery mode shapes the acquisition and structure of the initial microbiota across multiple body habitats in newborns." *Proceedings of the National Academy Sciences* 107, no. 26 (2010):11971–5.

17. Lotz, M., D. Gütle, S. Walther, S. Ménard, C. Bogdan, and M. W. Homef. "Postnatal acquisition of endotoxin tolerance in intestinal epithelial cells." *Journal of Experimental Medicine* 203, no. 4 (2006): 973–84.

18. Neu, J. and J. Rushing. "Cesarean versus vaginal delivery: long-term infant outcomes and the hygiene hypothesis." *Clinics in Perinatology* 38, no. 2 (2011): 321–31.

19. Thanabalasuriar, A. and P. Kubes. "Neonates, Antibiotics and the Microbiome." *Nature Medicine* 20, no. 5 (2014): 469–70.

20. Thysen, A. H., J. M. Larsen, M. A. Rasmussen, J. Stokholm, K. Bønnelykke, H. Bisgaard, and S. Brix. "Prelabor cesarean section bypasses natural immune cell maturation." *Journal of Allergy and Clinical Immunology* 136, no. 4 (2015): 1123–1125 e6.

Chapter 5

21. Arroyo, R., V. Martín, A. Maldonado, E. Jiménez, L. Fernández, and J. M. Rodríguez. "Treatment of infectious mastitis during lactation: antibiotics versus oral administration of Lactobacilli isolated from breast milk." *Clinical Infectious Diseases* 50, no. 12 (2010): 1551–8.

22. Bäckhed, F., J. Roswall, Y. Peng, Q. Feng, H. Jia, P. Kovatcheva-Datchary, Y. Li, Y. Xia, H. Xie, H. Zhong, M. Khan, J. Zhang, J. Li,

L. Xiao, J. Al-Aama, D. Zhang, Y. Lee, D. Kotowska, C. Colding, V. Tremaroli, Y. Yin, S. Bergman, X. Xu, L. Madsen, K. Kristiansen, J. Dahlgren, and J. Wang . "Dynamics and Stabilization of the Human Gut Microbiome during the First Year of Life." *Cell Host & Microbe* 17, no. 5 (2015): 690–703.

23. Cabrera-Rubio, R., M. C. Collado, K. Laitinen, S. Salminen, E. Isolauri, and A. Mira. "The human milk microbiome changes over lactation and is shaped by maternal weight and mode of delivery." *American Journal of Clinical Nutrition* 96, no. 3 (2012): 544–51.

24. Nylund, L., R. Satokari, S. Salminen, and W. M. de Vos . "Intestinal Microbiota During Early Life—Impact on Health and Disease." *Proceedings of the Nutrition Society*. 73, no. 4 (2014): 457–69.

25. Rautava, S., R. Luoto, S. Salminen, and E. Isolauri . "Microbial contact during pregnancy, intestinal colonization and human disease." *Nature Reviews Gastroenterology & Hepatology* 9, no. 10 (2012): 565–76.

Chapter 6

26. Du Toit, G., G. Roberts, P. H. Sayre, H. T. Bahnson, S. Radulovic, A. F. Santos, H. A. Brough, D. Phippard, M. Basting, M. Feeney, V. Turcanu, M. L. Sever, M. Gomez-Lorenzo, M. Plaut, and G. Lack. "Randomized trial of peanut consumption in infants at risk for peanut allergy." *New England Journal of Medicine* 372, no. 9 (2015): 803–13.

27. Krebs, N. F. and K. M. Hambidge. "Complementary feeding: clinically relevant factors affecting timing and composition." *American Journal of Clinical Nutrition* 85, no. 2 (2007): 639S–645S.

28. Parnell, J. A. and R. A. Reimer. "Prebiotic fiber modulation of the gut microbiota improves risk factors for obesity and the metabolic syndrome." *Gut Microbes* 3, no. 1 (2012): 29–34.

29. Prescott, S. and A. Nowak-Węgrzyn. "Strategies to prevent or reduce allergic disease." *Annals of Nutrition and Metabolism* 59, suppl. 1 (2011): 28–42.

30. Sonnenburg, E. D. and J. L. Sonnenburg. "Starving our microbial self: the deleterious consequences of a diet deficient in microbiota-accessible carbohydrates." *Cell Metabolism* 20, no. 5 (2014): 779–86.

Chapter 7

31. Dellit, T. H., R. C. Owens, J. E. McGowan, D. N. Gerding, R. A. Weinstein, J. P. Burke, W. C. Huskins, D. L. Paterson, N. O. Fishman, C. F. Carpenter, P. J. Brennan, M. Billeter, and T. M. Hooton. "Infectious Diseases Society of America and the Society for Healthcare Epidemiology of America guidelines for developing an institutional program to enhance antimicrobial stewardship." *Clinical Infectious Diseases* 44, no. 2 (2007): 159–77.

32. Dethlefsen, L., S. Huse, M. L. Sogin, and D. A. Relma. "The pervasive effects of an antibiotic on the human gut microbiota, as revealed by deep 16S rRNA sequencing." *PLOS Biology* 6, no. 11 (2008): e280.

33. Hempel, S., S. J. Newberry, A. R. Maher, Z. Wang, J. N. V. Miles, R. Shanman, B. Johnsen, and P. G. Shekelle. "Probiotics for the Prevention and Treatment of Antibiotic-Associated Diarrhea: A Systematic Review and Meta-analysis." *JAMA* 307, no. 18 (2012): 1959–69.

34. Jernberg, C., S. Lofmark, C. Edlund, and J. K. Jansson. "Long-term impacts of antibiotic exposure on the human intestinal microbiota." *Microbiology* 156, pt. 11 (2010): 3216–23.

35. Marra, F., C. A. Marra, K. Richardson, L. D. Lynd, A. Kozyrskyj, D. M. Patrick, W. R. Bowie, and J. M. FitzGerald . "Antibiotic use in children is associated with increased risk of asthma." *Pediatrics* 123, no. 3 (2009): 1003–10.

36. Van Boeckel, T. P., C. Brower, M. Gilbert, B. T. Grenfell, S. A. Levin, T. P. Robinson, A. Teillant, and R. Laxminarayan. "Global trends in antimicrobial use in food animals." *Proceedings of the National Academy of Sciences* 112, no. 18 (2015): 5649–54.

37. Van Boeckel, T. P., S. Gandra, A. Ashok, Q. Caudron, B. T.

Grenfell, S. A. Levin, and R. Laxminarayan. "Global antibiotic consumption 2000 to 2010: an analysis of national pharmaceutical sales data." *The Lancet Infectious Diseases* 14, no. 8 (2014): 742–50.

Chapter 8

38. Azad, M. B., T. Konya, H. Maughan, D. S. Guttman, C. J. Field, M. R. Sears, A. B. Becker, J. A. Scott, and A. L. Kozyrskyj. "Infant gut microbiota and the hygiene hypothesis of allergic disease: impact of household pets and siblings on microbiota composition and diversity." *Allergy, Asthma & Clinical Immunology* 9, no. 1 (2013): 15.

39. Fujimura, K. E., T. Demoor, M. Rauch, A. A. Faruqi, S. Jang, C. C. Johnson, H. A. Boushey, E. Zoratti, D. Ownby, N. W. Lukacs, and S. V. Lynch. "House dust exposure mediates gut microbiome Lactobacillus enrichment and airway immune defense against allergens and virus infection." *Proceedings of the National Academy of Sciences* 111, no. 2 (2013): 805–10.

40. Pelucchi, C., C. Galeone, J. Bach, C. La Vecchia, and L. Chatenoud. "Pet exposure and risk of atopic dermatitis at the pediatric age: A meta-analysis of birth cohort studies." *Journal of Allergy and Clinical Immunology* 132 no. 3 (2013): 616–622 e7.

Chapter 9

41. Cherednichenko, G., R. Zhang, R. A. Bannister, V. Timofeyev, N. Li, E. B. Fritsch, W. Feng, G. C. Barrientos, N. H. Schebb, B. D. Hammock, K. G. Beam, N. Chiamvimonvat, and I. N. Pessah. "Triclosan impairs excitation-contraction coupling and Ca2+ dynamics in striated muscle." *Proceedings of the National Academy of Sciences* 109, no. 35 (2012): 14158–63.

42. Hesselmar, B., A. Hicke-Roberts, and G. Wennergren, "Allergy in Children in Hand Versus Machine Dishwashing." *Pediatrics* 135, no. 3 (2015): e590–7.

43. Hesselmar, B., F. Sjoberg, R. Saalman, N. Aberg, I. Adlerberth, and A. E. Wold. "Pacifier Cleaning Practices and Risk of Allergy Development." *Pediatrics* 131, no. 6 (2013): e1829–37.

44. Tung, J., L. B. Barreiro, M. B. Burns, J. Grenier, J. Lynch, L. E. Grieneisen, J. Altmann, S. C. Alberts, R. Blekhman, and E. A. Archie. "Social networks predict gut microbiome composition in wild baboons." *Elife* 4 (2015).

Chapter 10

45. Cox, L. M., S. Yamanishi, J. Sohn, A. V. Alekseyenko, J. M. Leung, I. Cho, S. G. Kim, H. Li, Z. Gao, D. Mahana, J. Zárate Rodriguez, A. Rogers, N. Robine, P. Loke, and M. Blaser. "Altering the Intestinal Microbiota during a Critical Developmental Window Has Lasting Metabolic Consequences." *Cell* 158, no. 4 (2014): 705–21.

46. Kleiman, S. C., H. J. Watson, E. C. Bulik-Sullivan, E. Y. Huh, L. M. Tarantino, C. M. Bulik, and I. M. Carroll. "The Intestinal Microbiota in Acute Anorexia Nervosa and During Renourishment." *Psychosomatic Medicine* 77, no. 9 (2015): 969–81.

47. Magrone, T. and E. Jirillo. "Childhood Obesity: Immune Response and Nutritional Approaches." *Frontiers in Immunology* 6 (2015): 76.

48. Park, S. and J. H. Bae. "Probiotics for weight loss: a systematic review and meta-analysis." *Nutrition Research* 35, no. 7 (2015): 566–75.

49. Turnbaugh, P. J., F. Bäckhed, L. Fulton, and J. I. Gordon. "Diet-Induced Obesity Is Linked to Marked but Reversible Alterations in the Mouse Distal Gut Microbiome." *Cell Host & Microbe* 3, no. 4 (2008): 213–23.

50. Turnbaugh, P. J., R. E. Ley, M. A. Mahowald, V. Magrini, E. R. Mardis, and J. I. Gordon. "An obesity-associated gut microbiome with increased capacity for energy harvest." *Nature* 444, no. 7122 (2006): 1027–31.

Chapter 11

51. Hartstra, A. V., K. E. C. Bouter, F. Bäckhed, and M. Nieuwdorp. "Insights Into the Role of the Microbiome in Obesity and Type 2 Diabetes." *Diabetes Care* 38, no. 1 (2015): 159–65.

52. Hu, C., F. S. Wong, and L. Wen, "Type 1 diabetes and gut microbiota: Friend or foe?" *Pharmacological Research* 98 (2015): 9–15.

53. Karlsson, F. H., V. Tremaroli, I. Nookaew, G. Bergström, C. J. Behre, B. Fagerberg, J. Nielsen, and F. Bäckhed., "Gut metagenome in European women with normal, impaired and diabetic glucose control." *Nature* 498, no. 7452 (2013): 99–103.

54. Qin, J., Y. Li, Z. Cai, S. Li, J. Zhu, F. Zhang, S. Liang, W. Zhang, Y. Guan, D. Shen, Y. Peng, D. Zhang, Z. Jie, W. Wu, Y. Qin, W. Xue, J. Li, L. Han, D. Lu, P. Wu, Y. Dai, X. Sun, Z. Li, A. Tang, S. Zhong, X. Li, W. Chen, R. Xu, M. Wang, Q. Feng, M. Gong, J. Yu, Y. Zhang, M. Zhang, T. Hansen, G. Sanchez, J. Raes, G. Falony, S. Okuda, M. Almeida, E. LeChatelier, P. Renault, N. Pons, J. Batto, Z. Zhang, H. Chen, R. Yang, W. Zheng, S. Li, H. Yang, J. Wang, S. D. Ehrlich, R. Nielsen, O. Pedersen, K. Kristiansen, and J. Wang. "A metagenome-wide association study of gut microbiota in type 2 diabetes." *Nature* 490, no. 7418 (2012): 55–60.

55. Vrieze, A., E. van Nood, F. Holleman, J. Salojärvi, R. S. Kootte, J. F. W. M. Bartelsman, G. M. Dallinga-Thie, M. T. Ackermans, M. J. Serlie, R. Oozeer, M. Derrien, A. Druesne, J. E. van Hylckama Vlieg, V. W. Bloks, A. K. Groen, H. G. Heilig, E. G. Zoetendal, E. S. Stroes, W. M. de Vos, J. B. Hoekstra, and M. Nieuwdorp. "Transfer of Intestinal Microbiota from Lean Donors Increases Insulin Sensitivity in Individuals with Metabolic Syndrome." *Gastroenterology* 143, no. 4 (2012): 913–6.e7.

Chapter 12

56. Collins, S. M. "A role for the gut microbiota in IBS." *Nature Reviews Gastroenterology & Hepatology* 11, no. 8 (2014): 497–505.

57. Colman, R. J. and D. T. Rubin. "Fecal microbiota transplantation as therapy for inflammatory bowel disease: A systematic review and meta-analysis." *Journal of Crohn's and Colitis* 8, no. 12 (2014): 1569–81.

58. de Sousa Moraes, L. F., L. M. Grzeskowiak, T. F. de Sales Teixeira, and M. D. C. Gouveia Peluzio. "Intestinal Microbiota and Probiotics in Celiac Disease." Clinical Microbiology Reviews 27, no. 3 (2014): 482–9.

59. de Weerth, C., S. Fuentes, and W. M. de Vos. "Crying in infants: on the possible role of intestinal microbiota in the development of colic." *Gut Microbes* 4, no. 5 (2013): 416–21.

60. de Weerth, C., S. Fuentes, P. Puylaert, and W. M. de Vos. "Intestinal Microbiota of Infants with Colic: Development and Specific Signatures." *PEDIATRICS* 131, no. 2 (2013): e550–8.

61. Gevers, D., S. Kugathasan, L. A. Denson, Y. Vázquez-Baeza, W. Van Treuren, B. Ren, E. Schwager, D. Knights, S. Song, M. Yassour, X. Morgan, A. Kostic, C. Luo, A. González, D. McDonald, Y. Haberman, T. Walters, S. Baker, J. Rosh, M. Stephens, M. Heyman, J. Markowitz, R. Baldassano, A. Griffiths, F. Sylvester, D. Mack, S. Kim, W. Crandall, J. Hyams, C. Huttenhower, R. Knight, and R. Xavier. "The Treatment-Naive Microbiome in New-Onset Crohn's Disease." *Cell Host & Microbe* 15, no. 3 (2014): 382–92.

62. Wacklin, P., P. Laurikka, K. Lindfors, P. Collin, T. Salmi, M. Lähdeaho, P. Saavalainen, M. Mäki, J. Mättö, K. Kurppa, and K. Kaukinen. "Altered Duodenal Microbiota Composition in Celiac Disease Patients Suffering from Persistent Symptoms on a Long-Term Gluten-Free Diet." *American Journal of Gastroenterology* 109, no. 12 (2014): 1933–41.

Chapter 13

63. Arrieta, M. C. and B. Finlay. "The intestinal microbiota and allergic asthma." *Journal of Infection* 69, suppl. 1 (2014): S53–5.

64. Arrieta, M. C., L. T. Stiemsma, P. A. Dimitriu, L. Thorson, S. Russell, S. Yurist-Doutsch, B. Kuzeljevic, M. J. Gold, H. M. Britton, D. L. Lefebvre, P. Subbarao, P. Mandhane, A. Becker, K. M. McNagny, M. R. Sears, T. Kollmann, W. W. Mohn, S. E. Turvey, and B. Finlay. "Early infancy microbial and metabolic alterations affect risk of childhood asthma." *Science Translational Medicine* 7, no. 307 (2015): 307ra152.

65. Holbreich M., M. Stein, R. Anderson, N. Metwali, P. S. Thorne, D. Vercelli, E. von Mutius, and C. Ober. "Allergic Sensitization and Enviromental Exposures in Amish and Hutterite Children." *Journal of Allergy and Clinical Immunology* 133, no. 2 (2014): AB13.

66. Ly, N. P., A. Litonjua, D. R. Gold, and J. C. Celedón. "Gut microbiota, probiotics, and vitamin D: Interrelated exposures influencing allergy, asthma, and obesity?" *Journal of Allergy and Clinical Immunology* 127, no. 5 (2011): 1087–94; quiz 1095–6.

67. Olszak, T., D. An, S. Zeissig, M. P. Vera, J. Richter, A. Franke, J. N. Glickman, R. Siebert, R. M. Baron, D. L. Kasper, and R. S. Blumberg. "Microbial Exposure During Early Life Has Persistent Effects on Natural Killer T Cell Function." *Science* 336, no. 6080 (2012): 489–93.

68. Russell, S. L., M. J. Gold, M. Hartmann, B. P. Willing, L. Thorson, M. Wlodarska, N. Gill, M. Blanchet, W. W. Mohn, K. M. McNagny, and B. B. Finlay. "Early life antibiotic-driven changes in microbiota enhance susceptibility to allergic asthma." *EMBO Reports* 13, no. 5 (2012): 440–7.

69. Schaub, B., J. Liu, S. Höppler, I. Schleich, J. Huehn, S. Olek, G. Wieczorek, S. Illi, and E. von Mutius. "Maternal farm exposure modulates neonatal immune mechanisms through regulatory T cells." *Journal of Allergy and Clinical Immunology* 123, no. 4 (2009): 774–82.e5.

70. Sharief, S., S. Jariwala, J. Kumar, P. Muntner, and M. L. Melamed. "Vitamin D levels and food and environmental allergies in the United States: Results from the National Health and Nutrition Examination Survey 2005–2006." *Journal of Allergy and Clinical Immunology* 127, no. 5 (2011): 1195–202.

Chapter 14

71. Braniste, V., M. Al-Asmakh, C. Kowal, F. Anuar, A. Abbaspour, M. Toth, A. Korecka, N. Bakocevic, L. G. Ng, P. Kundu, B. Gulyas, C. Halldin, K. Hultenby, H. Nilsson, H. Hebert, B. T. Volpe, B. Diamond, and S. Pettersson. "The gut microbiota influences blood-brain barrier permeability in mice." *Science Translational Medicine* 6, no. 263 (2014): 263ra158.

72. Hsiao, E. Y., S. W. McBride, S. Hsien, G. Sharon, E. R. Hyde, T. McCue, J. A. Codelli, J. Chow, S. Reisman, J. Petrosino, P. Patterson, and S. Mazmanian. "Microbiota modulate behavioral and physiological abnormalities associated with neurodevelopmental disorders." *Cell* 155, no. 7 (2013): 1451–63.

73. Pärtty, A., M. Kalliomäki, P. Wacklin, S. Salminen, and E. Isolauri. "A possible link between early probiotic intervention and the risk of neuropsychiatric disorders later in childhood: a randomized trial." *Pediatric Research* 77, no. 6 (2015): 823–8.

74. Petra, A. I., S. Panagiotidou, E. Hatziagelaki, J. M. Stewart, P. Conti, and T. C. Theoharides. "Gut-Microbiota-Brain Axis and Its Effect on Neuropsychiatric Disorders with Suspected Immune Dysregulation." *Clinical Therapeutics* 37, no. 5 (2015): 984–95.

Chapter 15

75. Ang, L., S. Arboleya, G. Lihua, Y. Chuihui, Q. Nan, M. Suarez, G. Solís, C. G. de los Reyes-Gavilán, and M. Gueimonde. "The establishment of the infant intestinal microbiome is not affected by rotavirus vaccination." *Scientific Reports* 4 (2014): 7417.

76. Kimmel, S. R., "Vaccine Adverse Events: Separating Myth from Reality." *American Family Physician* 66, no. 11 (2002): 2113–21.

77. Valdez, Y., E. M. Brown, and B. B. Finlay, "Influence of the microbiota on vaccine effectiveness." *Trends in Immunology* 35, no. 11 (2014): 526–37.

Chapter 16

78. Zeevi, D., T. Korem, N. Zmora, D. Israeli, D. Rothschild, A. Weinberger, O. Ben-Yacov, D. Lador, T. Avnit-Sagi, M. Lotan-Pompan, J. Suez, J. A. Mahdi, E. Matot, G. Malka, N. Kosower, M. Rein, G. Zilberman-Schapira, L. Dohnalová, M. Pevsner-Fischer, R. Bikovsky, Z. Halpern, E. Elinav, and E. Segal. "Personalized Nutrition by Prediction of Glycemic Responses." *Cell* 163, no. 5 (2015): 1079–1094.

Index

ABOUT THE AUTHOR

Not yet thirty years old, PETER SHEAHAN is globally recognized as a leading expert in workforce trends and generational change. He leads a multimillion-dollar consulting practice, attracting clients such as Coca-Cola, L'Oréal, and Ernst & Young. He lives in Sydney, Australia.

FL!P

How to Turn Everything You Know on Its Head— and Succeed Beyond Your Wildest Imaginings

PeterSheahan

HARPER

NEW YORK · LONDON · TORONTO · SYDNEY

TO ALL YOU FLIPSTARS OUT THERE TRYING SOMETHING NEW,
TAKING A RISK, AND PUTTING IT ALL ON THE LINE. HOLD THE NERVE!

HARPER

This book was originally published in Australia in 2007 by Random House Australia.

A hardcover edition of this book was published in 2008 by William Morrow, an imprint of HarperCollins Publishers.

FIRST HARPER PAPERBACK PUBLISHED 2009.

The Library of Congress has catalogued the hardcover edition as follows:
Sheahan, Peter.
    Flip : how to turn everything you know on its head—and succeed beyond your wildest imaginings / Peter Sheahan. — 1st ed.
            p. cm.
    Includes bibliographical references and index.
    ISBN 978-0-06-155895-5
    1. Strategic planning.   2. Management.   I. Title.
HD30.28.S425 2008
658.4'03—dc22

2008002805

ISBN 978-0-06-171963-9 (pbk.)

09 10 11 12 13 WBC/RRD 10 9 8 7 6 5 4 3 2 1

contents

Preface to the Paperback Edition

Flip was written before world credit markets froze in 2008. The ensuing economic uncertainty has made it even more vital to free yourself and your enterprise from business-as-usual ideas and behavior. The ability to Flip your thinking and your actions, adopting new worldviews, business models, and modes of collaboration, will always differentiate the big winners from the also-rans. In this environment it could mean the difference between surviving and not.

In September 2008, when the world's stock markets were in freefall, I caught up with one of my clients, a regional CEO of Apple. After some pointless chat, I asked him, "What is Apple going to do in response to the market turmoil and the economic crisis we are facing?"

He shot me a strange and confused glance in response. "What do you mean?" he asked.

"Well, with your new retail store, the new products that you have just launched in the midst of the most frightening economic times in recent memory, how are you going to deal with the crisis?"

"We are going to innovate through it!" he replied, with unmistakable conviction. He was not participating in some positive-thinking exercise—he was so certain that Apple would be fine that he went on to say he believed the company would come out of this crisis in better shape than they had entered it in. He said that the economic situation was a once-in-a-generation opportunity to sort the wheat from the chaff in the market.

Compare his response to that of the managing partner of a midtier accounting firm—let's call it Firm A—who I met about two months after my Apple conversation. Firm A had just fired a significant number of its staff. This had happened after a period of five years when all that the firm's leadership could talk about was how hard it was to attract and retain great accounting talent. I was so curious that I had to ask, "Is your business suffering?"

"No," he replied. "We are getting ready!"

"Getting ready for what?"

"For the downturn," he responded. "We have more work than ever, and it looks like our newly formed insolvency practice will be a hit, but you never know."

A short time afterward, I mentioned this to a partner at competitive Firm B, and their response was, "Do you have the contact details for the staff who were let go?" Now Firm A may have better short-term profits, but the current financial crisis will pass sooner or later. And when it does, Firm B will be eating Firm A for lunch.

Now I am not some Pollyanna who believes that the world is perfect and that we don't have to respond to market conditions. We do. But the most empowering mindset to adopt is the one that says, "We will innovate through this mess, remaining confident in the quality of our staff, products, and systems." Many markets in the world had good fundamentals yet managed to talk themselves into a recession.

More than ever, you need to have the mindset, flexibility, and courage to innovate and change not just your organization but yourself as well. You can't really do one without the other.

The financial crisis is among a number of major challenges whose solutions require innovative, game-changing thinking. That begins with the core idea of Flipping—developing entirely new models for thinking about your world and your customers. In response to financial uncertainty, consumers become more cautious about spending, but they still unconsciously make decisions based on their emotional state. Strategies for dealing with this kind of buying are outlined in "Fast, Good, Cheap: Pick Three—Then Add Something Extra" and "Absolutely, Positively Sweat the Small Stuff."

The growing distrust of both government and private-sector organizations will bring the importance of deep personal relationships to the forefront of brand strategies. Not to mention the need to resist the knee-jerk, short-term, shareholder-as-the-only-important-stakeholder kind of decisions that contributed to the financial crisis and ultimately led to thousands of people being put out of work. These decisions will affect the employer and the brands of companies for decades. Demographics will ensure that the "war for talent" and skills shortages that plagued business prior to the current economic challenges will return with more intensity than ever, and the best and brightest will remember which organizations demonstrated concern for their people and which did not. In other words, you need to remember that "Business is Personal."

How to identify and develop new products and services that stimulate demand, and better still, generate whole new business opportunities and markets, is the heart of "Mass-Market Success: Find It on the Fringe."

The organizational model that will emerge at the other end of this journey will inevitably be a much more agile and adaptive one than most enterprises now have. "To Get Control, Give It Up" explains how to acquire the collaborative and co-operative competencies that already define today's most successful organizations, which will be a necessary characteristic of future excellence.

Most important of all is that you can't let frozen capital markets freeze your behavior. You have to remain active and ramp up, not slow down, your experimenting. As a leader, investor, parent, or team member, it is so easy to be paralyzed by the fear brought on by uncertainty. You will need to act in spite of uncertainty, and without doubt those leaders and organizations that stay proactive will reap the greatest rewards in profit, brand reputation, and talent acquisition when overall economic conditions improve. A recently completed study by the Corporate Executive Board showed that the most profitable businesses in the long upswing that just ended had been the most forward-thinking and innovative in the preceding downswing.

As Yogi Berra said, "When you come to the fork in the road, take it!" I wish you strength and excitement on your journey, and I believe that the ideas presented in *Flip* will be of great value to you during the transformative years that we are beginning to experience.

introduction:
GETTING FLIPPED!

> **Flip:** A shift in mind-set and thinking; often a
> counterintuitive approach that reflects the hard reality
> of the business landscape as it is today, and not as it used
> to be.

Business today requires new perspectives on strategy, operations, customers, and staff. Most of all it requires a level of
flexibility that has previously been considered a weakness in
some organizations. To navigate in an increasingly complex
environment and zig when the competition are all zagging,
you must flip repeatedly between different perspectives and
paradigms. And you must do so in real time and on demand.

Flip provides a foundation and guidelines for developing and
employing that flexible mind-set. It stimulates a way of thinking—
an approach to business and life—that will remain relevant regardless of the changes that will occur at an increasingly rapid

pace in coming years. It does not replace one rigid set of rules with another that will soon be outmoded. To *flip* is to turn conventional wisdom on its head, whether it be the wisdom of 1987, 2007, or 2027.

This book will, I hope, be a tonic for many readers. Daily I meet people who are struggling to keep their heads above water, crippled by their desire to always be "right" and to have an answer about what will happen in the future. But you can't know this, nor do you need to. You just need to notice what is and isn't working, and then be willing to *flip*.

The complexity of the current business climate means there is no single right way forward. There are many ways, and the most successful companies and leaders will try them all at one time or another. They will keep flipping. This is what Toyota is doing with the Scion and with hybrid technology throughout the Toyota and Lexus brands. It is what Apple is doing with the iPod. It is what Richard Branson does in virtually every industry he touches. And it is what Wal-Mart is doing in becoming a leader in environmental sustainability. Flipping, you are about to discover, is what the world's most effective organizations and individuals do to distinguish themselves from the competition.

• • •

There has never been a more exciting time to be in business. Things are changing faster and faster, new opportunities and new markets are opening up every day. In order to profit from these new opportunities and markets you first need to understand exactly what is changing, and

MIND-SET FLEXIBILITY, NOT PROPRIETARY EXPERTISE OR RESOURCES, WILL DEFINE THE SUCCESSFUL BUSINESSES AND LEADERS OF THE FUTURE.

what impact this is having on your existing business. Then you need to *flip*. Flip from the perspective that has governed your business and competitive strategy into new perspectives and strategic directions tomorrow and the day after tomorrow. Mind-set flexibility, not proprietary expertise or resources, will define the successful businesses and leaders of the future.

This book will show you how seemingly upside-down thinking creates straight-up success. That thinking begins with understanding that we have seen the end of an *either/or* world. Conventional business models are built on choices between mutually exclusive options. That's a nonstarter in a world where few, if any, options are truly mutually exclusive. For the foreseeable future, you must *Think AND, Not OR*.

Consider the following statements:

- The world is getting smaller every day.
- The world is getting bigger every day.

Both statements are true. The world is getting smaller all the time because people, things, and ideas move from one place to another in greater and greater numbers at greater and greater speed. And the world is getting bigger all the time for the same reason; all those people, things, and ideas in motion create an ever-increasing number of possibilities and options. Concentrate on one of these statements at the expense of the other and you will miss crucial possibilities and options and cripple your decision-making ability.

IF YOU LEARN TO FLIP, YOU WILL FUTURE-PROOF YOUR ORGANIZATION AND YOUR CAREER.

The need to "Think AND, Not OR" will pop up throughout

the book. It is the overarching context for six major flips I will explore:

- *Action Creates Clarity*—to move forward you must act in spite of ambiguity. Your action will create the clarity you're looking for.
- To keep pace with rising expectations, you can't just be fast, good, or cheap, or even any two of these. Instead you must recognize that *Fast, Good, Cheap: Pick Three—Then Add Something Extra* has become the price of entry in every industry.
- To keep *ahead* of rising expectations and develop competitive advantage, you must *Absolutely, Positively Sweat the Small Stuff.*
- To satisfy customers' needs for engagement and contact— spiritual, emotional, physical—remember that business is not business, *Business Is Personal.* One thing that has not changed in a globalizing economy is that people want to do business with people they know, like, and trust.
- To win mass-market success, *Find It on the Fringe.* The way to separate yourself from the competitive herd is by being courageous and creating new market space.
- *To Get Control, Give It Up.* You cannot command and control customers or the talented staff you need to reach them. Instead you must empower others to create, dream, and believe for you.

Within each of the major flips, there will be more subtle ones. I hope as you read this book you will begin to see opportunities to flip in all parts of your business and life, and each chapter includes suggestions of things to do when you finish reading it. More than specific examples of upside-down thinking, *flipping* is a philosophy. To be a *flipstar* like Richard Branson, Steve Jobs, and Rupert Murdoch, to name a few, is to embrace the need for mind-set flexibility. To understand that what was right for yesterday may not be right for today, and likely will be dead wrong for tomorrow. If you learn to flip, however, you will future-proof your organization, your career, and ultimately your ability to live life to the fullest.

THE FOUR FORCES
OF CHANGE

I hope everyone knows that the world is changing. Pundits in every field talk incessantly about the constant change we are experiencing, to the point that "change agent" has become one of the biggest clichés of our time. But what specifically is changing? And why does it feel so pervasive?

Allow me to introduce you to the four forces of change that are completely redefining the way you compete in the marketplace, attract and reward staff, and even live your life. And then, through a series of flips, allow me to help you deal with these four forces of change. (The latter being more important.) This is not a book about the fact of change, it is a book about how to handle and profit from change. First, however, it is essential that we identify succinctly what is changing, because only then can we get more specific about what to do about it.

Here are the four forces of change I will be referring to throughout the book:

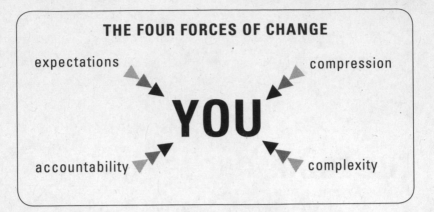

THE FOUR FORCES OF CHANGE

expectations   compression

YOU

accountability   complexity

1. increasing compression of time and space
2. increasing complexity
3. increasing transparency and accountability
4. increasing expectations on the part of everyone for everything

These forces are squeezing organizations and individuals alike. If you want to, you can see them as enemies, as the Four Horsemen of the Apocalypse that are ending business as you know it. Or you can welcome them as allies that are indeed changing the nature of business in challenging ways, but that also have the potential to accelerate your success and help you achieve competitive advantage.

None of the four forces is new. They've been around in one form or another at least as long as human beings have been creating complex societies. But there has been a dramatic shift in how the four forces affect our daily lives. They've never been

IF YOU AND YOUR COMPANY LEARN TO RIDE THE WAVE OF CHANGE, YOU STAND TO REAP MASSIVE WINDFALLS.

in as tight and immediate a feedback loop as they are today. That's why the pundits can't stop talking about change and change agents. That's why organizations and enterprises all over the world are freaking out about these shifts. And that's why you and your company must get on the front foot and learn to ride the wave of change—you stand to reap massive windfalls. Just think of Toyota and the car-industry-shifting Prius, or Apple and iTunes leading to the phenomenal sales of the iPod.

Let's take a look at these four forces.

1. INCREASING COMPRESSION OF TIME AND SPACE

Outside of science fiction and the thought experiments of theoretical physicists, it is not actually possible to compress time or space. However, our perceptions of both are certainly malleable.

Compression of time

Human beings have always been impatient. Today we expect things to happen faster than ever before. And not just a little bit faster, but over the last few years a lot faster. The quicker something can be done, the quicker we expect it to happen, whether it's the movement of goods by overnight courier companies such as DHL, FedEx, and UPS, the movement of people on flights from one side of the world to the other, or the movement of digital information from anywhere to anywhere via broadband Internet (something that merits fuller discussion below). They all feed our insatiable need for speed.

Staffan B. Linder's book *The Harried Leisure Class* notes that as economic growth and affluence increase, "the pace is quickening, and our lives are in fact becoming steadily more hectic." This has formed the basis of a commonly discussed sociological concept: as affluence increases, so does pressure around time. Ask yourself, are you feeling a little bit of pressure around time? Consider the following example.

I was discussing this idea of compression around time with the partners of a leading law firm. Afterward the CEO told me how only a decade or so ago as a young lawyer, he typically spent an hour dictating a letter to a client. Then he figured on half a day for the letter to be typed and returned to his in-box for signature before being put in the mail, two days for it to reach the recipient, two days for the recipient to draft and send a reply, and two more days for the reply to reach his desk— more than a week for a single exchange of business letters. Today, as CEO of his firm, he types and sends his own e-mails, or a short message text on his BlackBerry if he's traveling, and he expects an answer later the same day, if not within the hour.

And this CEO is not some "young twentysomething" exhibiting the impatience of youth. Increasingly we begin our workdays not fully rested, because we got to bed so late the night before, whether from trying to get overdue work done or to have a bit of social life in the midst of all our other commitments. When the alarm first goes off, we hit the snooze button and go into what a friend of mine calls "mathematician mode," calculating the absolute last possible minute we can get out of bed and not be late. Then we race to catch an express train or bus to the office. Whether it's a blessing or a curse, technology frees us from the need to interact with anyone as we board; we just insert our prepaid ticket in the slot. When we arrive at work we mill about restlessly, waiting for the express elevator.

Then we spend the day responding to the hectic demands of colleagues and superiors. We have two-minute noodles for lunch. And when we get home, we pop our instant dinner in the microwave and stand there thinking, *Come on now, I have not got all minute.*

An exaggeration, perhaps, but I'm sure most people will agree it's only a slight one.

Unless you want to go off into the bush and be a hermit, there is no escape from the nearly instant communication and feedback loops represented by e-mail, text messaging, and cell phones. Whether we're talking about countries, companies, or individuals, events that happen on the other side of the globe can and do have an immediate impact on our daily lives.

This is only half the story. It is not just that we want what we want faster, but that we change our minds about what we want more quickly. Consider that the average time from concept to product in the U.S. automobile industry is down from between five and seven years (about ten years ago) to around two years today. And what is ironic about this is that the U.S. automotive industry is considered to be among the least innovative and slowest to change on the globe. It is competing with companies such as Toyota, which brings entire vehicle ranges (Lexus and Scion) and new value propositions (hybrid engines) to market while most of their rivals are still trying to digest the fact that there might be a significant near-term profit opportunity in midpriced luxury cars, customizable cars for Generation Y, or eco-friendly engines.

> IF THE LATE TWENTIETH CENTURY WAS ABOUT DOING MORE WITH LESS, THEN THE EARLY TWENTY-FIRST CENTURY WILL BE ABOUT DOING MORE WITH LESS, FASTER!

Back in 1979, when the Walkman, not PlayStation, was its signature product, Sony went from product inception to product launch in under four months—remarkable for the time. Just recently, however, the PlayStation 3 cost Sony millions because of delays associated with its launch. It was shown to the public at the E3 games convention in May 2005, but didn't hit the shelves until November of the next year. By the time the PS 3 got to market, its competition, the Microsoft Xbox 360, had shipped almost 10 million units, and Sony, a company that was once famous for its speed to market and relentless pursuit of first-mover advantage, lost almost $2.3 billion because of the late entry.

In summary, increased affluence and rapidly developing communications technology are compressing our expectation around time. If the late twentieth century was about doing more with less, then the early twenty-first century will be about doing more with less, faster!

Compression of space

Compression of the way we view space is shrinking the world. People no longer see geographical distance as a barrier to the way they do business. The world is the new market, especially in light of an increasing number of international free-trade agreements.

Distances have always been relative to the time it takes to travel them. In our grandparents' youth, the other side of the world was weeks away. Now it is one day away by plane, or one second away given communications technology that makes it less and less likely that we actually need our flesh and bones to be on the other side of the globe.

My own business is an example of this. Daily, thousands of people from all over the planet log onto www.petersheahan.com and some buy products—some of which are digital and can be downloaded instantly—or subscribe to a free RSS feed that will help keep them on the cutting edge of new markets and trends. The point here is not to plug the "Peter Sheahan Live" section on my site (although I am glad to do so), but to point out that neither time nor geography poses any barrier to these transactions of value. I get visitors from countries I have never been to and make sales to people I have never met. I am able to complete transactions, even though neither I nor any other human is there to service the customer.

COMPRESSION OF DISTANCE MEANS THERE ARE NEW MARKETS TO BE SERVICED, AND NEW WAYS TO SERVICE EXISTING MARKETS.

Compression of distance means there are new markets to be serviced, and new ways to service existing markets. It also means there are more competitors—the most dangerous of whom may be a twenty-year-old at her computer in Sydney, San Jose, Seville, or Seoul, who in less than a decade could dominate your market.

Now, it is important to put the current status of this change in the proper context. Some economists argue that in many ways globalization is overstated. Consider that, for instance, of all the phone calls made in the world only 10 percent are international calls. And of all the investment in the world today, again only 10 percent is foreign (international) investment.[1]

But even if the direct international component of business holds steady at around 10 percent, we are now competing against international benchmarks. In an increasingly connected world, customers are increasingly exposed to global trends and fashions. The customer's sense of what the neighborhood business can do has irrevocably changed. Although people will always prefer to do business with people they know, like, and trust (see chapter 5, "Business Is Personal"), they expect those people to deliver at a global standard of excellence, not a local one. There is no escaping the need to position your business today to compete—in real time and on demand—in the increasingly globalized world of tomorrow. Flips are the future-focused way to achieve this.

2. INCREASING COMPLEXITY

Increasing compression of time and space produces increasing complexity. Businesses are being hit from all directions—from above, as they are saddled with dense and complex regulatory regimes; from inside, with the challenges posed by the adoption of sophisticated new technologies and the explosion of information networks; and from below, with the diversifying and intangible new demands of consumers.

COMPLEXITY BRINGS UNCERTAINTY AND COMPLEXITY THAT CAN PARALYZE INNOVATION AND POSITIVE ACTION.

The uncertainty and ambiguity this complexity brings can induce paralysis that prevents innovation and positive action, and may in turn force the downfall of once-great companies and careers.

Increasing complexity is being driven by, among others, the following six things:

- Rapidly interconnecting networks of ideas and people—from interstate highways, planes, and containerized shipping to the information superhighway of the Internet.

- Disruptive technology—innovations in product and process almost always have unintended consequences that challenge our ability to adapt, and reward those with the flexibility to flip into new modes of acting and thinking.
- Explosion of choice—in a globalizing economy, no one has a monopoly on any product or service for long, and the consumer's biggest problem is often choosing among apparently identical offerings.
- Increasingly intangible desires of the market—rising affluence shifts the business imperative from supplying customer needs to meeting customer desires for emotional fulfillment, no matter how mundane the product or service.
- Increased sophistication of technology, systems, and processes—complexity begets complexity.
- Legislation—whether it's financial transparency, safety, the environment, or human rights, the world's governments are regulating it.

Rapidly interconnecting networks of ideas and people

I've already mentioned increasing flows of people, goods, and data, but there's more to the story. Think of every person in the world as a node in a vast information system. Each node has different perspectives, ideas, and desires from the other nodes in the system. Some differences are slight; others are large. And a slight difference in one context may loom large in another.

Now connect those nodes by all the networks—physical and virtual—that link them together: text message, e-mail, landline phones, cell phones, express delivery, container shipping, and transportation networks for people comprising motor vehicles, trains, ships, and planes. These links crisscross the developed and the developing world, and more of them are overlapping all the time. Traffic on all these networks is constantly increasing, most notably in the form of digitized words, images, and numbers moving at the speed of light.

A senior executive at Google shared some interesting statistics with me when I spoke to a large group of managers at Google's headquarters in Mountain View, California. Did you know, for example, that 20 to 25 percent of daily searches on Google are unique? They have never been searched before— 25 percent! This is not so surprising when you realize that 25 to 30 percent of the Web is new *all the time*. We are generating content—opinions, survey results, perspectives, ideas, or just pointless garbage—so fast that at any one time more than one-quarter of the World Wide Web is brand-new. According to *Time* magazine, the world produced 161 exabytes (161 *billion* gigabytes) of digital information in 2006. That's more than three million times the amount of information contained in all books ever written.[2]

This ongoing, explosive content creation challenges us to assimilate greater and greater amounts of often conflicting information, increasing the complexity, ambiguity, and uncertainty in our lives. This rising noise puts a premium on what I call *confusion management,* which is without a doubt the most important asset for a leader today. Confusion management means dealing with ambiguity, contradiction, and uncertainty while still retaining the ability to function. More on how to do this later, but for now back to the challenge it represents. Separating the wheat from the chaff in the information pouring in on us, recognizing when two or more apparently conflicting ideas must be utilized in tandem, and accepting that new ambiguities will constantly arise are exactly what will distinguish winners from losers.

Disruptive technology

Add to this immense creation of knowledge (a generous label for much of what is online) the fact that the world has always been and always will be unpredictable because of unanticipated consequences. New technologies in particular have a history of creating unintended consequences, and nowadays new technologies enter the market faster and more frequently than ever before.

Take, for example, the diminishing impact of a thirty-second television commercial. Once a guaranteed way to drive sales, this form of media and advertising has become much less effective, especially with younger people, since the growth of the Internet, or more specifically of massive multiplayer online games, social networking sites, and greater access to broadband connections. Imagine what it is like for the forty-year-old ad account director advising the fifty-five-year-old consumer-goods brand manager that the market requires entirely new messages delivered in entirely new formats through entirely new media. You'll definitely have confusion and ambiguity in the mind of a once-unstoppable executive. Perhaps you *are* that forty-year-old ad executive, or your industry's equivalent. Oh, and just when you've worked out what that new message and media should be, the fickle consumer has moved on to even newer things.

Whenever technology changes rapidly, actions you took yesterday may have a different effect today. These unintended consequences make planning difficult, and they constitute yet more new information that eventually joins the crowded flow of data you must digest and evaluate. This book will help you do just that. I will speak in later chapters about the need to unlearn and relearn at a much faster rate in order to stay in step with not just this but the other three forces of change as well.

> WHENEVER TECHNOLOGY CHANGES RAPIDLY, ACTIONS YOU TOOK YESTERDAY MAY HAVE A DIFFERENT EFFECT TODAY.

Explosion of choice

More information, more knowledge, and more options ultimately mean more choices. Yet too much choice can actually paralyze. This is both a challenge and an opportunity.

The opportunity is for a business that can make itself "easier" for customers to choose and interact with. The challenge is that it is not just the customer faced with more choices, but the business as well. In an increasingly global economy with more and more sophisticated technology, there are not only more markets you can serve, but more ways you can serve them with a greater array of offerings.

"The paradox of choice" is now a fact of business throughout the world, as companies reexamine which products and services to specialize in, which markets to serve, and how best to design, create, and deliver a growing selection of offerings to increasingly broad and diverse markets.

Increasingly intangible desires of the market

Not only are the markets we serve more diverse and more demanding in terms of choice, they are also increasingly looking to intangible qualities to differentiate between one offering and another. The truth is most businesses are operating in oversupplied markets, where customers can choose between multiple products with the same functional value. In such markets, customers inevitably base the decision to buy one product over another on previously "superficial" features, a subject I'll explore in detail in chapter 4, "Absolutely, Positively Sweat the Small Stuff."

Increased sophistication of technology, systems, and processes

The technology, systems, and processes that you and your competitors use to access and service the market keep becoming more sophisticated. It is commonplace to buy things such as a server, laptop, or software, only to find that the tool requires an update—or even worse is completely obsolete—before you have even integrated it into your existing business processes and trained your team to use it.

To share a personal example, I have spent serious money and time configuring my own business systems so I could get the information I need from the database whenever and wherever I want. I can't tell you how challenging it has been to get members of my team to understand how the technology works and also to get me the exact information I need. Just when I thought I had my finger on it, online customer-relationship management (CRM) programs like Salesforce.com may have gathered enough momentum and critical mass to make my locally stored data a business process of the past.

This both excites me and hugely frustrates me. I'm sure you have had similar experiences.

Legislation

It is not just new technology, new choices, and empowered staff and consumers that are leading to increased complexity. It is legislation as well. Consider the following two examples.

Enacted in the aftermath of the Enron, WorldCom, and other business scandals, the Sarbanes-Oxley Act of 2002 (SOX), named for its key sponsors in the U.S. Congress, places new demands on corporate financial reporting. Because of the global reach of the U.S. economy, SOX has serious global impacts. For example, its strict guidelines on independent auditing of financial results have greatly complicated matters for both non-U.S. and U.S. companies, especially in terms of their relationships with the four major global accounting firms.

And this is to say nothing of the complexity of tax legislation around the world, especially for multinational companies with operations that span the globe. I read recently that the U.S. Internal Revenue Code is 7,500 pages and contains 3,500,000 words.

MANAGING CONFUSION IS USUALLY A CASE OF DOING MORE, NOT LESS, AT LEAST INITIALLY.

The second is a more grassroots example of a take-out restaurant I visited the day after attending a friend's wedding. Still recovering from the previous evening, I went to buy a kebab to try to settle my stomach. When I asked them to cut my kebab in half, they said they were not allowed to do so because of hygiene regulations. They even pointed out a sign they had posted to that effect. I told them they were off their rocker and wondered if it wasn't just an excuse to avoid extra work. But a week later I was doing some consulting for the Institute of Environmental Health and I related my experience to a former chief food inspector, who suspected, but could not say for sure, that cutting the kebab in half wasn't forbidden. Now, if a former chief food inspector doesn't know for sure, I obviously can't blame the restaurant proprietor for interpreting the law as he did.

All of this bureaucracy and complexity tends to lead to paralysis. In the face of the confusion, you do less. But taking less action is often counterproductive. Managing confusion is usually a case of doing more, not less, at least initially.

3. INCREASING TRANSPARENCY AND ACCOUNTABILITY

As my reference to Sarbanes-Oxley may suggest, the flip side (I couldn't resist) of the increasing volume and complexity of information flow is greater transparency and accountability. The digital communications revolution has put global information in the hands of literally billions of individuals, who then can share that information with one another at will. I have talked about the expanding exchange of knowledge and ideas, and how much additional complexity it creates for companies as they try to manage their databases. Even more important, information technology puts the power to obtain and share data in the hands of the individual.

The constant upswell in information, misinformation, complexity, ambiguity, and confusion thus does not increase opportunities to hide your mistakes and misdeeds. It actually has the opposite effect, subjecting you to increasing transparency and accountability.

You would think that people have enough to do, given how

busy everyone is, and how overloaded they are with information, without worrying themselves about your actions. Again the opposite seems to be true. There are always people ready to catch other people's mistakes with today's information technology. When they find mistakes, or what they think are mistakes, the very same technology allows them to spread their knowledge and opinions to literally the whole world. There is a good chance they are in the cubicle outside your office, bored out of their minds and looking for something more interesting to do.

Accountability is being forced onto businesses in three interconnected ways:

- *Top-down accountability.* This is where legislation such as Sarbanes-Oxley and government oversight are being introduced to make businesses more accountable for their behavior. In the near future, legislation on new behaviors such as carbon offset trading will shed an increasingly bright light on companies' environmental impacts.
- *Lateral accountability.* The existence of competitors in your marketplace gives customers the ability to "talk with their feet" when they don't feel as though you are meeting their increasingly intangible and constantly changing desires.
- *Bottom-up accountability.* A grassroots movement (be it word of mouth or mouse) can have a big impact on your reputation because of a positive or negative experience people have had with your brand.

Let me give you an example of these three elements playing together. A bottom-up movement has led to a growing awareness of issues related to climate change and the impact business has on the environment. This growing awareness creates an opportunity for a company to differentiate itself (or for a consumer to discriminate against a company) based on its level of "greenness." Take Westpac's "every generation should live better than the last" campaign and its joining nine other global banks in signing the "Equator Principles" in 2003, promising not to finance projects that endanger local communities or the environment. It was the only Australian bank to do so. A consumer may now keep competitors of Westpac accountable by not doing business with them, because they have not been as transparent and "green" in their behavior.

Or consider the example of McDonald's. Once upon a time if you had a good experience in a restaurant you might tell two or three people. If it was a bad experience you might tell ten or eleven. These days if you have a bad experience you can tell many times that number. You type an e-mail, put up a blog post, and click—a viral public relations attack!

I followed with interest the move by McDonald's to post a series of short videos on YouTube under the theme of "the McDonald's you don't know." On a recent visit to YouTube, I found the most popular of the McDonald's-produced videos had received 1,746 views. What intrigued me was that a video directly above this one had been viewed 387,747 times. It was not a video about how "great" the Golden Arches are, according to the company, rather it was a frame-by-frame video of how McDonald's had allegedly inserted its logo into the programming of the popular TV program *The Iron Chef.* This may sound perfectly normal, except that the alleged logo allegedly appeared for only one frame, long enough to be processed subconsciously but not long enough for the viewer to realize what has just happened, a very sneaky way of manipulating the viewer. It is an example of how companies can be held accountable in the public's mind for their actions—or even their alleged actions.

4. RISING EXPECTATIONS

The fourth force of change results from the other three and in turn feeds back into them: rising expectations for faster, better, cheaper products, for more varied options, and for greater transparency and flexibility in response to customer needs and wants. Let's consider the following product time line.

Twenty years ago it was standard to have two or three keys for your car—one to open the door and start the ignition, one to open the trunk, and perhaps another for your gas cap. This evolved into just one key, with buttons inside the car for access to the gas cap, trunk, and hood.

In an attempt to further improve the user experience, luxury-car manufacturers decided to save drivers the immense effort required to stick a key in a lock and offered remote keyless entry, allowing the driver to press a button to unlock the door. Then they would need to put that remote key into either a traditional ignition and turn the car on, or more recently hit the start button.

Now even this is too much, and you don't even have to take the key out of your pocket. The car automatically senses the key in proximity, and when you place your hands on the door handle the car unlocks instantly. Then, leaving the key in your purse or pocket, you hit the start button and the engine starts up. Not only that, but some leading cars will automatically reconfigure the seat and mirror settings based on the programmed owner of the key. To top things off, keyless ignition moved in the space of a few months from a luxury-car feature to one available on mass-market cars such as the Nissan Altima.

> A "SATISFIED NEED" NO LONGER MOTIVATES. ONCE A NEED IS MET WE MOVE UP OUR HIERARCHY AND START DESIRING MORE AND MORE OF LESS AND LESS PRACTI-CAL THINGS. THINGS THAT WERE ONCE A DESIRE RAP-IDLY BECOME A NECESSITY.

What's next is guesswork, but perhaps it will be fingerprint or voice recognition only, no need for a key. Or, wackier still, your car will know what you smell like and open and start upon sensing your uniqueness. (That is probably not a visualization you needed just now.)

The point is that what we are satisfied with today, we will not be satisfied with tomorrow. A "satisfied need" no longer motivates. We move endlessly up our hierarchy of needs and start desiring more and more of less and less practical things. Things that were once a desire rapidly become a necessity. Or, in business terms, features that once differentiated your product in the market fast become the price of entry, as competitors rip off, copy, and enhance your own innovations.

In autumn of 2006 the Pew Research Center, a nonprofit, nonpartisan "think tank," reported on the percentage of people who consider various items a necessity in life. Not nice to have, but a necessity. Here are a few selected items from that report:

- car, 91 percent
- home air-conditioning, 70 percent
- cell phone, 49 percent
- television, 64 percent
- cable television, 33 percent
- high-speed Internet, 29 percent
- flat-screen television, 5 percent
- iPod, 4 percent[3]

High-speed Internet is barely a decade old, and it has only had a critical mass of users for less than five years, yet already just under 30 percent of people surveyed in a random sample said it was a necessity for their life. At first glance you may think the 5 percent who call a flat-screen television a life necessity should join the real world, but if you do this survey again in five years, as no doubt Pew will, it will be more like 25 percent saying a flat-screen television is essential, and 5 percent saying one in each room. Expectations keep rising higher and higher.

The way you treated your customers and staff up until today will not be good enough tomorrow. You need to keep getting better, and with the compression of time you had better start getting better a whole lot faster, too.

In closing this chapter, I will say again, *change is not new*. Hopefully now you have a clear understanding of exactly what is changing, and are beginning to see the impacts these changes are having on your business. Within this context, let's now explore the *flips* and how to turn them to your advantage.

FIVE THINGS TO DO NOW

1. Hold a solo brainstorming session and write a list of all the ways the compression of time and space affects your business.

2. Think about complexity. What impact does it have on your business? More important, what impact does it have on your customers? Now think about strategy: How can you alleviate some of that complexity by simplifying existing services, or offering entirely new products and services?

3. Nominate someone from your team to become a red-tape renegade. Give them six months to cut as much bureaucracy and complexity from your internal communications and the interactions you have with your market as possible.

4. Ask yourself, "What do we currently do for our customers as an added bonus that may become the 'price of entry' in our market in the next few months or years?" Start thinking about how you can add new value to the same market.

5. Present the findings of the above four activities to your boss, or better still the executive team, and make some recommendations for the business.

ACTION CREATES
CLARITY

There is nothing more important in business today than an action orientation. If you take nothing else from this book, take this: you can't plan your way to greatness. There is nothing more valuable for your business (and for your life and career) than to do away with your commitment to microplanning everything and to let loose with some bold and courageous action.

This does not mean planning and strategizing are unimportant. Of course, having a plan remains vitally important. Although some pundits have declared "the end of strategy," I think that is far from the truth. But the big changes happening in the world mean that your plans must be shorter and more flexible. They must be built on the fundamentals of the most impressive human characteristic, behavioral flexibility. Most important, you must let your strategy and vision be refined, crystallized, and improved by the feedback you get from the actions that you take.

As you read your way through this book, you're going to be confronted with some pretty challenging suggestions about the things you must do to stay competitive in the face of the changes that are occurring in the market place. The overarching lesson from the ferocious rapidity of all of these changes is that you can't possibly understand them *all* before they happen, nor can you predict precisely what those changes might look like tomorrow.

So the worst thing you can possibly do is sit down and try to factor every suggestion and every change into your next project plan. It would be time poorly spent. The *best* reaction you could have is to put this book down and *do something*. Trust me when I say the feedback you get will be more valuable than any plan you ever fashion.

Consider this book itself as a case study. As I was struggling to gain traction with this book, a mentor asked me, "Pete, how's *Flip* coming along?"

"Great," I replied.

"How many words have you written?"

"None. I am still doing research."

"How long have you been talking about this book, two years? Isn't one of the chapters about how action creates clarity, or something like that?"

"What's your point?" I asked.

"Nothing really . . ."

It is always a little aggravating when someone gives you your own advice. We teach what we most need to know, the saying goes. I had taken two years to get to this point. Thorough and well researched, perhaps; slow and delayed, *yes*! I had employed a full-time researcher, and even sponsored a Ph.D. project on related themes. Despite all of this work, until my mentor's question I had nothing to show for it.

Perhaps this sounds like the logical way to approach writing a book or accomplishing any major task. Given that this is my fifth book, I would like to suggest that it is not. It is in fact the opposite. Your best work does not happen when you are planning. Your best work happens when you are in the flow. On a plane, in a coffee shop, or locked in your office late at night. It happens when you are taking action. Sure, you need planning time, but at some point you have to stop thinking, stop planning, and just *do something*. Anything!

Counterintuitively, the clarity I was so desperately trying to find before I started writing came only *after* I started writing. The point is, of course, that only through taking action, and the more the better, I say, will your strategy reveal itself (and rarely in one fell swoop) and the clarity you seek will be gained. Not only is the potential upside of this action orientation great, the downside of overplanning and overanalyzing can be worse than merely being confused. In a market being squeezed by constantly rising expectations, less client and customer loyalty, and a compression of time, distance, and required effort, inaction means you will likely miss the opportunity you were planning to exploit.

AT SOME POINT YOU HAVE TO STOP THINKING, STOP PLANNING, AND JUST DO SOMETHING. ANYTHING!

What are you procrastinating about, as you plan and plan and plan?

It is like the young couple (or singles nowadays) saving for their first house during a property price boom. House prices keep increasing faster than their ability to put together a sizable down payment. They effectively "save" themselves out of the market. The longer they "save" the further they are from their goal. A *flip,* no less. It can be the same in business. When we delay our action we deny ourselves the intensely valuable feedback that comes from putting the product to the test in front of a real consumer (or employee), who spends real money, and uses it in real life situations. Life and business move and change so fast that the longer we procrastinate the more we put ourselves at a competitive disadvantage.

It is small wonder that David Vice, CEO of Northern Telecom, says, "In the future, there will be two kinds of companies: the quick, and the dead."

What about you? Have you ever had an idea that someone else made a squillion on? You know, you had it in the back of your mind, shared it with friends, even made some preliminary notes about it, only to mess around thinking about it for so long that one day you read a two-page profile in a magazine about someone half your age who did what you wanted to do and made millions doing it.

Or worse, have you ever knocked down an idea that someone in your team came up with, by saying it was "a good idea, but the market would never want it or be prepared to pay for it," only to see your competitor nail you to the wall with it the following year?

You get the drift. If you want to develop a better business strategy, and a more compelling and achievable vision, then act first, and plan later. *Action creates clarity!* If you want a successful career, do the same. Stop trying to find the perfect answer, the perfect job, or the perfect product or business idea. Move! Do something! Anything! The action you take will generate clarity. From the clarity you will gain clearer vision on what you want to achieve, and the feedback you will receive will let you rejig your strategy to focus on the activities that are actually working as opposed to what you think might work. I call this "strategy on the go."

STRATEGY ON THE GO!

Strategy on the go is not the same as having no strategy at all. Dispensing with strategy altogether would be outright foolish, and even though some pundits have declared the end of strategy, this is not what I think nor is it what I am suggesting. I am, however, suggesting that we need a newer, more modern approach to strategy, an approach built on the fundamentals of behavioral flexibility and rapid decision making. It is time we let go of our obsession with detailed strategy, built on time frames of five, ten, or even twenty years. It is no longer possible to begin with point B in mind and reverse-engineer the result until you get back to point A with a detailed, step-by-step plan for achieving what you want.

> YOU NEED TO HAVE A BROAD VISION THAT COMPELS YOU TOWARD A BETTER FUTURE.

Instead, you need to have a broad vision, or what I call a trajectory, that compels you toward a better future. It should be flexible enough to absorb changes in market conditions and completely new technologies and products. The key is to map out how you are going to get there only in the broadest of strokes.

Imagine telling this to some of the Japanese companies. To

think that some of them still speak of fifty-year plans! There is a good chance that many of us will be dead in fifty years. How on earth are we supposed to know what the marketplace will want in fifty years, or what technology will be available, let alone what it will mean for our industry or our companies? Don't get too excited right now, as you laugh off the fifty-year plans, knowing you have a rock-solid five-year strategy. Perhaps even your five-year strategy is a little absurd. First, it is not rock solid. Second, in the current business climate a five-year plan could be just as dysfunctional as a fifty-year plan.

Meg Whitman of eBay could not have said it better when she observed, "Forget about five-year plans, we're working on five-day plans here." It's not that Whitman doesn't look five years into the future. She didn't pay $2.6 billion for Skype without thinking long term. But she and her colleagues are not following a detailed master plan; they're working out the plan as they go.

In an increasingly compressed and complex world with constantly shifting expectations, an obsession with planning and detail can be more of a hindrance than a help. It can wed your teams and business units to plans of action that are not working in the marketplace, and that are not reflective of how your business has changed. It is no different in your career. Be flexible in your approach, prepared to unlearn and let go of what no longer supports your ability to move toward your vision, and learn new skills and behaviors quickly that will take you to your goal.

Rupert Murdoch is doing the same. He paid a huge sum for MySpace without a clear and proven business model for how he would generate revenue from the site. MySpace gives us insight into a multinational company and its CEO taking action without a fully developed strategy. Rupert Murdoch is a flip-star! Here is what he said about MySpace and similar News Corporation ventures:

> The precise business model for sustained
> profitability from our digital investments is still
> uncertain at this point. Consequently, in some ways,
> we are embarking on a period of trial and error.
> —News Corporation Annual Report,
> December 2006

I have met Mr. Murdoch, having worked with News Corp. in the United States and having had the pleasure of dining with him in Los Angeles. The man is very impressive. He was a few days shy of his seventy-sixth birthday when we met, and he was as sharp as any twenty-six-year-old. He knew the MySpace purchase could be profitable, and the signs so far indicate that he is well on the way to being right, even if, as the quotation above shows, there are still some uncertainties.

News Corp.'s Fox Interactive Media division bought MySpace in July of 2005 for $580 million. Barely more than a year later, in August 2006, Google and News Corp. inked a deal under which Google will pay News Corp. at least $900 million, assuming Web traffic targets are met, to be the search and advertising provider for MySpace and some other Fox Interactive Media Web sites. At the time the deal was announced, MySpace was adding 250,000 users a day, suggesting that the Web traffic targets would have to be very high indeed to be out of reach. (In April 2007 they registered *thirty-one billion* unique page views.)

Flash-forward another seven months, to March 2007, when News Corp. and NBC Universal, the news and entertainment media division of General Electric, announced the creation of the largest Internet video-distribution network yet assembled. With initial distribution partners AOL, MSN, MySpace, and Yahoo, the network, hulu.com, will reach an unprecedented 96 percent of the U.S. Internet audience. And I am sure Internet portals in other countries will soon be added to the mix. At the time of the announcement, Cadbury Schweppes, Cisco Systems, Esurance, Intel, and General Motors had signed on as charter advertisers.

There's a flipstar time line for you. Spend roughly half a billion dollars with no definite idea of how to make it back, and eighteen months later monetize returns at roughly a billion dollars and counting.

The MySpace deals beautifully illustrate the overarching flip principle, "Think AND, Not OR." In the first place, the new video-distribution network signals that exploiting intellectual-property rights and giving the public free access to copyrighted materials on the Internet are not mutually exclusive concepts. Flexibility distinguishes flipstar organizations from their competitors. Instead of relying on business as usual in the television and movie business, News Corp. and NBC Universal are seizing the opportunity to make the Internet a fully fledged, advertiser-supported entertainment medium.

FLEXIBILITY DISTINGUISHES FLIPSTAR ORGANIZATIONS FROM THEIR COMPETITORS.

Second, note that in the August 2006 deal, Google and News Corp. are partners, whereas hulu.com, the News Corp./NBC

Universal video-distribution network, takes dead aim at Google's YouTube as a competitor. Likewise, outside the new Internet video network, NBC Universal and News Corp.'s Fox television, movie, and Internet divisions are themselves competitors rather than partners. Change partners and dance with your rival. It's obviously complicated and will no doubt sow confusion along the way, but Google, NBC Universal, and News Corp.—three of the most successful companies on the planet—recognize that the way forward is to "Think AND, Not OR," that "Action Creates Clarity," and that "confusion management" is now an inevitable cost of doing business. You can resent these facts if you like, or you can learn to love them and join the flipstars.

Take Burger King for another example. Burger King led the corporate charge into MySpace—a messy, ambiguous, and ultimately foreign situation—with action. They created a funny and personable profile, and the page took off. Burger King now has 134,000 "friends" (and probably a lot more by the time you read this), many of whom regularly post funny notes or even suggestions for burgers or promotions.

Fox itself is plunging in to promote the company's own products. Their brilliantly conceived MySpace profile for *X-Men: The Last Stand* has attracted almost three million friends. It is impossible to quantify how much the MySpace profile for *X-Men: The Last Stand* contributed to the movie's record-setting $120 million opening four-day holiday weekend release (the Memorial Day weekend) in the United States, but it surely didn't hurt. The previous Memorial Day weekend record gross was *The Lost World: Jurassic Park*'s $90.2 million in 1997. And *The Last Stand*'s Friday gross of $45.5 million trailed only 2005's *Star Wars: Episode III*'s record first-day take of $50 million.

MySpace is messy, but you need to learn to love the mess, to take some action without the clarity you desire. Others are choosing to sit on the sidelines and wait to see what, if anything, works.

Wired asked Rupert Murdoch how News Corp. was going to weather the cold dawn of a world in which traditional media are besieged by bottom-up, user-driven content: "'We'll figure it out,' he says, with a cheeky grin, scratching his head. 'You want to learn from MySpace,' he muses. 'Can you democratize newspapers, for instance? What does it mean for how we do sports or politics? I don't know—no one does. I just know we'll figure it out.'"[1]

This is what I am talking about. I love his new strategy, which strangely looks a little bit like: "I am not sure whether this will work or how I am going to go about it, but let's have a crack anyway. We have some bright people, we will pay to retain the bright people who started MySpace, and we will find out."

Even with the $900 million Google deal and the new Internet video-distribution network, it is still unclear whether News Corp.'s Internet properties will pay off as they should. But Rupert Murdoch and his colleagues act anyway. They learn to manage the confusion and accept the risk.

There's no detailed master plan. What News Corp. does have is a broad trajectory. They know the media business. They know that people, particularly the younger generations, are consuming media in new and different ways, and they know they want News Corp. to be a major player in the new media channels. MySpace is a leader of the so-called new media and Murdoch wants a piece of the action. Rather than try to reinvent the wheel and outspend some of the sites that have grown organically, he buys one of the most visited sites on earth. Then instead of doing what I reckon plenty of other media conglomerates would have done—namely, rebrand and integrate the Web site into its other media properties—he leaves it as is.

There's not a pixel of News Corp. presence on MySpace. "Obviously MySpace is a world unto itself," *Wired* quotes News Corp. COO Peter Chernin. "There's never been a second when we said, 'How do we put our stamp on it?' We'd be crazy to interfere."

It is possible that I may end up eating my words on this, and News Corporation shareholders eating their losses, but it is exactly the willingness to take risks based on a broad vision for the future, in spite of some ambiguity in the present, that the most successful leaders nurture in their executive teams. And you will need to embrace the same openness to risk in your career as well.

For another Aussie example, consider the Lonely Planet travel guidebooks. In 1973, after meeting on a park bench in Regent's Park, London, and later getting married, Maureen and Tony Wheeler spent their honeymoon traveling overland across Europe and Asia to settle in Australia. They weren't planning on writing a travel book, but when they told friends about their experiences, and these friends told their friends, they began to be besieged for advice. So they wrote up their experiences and self-published *Across Asia on the Cheap,* which they stapled together on their kitchen table in Sydney and sold for $1.80 a copy.

They sold 1,500 copies the first week, and a new business was born. For their follow-up guidebook, the Wheelers traveled throughout Southeast Asia in 1975. At this time the fires from the Vietnam War were still smoldering, and tourist travel in the region was almost unheard of. Yet *Southeast Asia on a Shoestring* proved another immediate success. Go forward thirty-plus years and the Lonely Planet guides have inspired imitators like the Rough Guides travel books and have carved out a large slice of the travel-book market. Lonely Planet now employs over five hundred people and has annual revenues of more than $100 million. In September 2007 the BBC's publishing division acquired it in a deal that put a lot of cash in the Wheelers' pockets (the exact sum was not disclosed) and still left them holding a 25 percent stake in the company.

Tony Wheeler is not a reckless man. Nor, with an MBA from the London School of Economics, is he the pot-smoking hippie some may expect. At the same time, says Tony (I've had the pleasure of meeting him, and he is an inspiring person), "We knew nothing about publishing." The Wheelers simply took action based on their own experiences, the eagerness of friends and acquaintances to learn from them, and their gut instincts.

Not that financial success came overnight. For a time they were so broke they would steam unpostmarked stamps off incoming mail and reuse them. That's creative recycling for you. This is the kind of hunger we need in our bigger organizations. A get-down-and-dirty-and-make-it-happen kind of attitude. Imagine what you could do with that desire and the money that so many of our bigger businesses have access to. Some months things were so bad that the Wheelers would go to dinner with their friends, pay by credit card, and collect their buddies' share of the bill in cash to get them through the month.

Go and find someone who is boot-strapping a new project or prototype in your business, and help get it off the ground. Tell their story to everyone in your company. Celebrate this kind of internal entrepreneurialism.

So many businesses have started just this way. Anita Roddick's Body Shop chain is another great example. This is how strategy gets done. Through action we get clarity.

If you have never watched the TV show *House,* then you should. It is a crash course in twenty-first-century strategy. Dr. House has a worldwide reputation as a brilliant diagnostician. In the medical profession, prescription in advance of diagnosis is generally considered a form of malpractice. Yet when he is confronted with patients whose ailments fit no obvious pattern, House boldly acts on the belief that prescription *is* diagnosis. He acts on his best hunch and, assuming the patients survive, adjusts their treatment according to their reaction to the first treatment.

Now I know it seems a little crazy to use a TV show as an example, but I can assure you the metaphor is powerful. Dr. House takes only the most complicated cases, for which he is the last resort. The more ambiguous and confusing the patient's symptoms, the more interested he is. He takes the seemingly unsolvable cases and solves them. He does it by acting on the best information he can get at the time of diagnosis,

GO AND FIND SOMEONE WHO IS BOOT-STRAPPING A NEW PROJECT OR PROTOTYPE IN YOUR BUSINESS, AND HELP GET IT OFF THE GROUND. TELL THEIR STORY TO EVERYONE IN YOUR COMPANY. CELEBRATE THIS KIND OF INTERNAL ENTREPRENEURIALISM.

without knowing for sure if he is right, and he treats the patient as though he is right. His reasoning is that if he is right then great, the patient will get better. And if he is wrong, not only will he now rule that educated guess off his list of possible illnesses, but the patient's response to the treatment will give him a better insight into what is wrong. Or said another way, action enhances clarity.

I was having a conversation recently with one of Australia's leading equity traders, who was mortified when I suggested that action creates clarity. As an equities trader, he always wants to make as informed a decision as possible before investing millions of dollars of clients' money. (Equally, despite using the *House* example above, there was no way I am going to suggest that physicians should go and prescribe drugs to patients on a whim.) However, there is still an element of doubt when he makes that investment. There is always a risk. He can use many inventive financial plays to mitigate that risk, but never completely. In a way he is taking action without complete clarity every day, even if he would never say this out loud.

You can't let the presence of risk, or the absence of clarity, prevent you from taking action. Whether that action be an investment, an attempt to ask someone out on a date, or more seriously, an attempt to cure a patient suffering what at first seems like an incurable disease.

DECISION MAKING ON THE MOVE

It may seem like a bit of a stretch, but human beings have been doing strategy on the go throughout the course of time. From as far back as studies have looked into, the early hunter-gatherer lifestyle conditioned human beings to be opportunistic foragers in a constantly changing environment. That required decision making on the move. Where is the best place to look for food today? How are we going to work together to find food as the weather changes day by day and season by season?

This lifestyle naturally stimulated our creativity and problem-solving skills, and as societies evolved and became more complex, these skills were honed in ever-more-challenging environments.

In this sense, it is very strange that we have become geared against, rather than toward, change. We arrange our organizations around command and control instead of around what has worked for most of time: change and adapt (*flipping*). Weekly I work with organizations with structures, reporting lines, and cultures that are against the very notion of flipping, and are for consistency and predictability. Even though at some level the organizations know this is dysfunctional—otherwise they wouldn't have asked me to consult with them—they resist hearing the fact and acting on it.

IT IS AS THOUGH THE PHENOMENAL SUCCESS OF BUSINESS IN THE LAST CENTURY, AND ESPECIALLY THE LAST FEW DECADES, HAS CREATED A SORT OF ARROGANCE THAT WE CAN PREDICT THE FUTURE.

Human success at adapting to changing conditions periodically lulls us into thinking that we've got it all figured out. Then we start developing and trusting long-term plans, which sooner or later—usually sooner—are exploded by a changing world. It is as though the phenomenal success of business in the last century, and especially the last few decades, has created a sort of arrogance that we can predict the future. I think it is fair to say we are increasingly feeling that confidence shaken as the four forces of change come to bear on us daily.

In what area of your business have you become overconfident? What do you assume to be immutable today that may become irrelevant tomorrow?

The truth is human beings are always much better off and much more productive when they abandon the desire to command and control things and accept the need to change and adapt to them. While the industrial age lent itself to systemizing and efficiency, that is not the case with the knowledge age, creative age, or—as I like to think of it—the relationship age. We think better on our feet, as Mihaly Csikszentmihalyi documented in his classic book *Flow: The Psychology of Optimal Experience,* and as Malcolm Gladwell also shows in his more recent examination of our capacity to make quick decisions in *Blink: The Power of Thinking Without Thinking.*

Of particular importance to this point are the studies Gladwell reports that demonstrate that we actually learn and adapt much faster than we think we do. We have a reliable ability to make decisions on the spot that are every bit as accurate as those over which we deliberate for months. We make *excellent* decisions quickly.

One study into the "adaptive unconscious," as it is known, used two decks of cards—one was stacked with cards that made it hard to win, one with cards that made it much more likely that you would win. Over time, test subjects gradually altered their behavior to draw more cards from the winning deck. But *more* interestingly, players actually altered their behavior *before* they reported that they had consciously detected a pattern. They were physically learning faster than they thought they were.

Gladwell argues, based on this and a whole host of other evidence, that "decisions made very quickly can be every bit as good as decisions made cautiously and deliberately." One of many possible examples of human beings' ability to make accurate judgments quickly is a 2006 study, is showing that people can accurately predict the result of political elections after seeing just ten seconds of footage of the candidates.

It is true; every action is a gamble, and that's scary. But like I said, we're evolved to thrive on taking lots of hunter-gatherer gambles every day. So stop complaining, because the time has come to . . .

BUILD A BRIDGE AND GET OVER IT!

We're all scared. Seriously. We do our best to pretend we are not, but even the most senior executives I meet are freaking out. They worry that new ideas might not take off. Will they meet budget? Will the stock market hammer that less-than-double-digit growth this year? Will they lose their jobs? And so on. This fear compounds, and is fueled by, the complexity of the world, especially as we try to plan further and further out into the future. This can often result in a paralysis of inaction.

Consider the photographic industry in recent times. It is well known that this industry has been thrown into turmoil by the digital electronics revolution. Recently I worked on the issue with the Photo Marketing Association of Australia, and Konica Minolta. Midway through a project for helping their retailers—mostly baby-boomer mom-and-pop operations—to promote and market digital products, Konica Minolta completely withdrew from the digital camera market globally, doing so literally overnight. One day we were working on a strategy of short-term actions to get better penetration in the market using their retail distribution network; the next day, nothing. Never mind, Sony stepped in to continue the work.

The problem was that Konica Minolta left its foray into the digital camera market too late, not releasing their first attempt, the Minolta D Image EX-1500, until 1998. They were aware of the technology, and even aware of competitors like Sony making gains in this market, but they waited as they tried to gauge the market better, searching for better information on consumers' responses to digital technology before they acted.

WHAT DECISION HAVE YOU BEEN PUTTING OFF? MAKE A DECISION *NOW*! TRUST YOUR INSTINCTS, AND GO WITH IT.

When they finally realized that digital technology was not only here to stay, but was already the dominant medium in consumer photography, it was too late. They had missed the early adopter's advantage. Instead of acting, they wasted time planning a response they never made to the biggest market opportunity to hit their industry in decades.

What decision have you been putting off? Make a decision *now*! Trust your instincts, and go with it.

Kodak and Sony were both pioneers of digital technology, dating as far back as the mid-1970s, but it was not until the mid-1990s that digital cameras met the market requirements of fast, good, and cheap at the price. Kodak started the digital trend with their DCS 100 camera in 1991, but they were almost crippled by the "disruptive" technology they introduced. Perhaps this is because the adoption of digital happened so fast that they were not ready for the change, because film was their primary product.

They seem to be back on track, and now have the number one U.S. market share in digital cameras and photo printers. They are still struggling to achieve their old levels of profitability, but at least they're on trend. The analog camera share of the market is fast falling to about 10 percent, and likely lower.

Sony got in fast enough, but only just, with their CyberShot in 1996. They had been playing in the digital space for two decades, and still they almost missed the boat. They had the same incomplete information and the same technological resources as Konica Minolta, and they faced the same ambiguity. Fortunately for employees and shareholders, Sony chose to act first and refine their planning second. Before you write Konica Minolta off, consider that they are now focusing on selling printing paper to be used for printing digital photographs, which to date is a very popular activity.

When digital cameras first came on the market, no one anticipated how quickly consumers would embrace them. But the nature of modern consumers is that they act *fast*. Either you move fast as well, or you lose.

After I spoke about the digital-camera revolution and its implications at a conference on behalf of my client Sony, a small-business owner named Jack approached me for some advice. Three years before, he had owned half a dozen photo-developing shops, with net positive cash flow of $250,000 each. On that basis he was making a lot of speculative investments and planning to retire in eight years.

When he came to me, Jack was down to one shop, which was barely breaking even. He no longer owned the freehold on the shop, but had been forced to sell it to service the debt on his speculative investments, which crashed when the technology stock bubble burst. To top it all off, his wife had left him.

As this example should make clear, when I talk about the need to stop planning and start acting, I'm not talking about acting blindly. Jack wanted to believe that his plan was foolproof, and he resisted looking at the accumulating evidence that it was not. While he was busy doing his wishful thinking, his smart competitors were beginning to offer digital photography lessons, digital photo frames, online photo-album and scrapbook design, and other value-added services, using market feedback to find out what customers wanted most. Also, making speculative investments is not really the same as strategy on the go.

After brainstorming with me, Jack has now begun to offer similar services. In effect he is transforming himself from a commoditized printer to a provider of customer experiences. He no longer sees his business as a simple and easily replicated transaction (printing photos for a fee) that the consumer could get down the road. He sees himself as a partner and mentor for his customers as they capture the memories of their lives. Through lessons and coaching, he is helping customers to adapt their own photography habits to the new digital world. His remaining shop is back on track, and he has a chance to return to the profitability he once enjoyed. If he'd had the guts to listen to the market and act on what it was telling him, instead of remaining loyal to an obsolete plan, he wouldn't have as much lost ground to make up. At any rate, I'm glad I could help him with his business.

Let's step back and look at the larger market again. When digital cameras arrived, most of the businesses that lived on selling traditional film and traditional film printing feared the end of printing photos and went into a state of denial. A few snapped out of denial and accepted reality fairly quickly, but most did not. Deep down Jack knew he was in denial, but he couldn't face up to his wishful thinking. He was hoping traditional film would hold on long enough for him to complete his detailed plan to retire in luxury in eight years.

In fact digital photography did not mean the end of printing photos at all. It meant that consumers no longer had to print their photos to look at them, but it also meant that consumers could easily take and store many more photos than ever before. The percentage of captured images that were printed might go down, but over the long term the total number of images printed would keep going up. One important reason for this is that most consumer photo printing is done by women, who are not content with storing pictures on a personal computer or the Web and who like to share prints with friends and family. The companies that acted in step with the market positioned themselves to profit from this, whether or not they saw the underlying dynamic clearly. The companies that bided their time to plan a response lost their place at the table.

THE ONES WHO WILL COME OUT ON TOP WILL BE THOSE WHO ACT IN SPITE OF THEIR FEAR.

Again I say: move! Do *something*. Whatever fear you are feeling, trust me, everyone else is freaking out about the same things. The ones who will come out on top will be those who act in spite of their fear. It will be those willing to expose themselves to the market (job or consumer market), be aware of the result, and adapt their behavior accordingly (think strategy on the go). I am constantly surprised at how few organizations want to do this.

AVOIDING REGRET

The saddest words in the English language are *if only*.

Ask yourself now: What opportunities are open to me, my career, my life, my business that are potential if-onlys tomorrow? What path of action do you think is worth taking a bet on now? Take it!

> ASK YOURSELF NOW: WHAT OPPORTUNITIES ARE OPEN TO ME, MY CAREER, MY LIFE, MY BUSINESS THAT ARE POTENTIAL IF-ONLYS TO-MORROW? WHAT PATH OF ACTION DO YOU THINK IS WORTH TAKING A BET ON NOW? TAKE IT!

You don't want to go through life continuing to pile up if-onlys. You want to say "if only" less and less, as you become smarter, more confident, and more successful. And the only way to do that is by getting comfortable with risk.

This takes on a new and even more potent twist when you delve into the psychology of the way humans process and experience regret. Many people will not act because they fear they will regret their action. Most people also operate under the assumption that they will regret foolish actions taken more than smart actions not taken. This is, interestingly, false.

Echoing commonsense wisdom, psychologist Daniel Gilbert, in *Stumbling on Happiness,* says, "In the long run, people of every age and in every walk of life seem to regret *not* having done things more than they regret things they *did*." This is borne out by a number of academic studies over time—and if you're interested, the most common regrets include "not going to college, not grasping profitable business opportunities and not spending enough time with family and friends."[2]

Despite this, the four driving forces of change seem to have made a lot of companies and CEOs risk averse. It is not surprising really. With the average tenure of a CEO these days barely five years, so many just try to get through it unscathed. The average tenure of a chief marketing officer is far less, and it is not surprising that this is the area most in need of some innovation. Paralyzed by fear of failure, risk-averse organizations and leaders try to plan their way to a secure future. The result is that they plan and procrastinate their way straight into the arms of the failure they want to avoid. Think of the major American motor companies for example. They failed to move toward small cars over the last several decades, and hybrid engines in the last decade or so, and watched both Japanese and European carmakers steal their market share.

Today's most successful organizations and people are risk happy. They embrace the messy ambiguity and confusion in the marketplace and the fact that you can't hide your mistakes, and they turn these forces into allies by continually taking chances and trying new things.

This earns them credit with customers, because in oversupplied markets, people want their brands to show leadership and courage. They love Nike for phasing out and updating an Air Max style before it's reached the height of its potential sales curve. They become even more loyal to Toyota's Scion when the company limits production in the face of increasing demand so that its hip young brand keeps its edge for the long term and doesn't get oversold in the short term. Meanwhile the business-as-usual brands like Dunlop and Ford can't even get customers' attention.

Risk-happy companies also attract the best staff. It used to be the case that companies that offered stable, long-term employment were more attractive. But these days, the average tenure of people aged twenty to twenty-seven at a single job is less than two years. Companies are finding that the best and the brightest young people are generally not interested in long-term stability and want to leave after a year because they're bored.

GE CEO Jeffrey Immelt worries that his company is not attracting enough young hunter-gatherers, because gas turbines are not as cool as iPods and search engines. "If we can attract the best twenty-two-year-olds," he has said, "then we can double, even triple in size. If not, then we are already too big." Not content with that, Immelt has challenged GE to achieve 8 percent organic growth, exactly the kind of action-oriented move that will keep GE an exciting

IS YOUR ORGANIZATION WORKING ON PROJECTS SO EXCITING THAT THE BEST AND BRIGHTEST IN YOUR INDUSTRY ARE BANGING THE DOOR DOWN TO COME AND WORK FOR YOU? IF NOT, WHY NOT?

place to work. It is not the action of a CEO who is afraid to fail, and GE's continued success is testament to the effectiveness of this approach.

Is your organization working on projects so exciting that the best and brightest in your industry are banging the door down to come and work for you? If not, why not? What could you get off the ground in the next thirty days that will excite not just the people who work for you but the talent that works for your competitors, too?

Do you think the people who are scrambling to work for Google (they get seven thousand unsolicited applications each day) want to settle into a predictable routine governed by a detailed, long-term plan? These are some of the brightest and most talented people in the world, and they are dying to work for Google precisely because it is impossible to predict what the company will look like in five years' time. Sure, Google has vision, but it acts on its vision in short-term installments, rolling out new concepts and products fast, abandoning what doesn't work, and moving on to something else in a heartbeat. At a recent conference in Los Angeles I got to spend some time with Laszlo Bock, the head of HR at Google. He was asked what he thought the Google workforce would look like in five years. He almost laughed as the question was being asked, and explained that their business changes so fast, as it adapts to the market, that they only really thought in detail about two years out from a workforce-planning point of view.

A risk-taking, action orientation isn't just a good business decision from an innovation and market-penetration point of view, it is also highly desirable from a staffing perspective. Ambiguity and confusion may scare you as a business leader, but bright and talented employees—the ones you really want and need—love the high-paced, high-energy environment that an action-oriented company creates.

THE SUREST ROUTE TO FAILURE IS NOT TO TAKE RISKS.

Risk-averse companies and people misunderstand one of the basic facts of life: the surest route to catastrophic failure is not to take any risks. Ice hockey great Wayne Gretzky famously said,

"I miss all of the shots I don't take." If you want to succeed at anything, you've got to take a lot of shots. You've got to throw plenty of mud against the wall and see what sticks and what doesn't. You've got to develop your risk tolerance by taking lots of small- to medium-size risks. In so doing, you develop your recovery ability, and you shrink risk. You also shrink your fear of failure and learn to make big risks manageable.

This is a key element of confusion management, the number one skill of a good leader in the flipped world. I am about to explain how to manage confusion, but first I must make a very important point.

It is vital that we do not confuse action creates clarity with speed to market. Speed to market can be very important, but putting something on the market and giving it to consumers before it is ready can have disastrous consequences—it can sometimes be enough to kill a company. This is not what I am trying to say. What this chapter is about is making sure you can have speed to market without compromising what you put out there.

For example, Atari was first in video games, but it was Nintendo, a fast follower, that survived to do battle with Sony's PlayStation and Microsoft's Xbox. Likewise Apple, a frequent flipstar example in these pages, brought the Newton PDA to market before it was ready, and saw the Palm Pilot become the first truly successful PDA.

CONFUSION MANAGEMENT

There is going to be more confusion in the business world
in the next ten years than in any decade, maybe in history.
—Steve Case, Former AOL Time Warner Chairman

I've already referred to confusion management. The best definition I have ever heard for confusion management came from a good friend and client of mine, Sheryle Moon, who at the time was a director of Manpower (the recruitment company), when she said the key to being a successful manager today is the ability to deal with ambiguity and still take effective action day to day. Here are my thoughts on what it takes to be a good confusion manager.

KEEP MOVING

It goes without saying that step one is to keep taking action. Action creates clarity, so take plenty of it. You may like to look at it in the following ways:

EXPERIMENTATION

Look at your work, your project, as an experiment. Allocate a reasonable amount of time each week to try new things. At Google, software engineers spend 20 percent of their time on projects of their own interest. Although the founders credit this idea to their experiences at Stanford, where Ph.D. students are encouraged to spend one day a week on something besides their officially sanctioned thesis topics, companies like 3M with their 15 percent rule have been doing the same for decades.

These ideas generally do not need to be approved by management. They are usually the ideas of the scientists themselves. From these little experiments came the Post-it notes at 3M, and Google News, Google Earth, and Froogle, among others, at Google.

Could your business implement something like 20 percent

time? What about one day a month dedicated to new-idea generation?

GUERRILLA TACTICS

I have met my share of potential flipstars who are caught inside very traditional and unexciting companies. They don't leave because they love the product they make or sell. My advice to them (and you) is to start little hidden experiments. Your company may not give you 20 percent time, but take some of it anyway. Use the company's resources to get a working prototype of the service or product you are envisioning up and ready for demonstration. You may get your butt kicked, but who cares. First, you probably don't even like your manager, and second, it is easier to apologize than to ask for permission.

DEDICATE ONE DAY A MONTH TO NEW-IDEA GENERATION.

These little behind-the-scenes activities are like guerrilla tactics. Here are some extra ideas for all of you hidden flipstars:

1. Recruit a champion. Find someone high in the organization who you think is a closet flipstar as well. Recruit them for your project. Stop them in the foyer, stalk them at lunch if you have to. Tell them what you are up to, tell them that it is undercover, and ask first that they make some suggestions on how to further the idea, and second that they help you to get your new idea a hearing at the level of the organization where things can actually change.

2. Recruit some helpers. If you are bored, I can guarantee some of your coworkers are, too. They probably don't have the rebellious energy required to start a guerrilla

project themselves, but I am willing to bet they would happily work with you on yours.

3. Make sure it works. Don't be all self-righteous just because you are doing something. Although I think you have a right to feel great about your risk taking and orientation to action—I have been writing about it nonstop in this chapter—it won't get you anywhere in your organization by itself. No one can argue with results. Get your prototype up and running, make sure it can work, crunch some numbers, and then show people. Not the other way around.

4. Move on. Your little experiments will often fail. Get over it. Start another one.

As Richard D'Aveni of Dartmouth University's Tuck School of Business puts it, "This is not the age of castles, moats and armor. It is rather an age of cunning, speed and surprise." Your career is no different. You can contemplate your navel all you wish, searching for your soul's purpose, the perfect job, the foolproof strategy, the most innovative product, or the grand unified theory of everything, but you won't find it there. You will find it by taking action, by trying something—damn near anything, I say. And the combination of wins and losses, successes and failures, and everything in between will deliver back to you the clarity you need to fine-tune those actions and strategies to make you more effective, and to generate the results you desire.

WHAT GUERRILLA PROJECT COULD YOU PERSONALLY KICK OFF? EVEN IF YOU ARE A VERY SENIOR EXEC, GET SOMETHING COOL STARTED. TODAY!

What guerrilla project could you personally kick off? Even if you are a very senior exec, get something cool started. Today!

DECIDE ON A TRAJECTORY

You really do need to have some idea of where you are going. We live in times of too much opportunity, rather than a time of not enough opportunity. The greatest challenge a flipstar will face will be to decide what to focus on. But decide you must.

WE LIVE IN TIMES OF TOO MUCH OPPORTUNITY, RATHER THAN A TIME OF NOT ENOUGH OPPORTUNITY.

The beauty of the trajectory model is it leaves plenty of room for serendipity. It is more about where you are heading, not necessarily exactly what you will be doing to get there. Being the leading bank will likely mean something vastly different in fifteen years from what it means today, but it is still a worthy goal. The best real estate company in twenty years might be one dominated by property management rather than by sales.

Cirque du Soleil has a trajectory that does not limit it to the production of circus events. Their trajectory is to "evoke, provoke, and invoke imagination." It sounds a little trite, I know, but they certainly have done this with their stage shows to date. This is not all they are doing. They are looking deeply into establishing Cirque du Soleil resorts, as well as staging stadium-size productions. I think there is huge potential for them in corporate training. This is on top of Cirque du Monde, which is a not-for-profit venture that travels into third-world countries and teaches juggling and other such imagination-building skills.

Cirque du Soleil's strategy would actually include all of the above items, but as you can see, their actions are quite varied, yet still fit into their broader vision.

While on the topic of imagination, I would like to encourage you to use some. It is the companies who in recent times had far-reaching imaginations that we celebrate today, companies that were able to see outside of their narrow sphere and imagine a bigger and brighter future. Think big! Dream even. Only this will inspire intelligent people to join you on your quest to create something that is bigger and better and faster than whatever is currently available.

Confusion management requires that we balance activity and big-picture thinking. Replace your detailed strategy for the long term with a basic trajectory, a direction you want to head in, within which you will have the flexibility and agility to change quickly as you respond to changes in the external and internal environment that will alter the landscape you will be competing in. And allow the ambiguity that comes with it to excite you rather than scare you.

Focus your strategy and planning on the shorter term, which I will outline in a moment.

I am using business-focused language here, but your career is exactly the same. I meet very few successful people who knew from a very early age what they wanted to do. Sure there are the Tiger Woods– and Eminem-style examples, but in reality they are few and far between. People may have known they wanted—for example—to be an entrepreneur, but did they know it would be in the specific industry they are in today? Probably not.

The chances are they would not even have been certain that they wanted to be an entrepreneur. In fact they may not even have known what the word meant when they got their first job. I certainly did not. I did not know that there were jobs even remotely like what I now do.

The trajectory is a powerful thing when the market changes in your direction. Google began at Stanford with Sergey Brin and his Ph.D. supervisor, Rajeev Motwani, as a data mining project. There is no way in the mid-1990s, when they were getting very interested in this topic and how it applied to the newborn Internet, that any of the MIDAS (Mining Data at Stanford) crew, which included among others Google cofounder Larry Page, could ever have imagined how valuable Internet search tools would become. As the project evolved and Brin and Page got more and more committed to it, they decided that no matter what, they would focus on better search functions. This remains at the core of the Google trajectory and all of their very powerful and valuable tools, and their skyrocketing revenues come from all things related to search.

Google could easily rest on the success of their AdWords business and choose to consolidate rather than aggressively pursue new applications for search technology. The millions of AdWords clicks, most only worth a couple of dollars each, are where Google gets almost all of its revenue—$3.66 billion a quarter and rising in 2007.

But instead of sitting back and counting the cash, Google keeps developing new opportunities and new services built around search capability.

They are also expanding into other areas: for example, taking Microsoft on in the business-software market, looking to replicate the success of Microsoft Office with a suite of online alternatives such as Google Spreadsheets under the banner of Google Apps.

CONNECT WITH THE FUNDAMENTALS

Some things never change. Seriously! In fact what we do actually changes very little over time. It is *how we do it* that changes far more often. Never lose sight of what you are trying to do. For example, the customer should be at the center of everything that you do. The new strategy for the Commonwealth Bank is built around this fundamental idea. Here is a thirty-thousand-plus-people organization that dominated the banking sector for many years. Under the leadership of David Murray, they embarked on a deep and widespread cost-cutting rampage. Most industry pundits agree that this was essential for the bank to generate better profitability and increased efficiency. The problem is they seem to have done so at the expense of quality customer service.

I don't think the banking sector is all that complicated. Sure the products, multitude of choices, and other such things can make it seem very complicated, but in reality all banks are offering the same sorts of products and services. The competitive advantage, then, will be very simple. Those who offer these products and services in the fastest, easiest, and, dare I say it, the most helpful and the friendliest way will win. Good customer service, among other things, is a fundamental in this industry. Stay connected to it, no matter what else changes. Sure, the way you deliver that customer service can change, using things like the Internet for example, and customer expectations of quality will continue to rise, but customer service will still remain a fundamental in the provision of banking products and services.

What are the fundamentals in your business? How are you performing in these areas? Are you staying abreast of new developments and rising expectations in these areas? If not, why not? What could you do now to get started?

WHAT ARE THE FUNDAMENTALS IN YOUR BUSINESS? HOW ARE YOU PERFORMING IN THESE AREAS? ARE YOU STAYING ABREAST OF NEW DEVELOPMENTS AND RISING EXPECTATIONS IN THESE AREAS? IF NOT, WHY NOT? WHAT COULD YOU DO NOW TO GET STARTED?

If you are running a large organization, you will obviously be delegating the accountability of the fundamentals to key people. Do so, but monitor them like a hawk.

Getting back to the Commonwealth Bank for a moment, their new CEO, Graeme Murray, who has successfully turned around the Commonwealth Bank's sister bank in New Zealand, ASB, and also Air New Zealand, has said publicly that business is much simpler than we make it out to be. There are some basics that are not negotiable and you must execute on these flawlessly. I am backing Murray to do this successfully at the Commonwealth Bank.

LOOK, LISTEN, AND UNLEARN

It is essential that we keep our eyes and ears open. The reason people and the businesses they run fall behind is that we as individuals are not paying attention to the changes taking place. We are not looking for them, we are not reading about them, and sometimes we notice them but refuse to listen to the lessons they give. Sure, you can trundle out any one of hundreds of excuses about being too busy, underresourced, or restricted by upper management (or the board and shareholders if you are upper management), but in reality these are all just excuses.

To be a flipstar you need to have high sensory acuity. That is, you must notice the nuances of change that occur around you daily and see what is happening. You must begin to act on the fact that more of your customers mention they visited your Web site before calling you, or that increasingly people get confused when you tell them what you do. The point is you won't be able to change and adapt if you do not know what is going on.

If you take repeated action that is not pushing in the right direction and you are not using this feedback to infer how you may need to change, then this is not intelligent behavior. It has been referred to colloquially as the definition of insanity; that is, doing the same thing over and over again and expecting a different result. Not all feedback is good in the first instance. The faster you move through denial, through believing that all is rosy when it is actually not, and make some changes, the better. First, you have to notice that all is not rosy.

And it is not just noticing things about your individual business and industry. It is said that the disposable aluminum drum invented by Canon that revolutionized the copier market was inspired by a can of beer. I know, I know. That particular engineer really needs to get a life. Thinking about photocopiers while having a beer. But in reality, this is how many such things actually happen.

Step one—keep moving. This will create plenty of feedback for you to observe. Go beyond this and look to what your competitors have done that has and has not worked.

Look to other industries for what they have done. How could you do something similar? I was with a large health care provider recently. The biggest challenge they have is finding nursing staff for their many hospitals nationwide. Skills shortages are rampant, and filling rosters is a nightmare. One of the members of the board had been a consultant for an airline and was responsible for implementing systems that would allow them to get the maximum out of their staff while cutting their numbers (and costs).

She was able to explain the rostering system the airline used—which mostly consisted of a staff bidding system, with the hard-to-fill shifts paying more money, and the traditionally desirable ones less—and show that it could meet the health company's needs. The board and senior managers adopted the same approach across multiple sites and are saving tens of thousands of dollars every month in a tight labor market. This is an example of listening to someone with the right expertise.

I would like to reiterate here that no matter how much information you gather, you will still be confused. In fact, the more you discover, the more confused you may become. Again, celebrate the confusion. Rather than dig deeper and deeper and run the risk of growing more paralyzed, revel in the ignorance and see it as an opportunity to try something new.

A groundbreaking piece of work by David Gray at BCG Strategy Group, published in the *Harvard Business Review,* argues that embracing "nescience" (the opposite of knowledge) is actually enormously beneficial. Too much knowledge can constrain thinking and limit you to acting within the confines of what you "know" to be true.

WHAT KNOWLEDGE OR PRACTICES DO YOU HOLD ON TO THAT ARE NO LONGER EMPOWERING? WHAT BEHAVIOR THAT ONCE DROVE YOUR SUCCESS DO YOU AND YOUR TEAM NEED TO UNLEARN?

The beauty of being a flipstar is you can always do the opposite of what the world suggests. People say knowledge is power. Perhaps knowledge is not power. Maybe ignorance is. David Gray even mourns the loss of ignorance:

> Unlike knowledge, which is infinitely reusable,
> ignorance is a one shot deal: Once it has been
> displaced by knowledge, it can be hard to get back.
> And after it's gone, we are more apt to follow well
> worn paths to find answers than to exert our sense
> of what we don't know in order to probe new
> options. Knowledge can stand in the way of
> innovation. Solved problems tend to stay solved—
> sometimes disastrously so.[3]

Two thousand years ago, people "knew" the universe was created in a week. A thousand years ago, humans "knew" the sun moved around the earth. Five hundred years ago, people "knew" the earth was flat. Imagine what you will "know" tomorrow.

What knowledge or practices do you hold on to that are no longer empowering? What behavior that once drove your success do you and your team need to unlearn?

MAKE UP YOUR OWN MIND

The most powerful way to figure out the right answer is to ask yourself what you think. I know this sounds ridiculously simple, but I can assure you it is very powerful. Ask yourself: What do I think? You will come up with new answers—answers that make sense, and that are based on all the experience that has brought you to where you are today.

Great leaders and managers must be prepared to make decisions about their business, and for that matter their lives. Often these decisions will be based as much on intuition and educated guesswork as on predictable data and knowledge. The flipstar makes the decision anyway. Why?

- *Decisions lead to action.* When you finally make up your mind about something, it usually leads to action. There is no need at this point to say why this is a good thing.
- *Decisions create momentum.* The action that follows your decision will give you the clarity that was preventing you from making the decision in the first place, and now you are off on a positive upward spiral. Action, clarity, confidence, decision, action, and so on.
- *Decisions create confidence.* The decision gives not just you a sense of confidence but also those around you. If you are to get your team or, if you are a CEO, your whole bloody company moving in the direction of your stated trajectory, you had better instill some confidence. Decisions create that confidence.

In reality you have no choice but to make up your mind. If you ponder too long you will become irrelevant faster and faster. Sure, a decision usually requires some change and some risk, but the risk of not deciding and not taking action is, of course, much higher.

To be a flipstar you will need some degree of delusional self-confidence, a willingness to believe in your opinions and ideas even if the whole world says they are crazy. John Chambers, CEO of Cisco Systems, says, "You have to have the courage of your conviction."

Let's get real about it. We spend so much of our lives pretending we know what we are talking about, trying to appear in control in front of those we are supposed to lead. But as I want to emphasize again, the reality is we are all throwing mud at the wall and hoping some of it sticks. Business texts and conventional wisdom suggest this is reckless and unintelligent behavior. I say it is just the reality of a global business world.

Now, don't stop pretending. This is an essential part of what you as a leader need to do. You must instill a sense of optimism in those you attempt to lead. Your team and colleagues will be attracted to confidence. An air of certainty must surround the leader. You can be honest about the confusion in the market, but you must remain certain that the business can pull through it, and that you are the one to lead them through.

I would suggest that without this blind faith it would become impossible to lead. The belief that tomorrow can be better than today makes the pain of the hard work bearable. The belief that your product is going to be a hit makes the late nights and inevitable setbacks tolerable as the promise of something great tomorrow keeps you going.

You must make a decision, and *move*.

GO NUTS!

By this point you should have a good idea of what is working and what is not. If you have been continually taking action, observing the results, and taking new action based on that feedback, you are ready to make a decision and bet the bank on it. There comes a time when your little experiments need to become full-scale attacks.

Whether you choose to do this through rapid evolutions or full-scale revolution is up to you. Either way, a rapid, full-blown effort is the only way change happens.

Bain & Co. conducted a study of twenty-one recent "corporate transformations" (large-scale management consulting projects involving massive company overhaul, mostly the CEO firing all of upper management). In every case where the transformations led to positive benefits for the company, the changes happened in two years or less. For the companies that changed fast and enjoyed quick and tangible results, their stock prices rose on average 250 percent a year as they revived.[4]

This kind of commitment creates energy. And energy is what we need. An apathetic staff is worse than an angry staff. An angry staff can have its energy redirected when an organization is ready to move in a positive new direction. Apathy is far harder to budge. Consider from the same study quoted above that 90 percent of people with heart problems that required invasive heart surgery never change their behavior postsurgery, even though it is likely to kill them. The same lazy habits that got them into this mess are likely to keep them in this mess.

> BUSINESS IS NO DIFFERENT FROM A KINDERGARTEN PLAYGROUND. IN ORDER TO PROGRESS ALONG THE MONKEY BARS, YOU NEED TO LET GO.

IBM has not treated its reinvention as a global services company lightly. They actually sold an important part of their hardware business—their entire PC division, including the ThinkPad laptops—to Lenovo in

China. Although IBM continues to manufacture mainframe computers (the latest is the Enterprise z9), jettisoning the part of the brand that had become most familiar to ordinary consumers was not a half measure. Of $91.4 billion in 2006 revenues, IBM got $20 billion from its systems-and-technology division, which manufactures mainframe computers and designs server processors and computer chips, and $47 billion from services. Who knows, maybe one day Apple will execute a similar flip and become a full-scale music business.

Business is no different from a kindergarten playground. In order to progress along the monkey bars, you need to let go.

RAPIDLY DEVISE SHORT-TERM PLANS

Strategy on the go requires that you do have a strategy—it just requires that it be shorter term not longer term, and that you allow yourself the requisite degree of flexibility within your longer-term plans.

This is a vital point. Long-term planning is built on your capacity to accurately predict not only the environment of the future but also the outcomes of each of your steps along the way. The problem we face is that the four compounding forces of the changing world make long-term prediction nigh on impossible. Prediction, as an activity, is getting harder and harder.

Consider this point, made by Daniel Gilbert: "All brains—human brains, chimpanzee brains, even ordinary food-burying squirrel brains—make predictions about the *immediate, local, personal future*. They do this by using information about current events ('I smell something') and past events ('Last time I smelled this smell, something big tried to eat me') to anticipate what will happen next ('A big thing is about to———')."[5]

But the more complex a system becomes, the less likely it is that past results are going to be accurate predictors of future outcomes. Think about this in terms of something we try to predict all the time: You can predict what the weather in your immediate vicinity will be like in the next ten minutes, but it becomes much harder to make an accurate prediction if you expand the time span and you increase the complexity of the system.

As both time span and complexity increase so does the margin for error when predicting the future. The preceding steps are an attempt to minimize the complexity you face, so add to that a reduction in the time span for which you are attempting to plan and you will reduce your margin of error.

This kind of short-term planning approach requires genuine speed. You must be able to devise and communicate your plan rapidly, and repeatedly, as you go.

CHANGE AND ADAPT

At a conference recently I heard the mountain climber Tim Macartney-Snape speak. He is one of the only people in history to climb Mount Everest without the assistance of oxygen. He said that the hardest thing about climbing Everest, apart from the lack of oxygen and the cold, is that the mountain "makes its own weather." Apparently, even the best climbing technology can't predict the weather patterns at that altitude. It reminded me of the marketplace. It makes its own weather; you either adapt and flip to stay on the mountain or you resist and get blown off it.

At the end of the day, you need to be flexible. You must be able to change and adapt your actions and strategy in order to more effectively propel your business, your career, and your life in the direction you desire, not just in little ways but in big ways. Sure, you may have a whole lot of personal or business capital invested in doing business a certain way, but if the world changes, so must you. This is true regardless of the infrastructure you have in place, the degrees you have, or the business models that have made you profitable.

When I first started my business, I wanted to work with students in schools. I wanted to teach them how to make a smooth transition from the classroom to work, and to give them the inside scoop about résumés, interview skills, employment expectations, and—most of all—about the changing career landscape. I thought a one-week course would be perfect. How wrong I was.

The mistake was not in the idea. There was a lot of support for the content I was intending to teach. It was the format of a one-week course that was wrong. But even this was not the mistake. There is nothing wrong with "guessing" what will work. In many ways this is what we do when we strategize (we hate to admit it, but it is all just educated guessing). The mistake was spending a year perfecting my guess before I opened myself up to the feedback of the market.

Month after month I worked on what I would say, on what day in the one-week program I would say it, and what activities I would get the students to engage in to drive their learning. Then, in less than a week, the market made it very clear that this was not what the schools could use. They could not afford to release

ACTION MUST HAPPEN NOW—RISKS MUST BE TAKEN: CALCULATED, INTELLIGENT, AND HIGH-PAYOFF RISKS.

students from class for that long. Even though it would provide a good learning opportunity, it was simply not practical. They wanted short (as quick as one hour), sharp sessions with simple take-home tip sheets for their students. Had I started earlier, had I committed before I was ready, had I acted in spite of my early confusion, I would have gained more clarity faster. I would have saved a year!

Action must happen now—risks must be taken: calculated, intelligent, and high-payoff risks, but risks nonetheless. You can change and adapt. You must believe it is possible.

FAST, GOOD, CHEAP: PICK THREE—THEN ADD SOMETHING EXTRA

A friend of mine had a successful business in Sydney, manufacturing point-of-sale materials and displays out of Plexiglas. Walk into a shoe store looking for Nikes or Reeboks anywhere in Australia, or a consumer electronics store looking for a new camera, and you would likely find a product in that store made by my mate. At least that was the case a decade or so ago.

He had a good thing going until one day when everything started to change. That's when Plexiglas made in China first began to trickle into the market, and then suddenly flooded in.

As I sat with him having a coffee in his office toward the end of this tumultuous period, he said to me, "Peter, I just don't get it. What am I doing wrong?"

He probably meant it as a rhetorical question. But I stood

up and walked to the wall behind him, where he pinned up family photos, *Far Side* cartoons, inspirational quotes, and that kind of jazz. In the middle was a piece of traditional workplace wisdom, a sign that said in block capital letters, FAST, GOOD, CHEAP: PICK 2.

My friend thought the sign was funny, but it also expressed a business philosophy that had served him well over the years. Perfection's not possible, and as Mick Jagger sang, you can't always get what you want. If you try hard, you can get what you need, so long as you don't ask for too much.

I took a red marker, crossed out the "2" and replaced it with a "3." That was my way of saying, "Getting what you need is yesterday. Today is getting what you want. And too much is never enough."

My friend had built his business by making good-quality products and responding promptly to his customers' needs. Compared to the Chinese imports, his products were not cheap, but he thought he had a quality advantage and could match anyone on efficiency and speed of delivery (at least initially anyway). As it turns out, many of my friend's customers thought his quality advantage was superfluous. The Chinese goods were made well enough to last through their normal span of use in a fast-changing sales environment, where customers are eager for the next new thing, and merchandisers replace their displays and point-of-sale materials frequently.

Behind my friend's wrong assumption about his quality advantage was a mistake about time and the pace of market change. It was two errors, really. One mistake was thinking the greater durability of his product was actually an advantage for most of his customers, rather than being increasingly irrelevant given how fast market trends change. The second mistake was in thinking he was a fast enough manufacturer and supplier to hold his market share. The Chinese manufacturers turned out to be much faster than he was as well as much cheaper, and were good

GETTING WHAT YOU NEED IS YESTERDAY. TODAY IS GETTING WHAT YOU WANT. AND TOO MUCH IS NEVER ENOUGH.

enough on quality to rapidly erode his market share. Not only that, but over time the Chinese imports arguably began to match the quality on which he once built his market position.

My friend's mistake wasn't unusual. It was business as usual. The standard advice in business textbooks is that you can't lead your league on fast, good, and cheap, so you should concentrate your efforts to excel and establish competitive advantage in one category, be average or good in a second category, and don't worry too much about lagging in the third category.

But remember, today's customers increasingly "Think AND, Not OR." In a world of global oversupply, global underdemand, and nonstop technological change, they get more of what they want faster and faster all the time. In the case of some of the most profitable businesses of our time (think of successful product launches such as the Apple iPod, Nintendo Wii, Toyota Scion, or the latest Nike shoe), customers get more of what they want before they know they want it.

The four forces of change I discussed in chapter 1—especially the fourth: rising expectations—are the cause of this change. The bottom line, as you can see from my mate's example, is that the more variety, the better quality, and the faster service customers enjoy, and the more comparative information that is available to them about you versus the competition, the more finicky, demanding, and impatient they become. In that context, competitive advantages quickly turn into competitive necessities that rivals can copy and adapt for themselves.

COMPETITIVE ADVANTAGES QUICKLY TURN INTO COMPETITIVE NECESSITIES THAT RIVALS CAN COPY AND ADAPT FOR THEMSELVES.

For you as a businessperson, that situation desperately requires you to flip on its head any notion that fast, good, cheap: pick two is sufficient in today's market. Genuine competitive advantage requires a minimum of all three, plus something above and beyond these necessities.

In order to really grasp this flip and implement it into your day-to-day operations you must understand:

- To compete in any market, being fast, good, and cheap is the price of entry. To achieve competitive advantage, you must lead the league in at least one category and be industry standard in the remainder.
- You must commit to a perpetual cycle of innovation. Fast today is not fast enough for tomorrow. Good today is not good enough for tomorrow. Cheap today is not cheap enough for tomorrow.
- Customer satisfaction sucks! If you're ever going to move beyond satisfying customer needs to fulfilling customer wants, which is where the big profits are, you've got to be fast, good, cheap, and *more*!

When I tell executives, "Customer satisfaction sucks!," they look at me like I'm crazy. If anything is accepted business wisdom today, it is that customer satisfaction is paramount.

Yes and no. Bear with me while I split a very important hair.

My cable-television provider recently called to ask if I was satisfied with my new package. Well, the service does what they promised it would do, but I'm not going out of my way to tell people how great it is. Instead of five television channels with nothing on that I want to watch, there are more than a hundred. I need digital-television service if I'm going to get the most out of my high-definition flat-panel television. But I don't want lots of channels. I want one that shows what I want to watch.

Lots of channels satisfied my need, but they didn't fulfill my want. The cable-television industry as a whole is a classic example of company needs and wants being out of alignment with customer needs and wants. New digital-video services such as TiVo are going some of the way to addressing this disparity, as are cable-television packages offering you the ability to choose your dominant programming such as baseball, movies, or some other genre that interests you. The programmers, the television networks, and the cable-television providers prefer to aggregate hundreds of channels into tiered services that only seem to fulfill customers' desire to watch what they choose when they choose.

That's the subscription model the industry wants. And some customers are probably perfectly happy with it. But other customers aren't going to be happy until they get à la carte pricing that enables them to pick and choose only the channels they really want to watch, or perhaps until they can freely choose high-definition video from millions of sources streaming over the Internet. Over the next few years, as YouTube continues to grow and more households opt for digital-video recorders like TiVo, I suspect many customers will want to program their own television viewing rather than settle for the cable-television industry's programming packages.

This will, of course, happen not only as technology gives customers more choices outside the traditional television providers' control, but also as peer-to-peer networks allow users to get recommendations from people they know and trust so they can sort through the limitless choices efficiently. The Internet video distribution network from NBC Universal and News Corporation, in collaboration with AOL, MSN, MySpace, and Yahoo, anticipates that future and represents a savvy response to the popularity of YouTube.

The NBC Universal–News Corp. network is a brilliant flip on the entertainment business as usual, and I have more to say about it in chapter 7, "To Get Control, Give It Up." The general point I want to make here is that customer satisfaction measures your ability to deliver what you have conditioned the customer to expect, not what he or she really wants. And that leaves you vulnerable to a competitor who will supply that want with a positive, memorable, mind-blowing customer experience, something that exponentially raises the standard on fast, good, and cheap, and instantly creates a whole new level of customer expectation.

GREAT YESTERDAY, GOOD TODAY, BAD TOMORROW

Once you recognize the "Fast, Good, Cheap: Pick Three—Then Add Something Extra" dynamic at work, you begin to notice it everywhere. I want to point you to some of the best current examples of it among the world's leading companies. But first let me tell you what happened to my friend's Plexiglas business.

When I crossed out the "2" in his FAST, GOOD, CHEAP: PICK 2 sign and replaced it with a "3," my friend got it instantly. He had been flipped, and he knew it. So without any detailed plan (he is the kind of person who showed me that action creates clarity), he began to act differently.

He stopped trying to stem the losses from his broad old customer base, and he began to concentrate on serving only the most profitable customers. This quickly led him to a much more specialized focus serving specialty and high-end retailers and letting the Chinese imports have the mass market. In effect, he evolved his business model via action, without too much planning (he was desperate), to one in which key accounts were all that mattered. This is a strategy that Noel Capon of Columbia Business School has shown to be a crucial component of success for many of the world's top companies, such as IBM, Xerox, and Citibank (see Professor Capon's *Key Account Management and Planning* for a great primer on this subject).

In my friend's Plexiglas business, the result has been to boost margins. He can now continue to charge more for his products than the Chinese imports cost, while delivering faster service and better quality than ever before. This may sound like he has picked *fast* and *good,* but still not *cheap.* No, he has picked cheap—cheap at the price. The customer's sense of a product's cheapness can be expressed by the ratio of value to price. Customers feel a product is cheap at the price, not when they pay the lowest possible price, but when they feel like they received great value and a good deal on all measures, including price.

My friend's customers pay a premium for his products, but they get exactly what they want. What they need is just a properly sized and shaped piece of Plexiglas. What they want is getting that from someone with a deep understanding of their needs and the ability to fine-tune the product offering to their specifications. Together these factors make my friend's premium pricing a bargain in the minds of his most profitable customers. He is indeed fast, good, and cheap.

His next challenge, if he really wants to be a flipstar, is to "Think AND, Not OR," and to work on selling a large volume of high-margin products. Because who says there needs to be a trade-off? Apple, after all, sold millions of iPods at a 30 percent price premium to other MP3 players. How? Well, as you will learn in chapter 4, "Absolutely, Positively Sweat the Small Stuff," $400 is a bargain if the item makes you feel cool.

Before moving on, I would like to give you an excellent example of how "Fast, Good, Cheap: Pick 2" has become "Fast, Good, Cheap: Pick Three—Then Add Something Extra." The original brand identity of McDonald's was fast and cheap. Of course the food had to be good enough to satisfy customers' needs for a meal, but it also had to be served in good enough surroundings, which simply meant that the restaurants were clean and neat, and it had to be available at lunch and dinner. For a long time that was all McDonald's needed to be, and it grew to become one of the world's dominant brands simply by adding more and more locations in the United States and other countries.

CUSTOMERS FEEL A PRODUCT IS CHEAP AT THE PRICE, NOT WHEN THEY PAY THE LOWEST POSSIBLE PRICE, BUT WHEN THEY FEEL THEY RECEIVED GREAT VALUE AND A GOOD DEAL ON ALL MEASURES, INCLUDING PRICE.

In 2002, however, McDonald's posted its first-ever quarterly loss because of declining margins and year-to-year same-store sales. Since then, the company has pursued a strategy of "better, not just bigger," adding only fifty to a hundred restaurants per year in the United States compared to hundreds a year in the 1990s. McDonald's looked at a new competitive landscape that included Starbucks, Burger King, Wendy's, and Dunkin' Donuts, among others, and it looked at changing customer behaviors, especially eating more meals on the run outside traditional mealtimes. Then McDonald's responded aggressively by encouraging franchisees to stay open later than the traditional 11 P.M. closing time (in 2002 only .5 percent of McDonald's restaurants in the United States were open 24/7; in 2007 that figure was 40 percent and rising) and to redecorate restaurants (many now include casual seating areas with flat-screen televisions as well as playrooms for children and other amenities). The company also increased investments in its test kitchens to develop new customer favorites like the McGriddle, first offered in 2003, and the Snack Wrap, first offered in 2006. And to position itself against Starbucks in particular as the customer's "third place" of choice, besides home and work or school, it launched the McCafé concept, which is proving successful both in the United States and foreign markets.

The result of recognizing that being fast, good, and cheap means meeting constantly rising customer expectations is that the first-ever quarterly loss in 2002 may be the only one McDonald's ever experiences. It is enjoying an unbroken streak of sales increases, and profit margins are up for both the company and its franchisees.[1]

McDonald's showed great flexibility in changing the hours of operation, décor, and menu of its U.S. restaurants. It showed equal flexibility in adapting its rebound tactics to different regional markets around the world. To remain fast, good, and cheap in the perceptions of Australian consumers, for example, McDonald's had to convince them that its version of "good" included healthy food. The breakthrough came with "Salads Plus," which drove hundreds of thousands of new customers into McDonald's restaurants in Australia, and increased same store sales nearly $1 million per year in many instances. Of course, the new customers did not eat only salads. They also bought lots of soft drinks and fries and burgers. More recently, in addition to introducing the McCafé concept that is also working in the United States and other markets, McDonald's in Australia has secured the Heart Foundation of Australia's seal of approval on a number of menu offerings. All in all, McDonald's regional market flexibility is a superb strategy from a very proactive company and a model for other global brands.

Throughout its lackluster period, McDonald's never lost sight of the need to stay fast. In fact, McDonald's has always been the fastest restaurant company in the fast-food business. As I said, to achieve competitive advantage, to have the best market share and the best profitability in your market segment like McDonald's, you have to lead the league on at least one of fast, good, and cheap, and be industry standard in the others. What McDonald's has successfully done is come up to standard as consumers' definition of "good" changed, while still remaining fast enough and cheap enough.

Another example of an industry constantly hammered by the need to be faster, better, and cheaper is the automotive fuel industry, especially in the context of dramatic changes in automotive technology. Traditionally the gas station was also a service station, which sought extra profit margin and brand loyalty by offering car maintenance and repairs as well as gas. But today's cars are more reliable than older cars, and they're filled with computer chips for diagnostics and vehicle systems control that the average service station can't fix.

In other words, the service-station business is obsolete, or quickly becoming so. That has led to a new business model in which gas stations are linked with convenience stores or other retail operations that sell drinks, snacks, and basic grocery and impulse items. The profit margin on convenience-store items is much greater than that on gas. According to A. C. Nielsen, 25 percent of household grocery purchases are "convenience" purchases and are not price driven.

For example, ExxonMobil created a new franchise of On the Run stores behind the gas pumps at Exxon and Mobil stations, and the other major gas brands made similar moves. The popularity of gas stations in combination with convenience stores inevitably attracted the interest of retailers that had never sold gas. In effect, fuel sales, although modestly profitable in themselves, served as a "loss leader" to drive higher-margin retail sales. For the same reason, there are now gas pumps at many Wal-Mart locations in the United States.

Well-established retailers such as Wal-Mart can make low-priced fuel sales profitable both because the gas-buying driver who stops to fill up at a Wal-Mart gas station is likely to buy other things at their stores, and because they are big and powerful enough to negotiate favorable pricing from gas wholesalers and refineries. Discount gas prices complement their existing reputation for delivering good value to customers.

Let's see what sense "Fast, Good, Cheap: Pick Three—Then

Add Something Extra" can make of this. Filling up the tank takes the same amount of time wherever you do it. Maybe you save a little time by stopping at the first station you see, or maybe you lose a little time by looking for a favored brand or a cheaper price. You may have a favorite brand, but the reason won't be because you like their gasoline. For the vast majority of drivers, there is no quality difference between different brands of gas. Assuming equivalent octane ratings, one brand is as functionally good as any other brand in the customer's eyes.

So if you're in the business of selling gas to drivers, how are you going to differentiate your brand from other brands, when one gas-buying experience is more or less as fast, good, and cheap as another? As these examples show, you "Think AND, Not OR." You combine two previously distinct retail sales categories into a new value proposition that offers customers who feel intense time and opportunity pressure throughout their lives a faster, better, cheaper way to buy gas and other daily essentials at one go. It's a value proposition tailored to increasingly affluent customers who rate the low gas price per gallon and time savings plus extra cost per item of quickly buying a few necessities and impulse items as better overall than taking the time to drive to a major grocery store, park in a large parking lot, walk from the car and through the aisles to find what they want, pay for their purchases, and walk back to their car.

FAST TODAY IS NOT FAST ENOUGH FOR TOMORROW, GOOD TODAY IS NOT GOOD ENOUGH FOR TOMORROW, AND CHEAP TODAY IS NOT CHEAP ENOUGH FOR TOMORROW.

The companies selling gas to drivers have to meet the challenge of "Fast, Good, Cheap: Pick Three—Then Add Something Extra" on two levels: the gas itself, and the gas in combination with a higher-margin sales channel. Within this mix, they can emphasize fast, good, or cheap, but to stay in the game they also have to be industry standard on the other two. Wal-Mart may emphasize cheap, but they must also be fast and good enough to keep attracting a profitable share of customers. Likewise, ExxonMobil's highly profitable On the Run stores may emphasize convenience, but if they seek too high a profit margin on things like milk that are available in many other places, they won't be able to sustain and increase their market share.

But remember that "Fast, Good, Cheap: Pick Three—Then Add Something Extra" is both a moving target and table stakes, the price of entry into any market. To build significant competitive advantage, you've got to offer something else, an X factor that will really differentiate you positively in customers' eyes. That's the subject of the next chapter, "Absolutely, Positively Sweat the Small Stuff."

As both the McDonald's and consumer gasoline examples show, fast today is not fast enough for tomorrow, good today is not good enough for tomorrow, and cheap today is not cheap enough for tomorrow. Three areas where we can see this are fashion, cars, and telecommunications.

FAST IS NOT FAST ENOUGH

We always notice when products and services take an increasing share of our wallet. But the increasing time pressure we all feel today makes us increasingly sensitive to the time that products and services take. We also have less and less patience for anything other than instant delivery of those products and services.

We repetitively hit the "close" button on elevators, when no one is racing to join us inside, because the five seconds it takes for the average elevator door to close is unbearably, excruciatingly long. We do this even though we know that the elevator system will not respond any quicker, no matter how many times we press the button.

We do it because technology is speeding up the world, and making just about everything but the closing of elevator doors happen faster and faster all the time. The result is that, especially for the youngest customers (and the youngest staff, too, when it comes to the pace of their careers), no demand on speed of delivery feels unrealistic or exorbitant.

Few industries operate at as rapid a pace as the retail clothing business. As fashions constantly change, the ones who profit are the ones who can quickly and efficiently provide styles, colors, and fabrics that appeal to the key demographic of young customers, who in turn set the trends for other demographic groups. Three retail clothing chains—Zara, H&M, and Uniqlo—now perform that fast, good, cheap hat trick better than any others, and Zara is perhaps first among equals in delivering fresh, exciting apparel to young consumers.

The Zara chain is owned by the Spanish company Inditex, which has seventy thousand employees and counting (they added eleven thousand employees in 2006), 3,131 stores in sixty-four countries and counting (they added 439 stores in 2006), and net 2006 sales of 8.1 billion euros ($12 billion; an increase of 22 percent over the prior year). Practicing what it calls a "fast fashion" system, Zara can design and distribute a fashion forward garment in fifteen days. Some Zara styles resemble the latest couture offerings, albeit in less expensive fabrics. Others beat the luxury fashion houses to market with Zara designers' fresh takes on the clothing trends of urban youth around the world.

Equally significant is the number of styles and variations Zara retails every year. Three teams of designers for women's, men's, and children's lines generate forty thousand or more designs a year, and about ten thousand of these make it into actual production of five to seven sizes and five to six colors per garment. That means Zara's supply-chain management must smoothly handle around three hundred thousand new stock-keeping units (SKUs) per year.

The final flip—or should I say wrinkle?—is that Zara produces each garment in very limited quantities. Most clothing companies try to milk the most popular styles and sell them in high volume, which inevitably creates lags in inventory supply and turnover, and at the end of most selling seasons triggers unprofitable discounting to move inventory that no longer excites customers. Instead, Zara says, so to speak, "We love stock-outs" (the retail term for being out of stock on a requested item).

The speed with which Zara changes garment styles and colors encourages impulse buying and more frequent store visits. Customers know that if they see something they like at Zara, they'd better buy it right then and there, because it won't be available later. They also

EVERYTHING ABOUT ZARA'S ORGANIZATION EXPRESSES THE BELIEF THAT THEY CAN NEVER BE FAST ENOUGH.

know that whenever they enter a Zara store, they're going to see new things. Thus, for example, Zara's London stores attract an average of seventeen store visits per unique customer per year, whereas their competitors attract an average of only four visits per unique customer per year. Zara's strategy makes its customers so curious to know what new clothes are on the racks that the company spends only 0.3 percent of sales on advertising versus 3 to 4 percent for most of the competition.

Everything about Zara's organization expresses the belief that they can never be fast enough, and that there is also no excuse for forgetting good and cheap while they're at it. Instead of isolating design, production, and marketing staff in separate silos, Zara's offices, stores, and other facilities are laid out to encourage the fast, free flow of information, with designers working in the midst of production and marketing so that feedback on new styles, production glitches, quality problems, and customer behavior becomes virtually immediate. This also sends a message to Zara's staff that no one is "cooler" than anyone else, or to put it another way, that everybody in the company is as cool as the design team.

Likewise, design and production move quickly thanks to an intensive use of computer-aided design (CAD) and just-in-time supply-chain management. The stores themselves are integrated into this blindingly fast feedback loop through daily PDA and weekly telephone communication on how customers are reacting to different offerings. The result is that whereas most competitors are hard-pressed to vary 20 percent of the order mix in any one selling season in response to customer behavior and other factors, Zara can adjust 40 to 50 percent of the order mix without strain.[2]

Zara's "fast fashion" system would break down if customers didn't think the clothes were of high enough quality or affordable enough, and the company could easily serve as an example of "good is not good enough" or "cheap is not cheap enough."

The same could be said of both H&M, which is based in Sweden and has 1,300 stores in twenty-nine countries, and Uniqlo, which is based in Japan and has more than 730 stores in Japan, China, South Korea, Hong Kong, the UK, and the United States.

In the "Fast, Good, Cheap: Pick Three—Then Add Something Extra" mix, H&M might be said to emphasize good. It delivers what the company Web site calls "discount high-end fashion" through special one-season-only collections from prominent designers such as Karl Lagerfeld and Stella McCartney, or in association with trendsetting personalities such as Madonna and Kylie Minogue.

Uniqlo (a name formed from the words *unique* and *clothes*) might be said to emphasize cheap. Despite a persistent economic downturn, between 1999 and 2002 Uniqlo opened two hundred new stores in Japan, selling "recession chic" so cheaply that some Japanese business commentators and many competitors accused it of causing deflation. (Sounds like sour grapes to me.)

All the same, "fast is not fast enough" rules the fashion industry. H&M's name-brand designer- and celebrity-connected offerings deliver up-to-the-minute, trendsetting quality at a discount price. And Uniqlo's business DNA derives from its parent company, which is named Fast Retailing Company, because young customers would never buy clothing that is not fashion forward, no matter how cheap it might be.

GOOD IS NOT GOOD ENOUGH

One of the great *flipstars* of modern business is Toyota. The only carmaker that is consistently gaining market share in all product categories in all market regions, Toyota has an enviable reputation for product quality. Scores of articles and several books have been written about the efficiency and speed of the Toyota Manufacturing System, which made just-in-time inventories a global business trend.

Toyota's quality advantage underpins a number of other strengths. It spends the least amount of time to make a vehicle of any carmaker and has been the least vulnerable to costly vehicle recalls (although it has had some lately). It also spends the least amount of promotion money per car sold, and its vehicles spend the least amount of time on dealers' lots. To top this all off, Toyota has the highest per-unit profitability of any volume or luxury-car maker.

Over the past few decades Toyota has shown a greater commitment than any other carmaker to the idea that good is not good enough, without ever forgetting the need to be fast enough and cheap enough in customers' eyes as well. Segment by segment, Toyota has leveraged its quality advantage from economy cars to midsize family sedans, luxury cars, sports cars, SUVs, and utility trucks. The most dramatic example of that remains the Lexus and its quick ascent to become the world's number one luxury car.

The more economy cars Toyota, Honda, and Nissan sold in the United States—the world's biggest car market—the more they wanted to expand the range of vehicles they sold there. All three sold luxury cars in Japan, and Honda was actually first to market a luxury car brand in the United States with its Acura range. There is no question that Acura has been a success, but there is also no question that Lexus quickly overtook it and left it a good distance behind.

When Toyota looked into entering the U.S. luxury-car

market, it saw two things. First, there was a gap in what might be called midprice luxury between normal cars and luxury automobiles from the likes of Jaguar and Mercedes-Benz. Second, most luxury automobiles were unreliable and very costly to maintain. Toyota therefore set out not only to build a luxury car of high quality at an attractive price, but also to deliver a total customer experience that would be second to none.[3]

After Toyota launched the Lexus in 1989, managers in the company followed up with all new buyers to confirm that they were happy with their cars and to offer to fix anything that was wrong. They not only fixed every mechanical problem, but they also returned the car freshly washed and with a full fuel tank. Toyota didn't rest with building a luxury car that matched or exceeded Mercedes-Benz and BMW in quality; they also made sure that their customer service was nothing short of mind-blowing.

Another Toyota flip of special interest here is that "Fast, Good, Cheap: Pick Three—Then Add Something Extra" does not necessarily mean lowest- or low-priced. Remember my friend the Plexiglas manufacturer. Cheap is not only an absolute measure of price, but a relative measure of value perceived and received. Toyota's products are cheap at the price, because a Lexus, say, works exactly as it should, retains value against the competition over its life span, and is supported by industry-leading customer service. Throughout its range, Toyota maintains a premium pricing advantage over its Japanese, Korean, European, Australian, and American competitors.

Having steadily grown to overtake DaimlerChrysler and Ford along the way, in the first quarter of 2007 Toyota displaced troubled General Motors as the world's number one carmaker. To remain at the top, it must continue to be fast, good, and cheap at the price in customers' eyes. In other words, it must continue to innovate, offering appealingly designed, user-friendly, and dependable vehicles.

CHEAP IS NOT CHEAP ENOUGH

Over the past two decades the telecommunications industry has alternately been the darling and the favorite whipping boy of the world's stock markets. Deregulation of telecom monopolies, mergers, and acquisitions, and the growth of cell-phone use have redrawn the telecom map several times. Throughout that process there has been steady downward pressure on pricing and margins.

The latest upheaval in the telephone business has come from the ability to bypass both landline and mobile networks and offer phone service via the Internet. Start-ups in the U.S. market such as Vonage, Internet service providers such as Earthlink, and cable-television providers such as Time Warner that offer broadband Internet connections through their cable networks have all gotten into the business of selling VoIP (Voice-over Internet Protocol) telephone service.

Although these companies have seized a new channel for reaching and serving customers, they are all still practicing business as usual, selling phone services for fees billed on a monthly basis. These services are fast and good, and can be much cheaper than those of the established phone companies (although the latter have fought back on price). Having tried VoIP, I am convinced that the service is almost cheap enough, but it is definitely not yet good enough for the low price to be compelling.

I tried VoIP in my business and found the quality substandard, unless I upgraded not only my plan but my broadband connection, too. Fast and cheap are not enough. The product also needs to be good if it is going to get mainstream support. From what I understand, big companies are getting excellent results with VoIP, but not without significant investment. One of the major advantages it offers is that it taps into existing network infrastructure and allows a lot more flexibility than traditional phone lines.

Vonage is the most prominent of the VoIP providers in the U.S. market. They seemed headed for as glorious an IPO and subsequent rise in share price as Google. But by the time Vonage did go public, the air was already leaking out of its balloon, thanks to an even bolder newcomer called Skype.

Skype's business model abandoned fast, good, cheap for instant, excellent, free at the entry level of service, which is all many customers will ever use. Two or more Skype users can communicate with crystal-clear digital sound quality on their broadband connections for absolutely no additional cost. And if they want to include someone on a traditional landline phone, that costs about two cents per minute. After eBay bought Skype for $2.6 billion, the two-cents-per-minute charge to call a normal phone was waived in North America for several months in order to build usage.

Instant, excellent, free: what kind of business model is that? But if you wonder how Skype will ever make enough profit for eBay to justify paying $2.6 billion for it, you're missing something. Sooner or later the technology that Skype used to create its service was bound to be exploited in a similar way by someone. Once that technology existed, the genie was out of the bottle. Not only that, but as with Google's purchase of You-Tube and News Corp.'s of MySpace, the value was in the network and the relationship the brands have built with their customer base. (More on this in chapter 5, "Business Is Personal.")

Skype's founders were willing to act first and strategize how to exploit the technology as they proceeded. The enthusiastic response of customers all over the world made Skype the gold standard of Internet calling and created several revenue streams, including calls between Skype users and regular landline and cell phones, business-tele- and videoconferencing, and business phone services. For eBay there is the chance not only to grow the Skype brand in those areas, but to grow its core e-auction and e-commerce business by plugging Skype telephony into the eBay network along with PayPal's financial transaction service.

No one can be certain yet if this is going to pay off at a level that justifies a $2.6 billion purchase price, but I love Skype for trying, and eBay for trying harder (or should I say for paying). An indicator that Skype will ultimately be profitable for eBay, in my view, is the growing universe of third-party products made specifically for use with Skype. Before the iPod there were many MP3 players on the market, but none generated the add-ons from third-party vendors that are now available for playing the iPod in a car, on a portable boom box or in a home entertainment system. Likewise there are other VoIP platforms, but only Skype is generating products from third-party vendors such as the Skype-ready WiFi phones from Philips, Netgear, and Belkin.

When eBay released its first-quarter numbers for 2007, first-quarter net revenues for the entire company rose 27 percent to a record $1.77 billion, and net income rose 52 percent to $377 million. First-quarter net revenues for Skype rose 123 percent to $79 million. CEO Meg Whitman said of Skype, "This is a very young business growing very fast."[4]

The picture admittedly looked less rosy on October 1, 2007, when eBay announced a $900 million one-time write-down on the $2.6 billion purchase of Skype. Critics of the company immediately released statements on eBay's overpaying for Skype. But there is a solid chance that eBay will eventually profit handsomely from Skype. Two weeks after eBay's announcement, MySpace and Skype formed an alliance that enables MySpace users to make phone calls and send instant messages via Skype. And at the end of October 2007, the British telecommunications provider 3 announced that it was releasing a Skype-ready cell phone for use in the UK, Australia, Denmark, Italy, and Hong Kong. These are not the signs of a failing brand. There are inevitably stumbles in any new venture, but on balance it looks like eBay has reason to feel good about Skype's long-term prospects.

Vonage has very clever ads that say it is "leading the Internet telephone revolution." But again, Vonage is a halfway revolutionary, trying to use a cheaper channel to conduct telephone business as usual. They are caught between the established phone companies' ability to match or nearly match them on price, on the one hand, and Skype's ability to profit by providing the same or better service for free, on the other. In Vonage's case, the market has responded by making its heralded IPO and subsequent share price lackluster, to say the least.

Skype's challenge to the telephone business as usual does

not mean that you have to give away your products and services for free, even at the entry level of a tier of products and services. My point is instead that Skype shows you can't take the customer value of your pricing for granted. If I'm going to be on one side of the change outlined above for the telecommunications industry, I want to be on Skype's side, not Vonage's. I want to be so fast, good, and cheap at the level that customers are uncomfortable with how much they are paying for something that they willingly pay me a premium for what I deliver on top of that. That's the formula eBay is following with Skype.

• • •

Sooner or later, somebody or something is bound to come along and yank the price floor out from under your entire industry. Flipstar Richard Branson has made a career doing just that. His most notable achievements in this regard are taking on the European, transatlantic, and Australian airline industries with his Virgin Express, Virgin Atlantic, and Virgin Blue airlines. The result was a substantial reduction in the price of airfares in all three markets. In August of 2007 Branson finally succeeded in launching Virgin America (for air travel within the United States) after hitting regulatory brick walls over the airline's ownership structure, which had to conform to U.S. laws that require U.S. airlines to be controlled by American citizens and at least 75 percent owned by American companies.

In any case, remember that cheap today is not cheap enough for tomorrow. Virgin Express

> SOONER OR LATER, SOMEBODY OR SOMETHING IS BOUND TO COME ALONG AND YANK THE PRICE FLOOR OUT FROM UNDER YOUR ENTIRE INDUSTRY.

opened the door for Ryan Air and other low-cost air-travel competitors in the European market. And Tiger Airways, backed by the founder of Ryan Air, and Jetstar are both giving Virgin Blue a run for the low-cost air traveler's dollar in Australasia, which has always been a very cutthroat market.

FAST, GOOD, CHEAP, AND MORE!

The price of entry in every market today is undeniably "Fast, Good, Cheap: Pick Three." But because a satisfied need no longer motivates and expectations keep rising, the price of entry will soon be "Fast, Good, Cheap: Pick Three—Then Add Something Extra."

The ante to get in the game keeps rising, the table stakes keep getting bigger, because of the feedback loop between increasing compression of time and space, increasing complexity and ambiguity, increasing transparency and accountability for actions that have to be performed in conditions of high uncertainty, and increasing customer expectations. You can't escape that challenge. You can only flip it to your advantage by meeting it sooner than your competition. You can attune your behavior to the psychology of human expectations in a time of constant technological development, or you can be left by the side of the road.

If "Fast, Good, Cheap: Pick Three" is not yet the price of entry in your market, it soon will be. And if you want to top the standings when that happens, you must not only ensure that you are fast enough, good enough, and cheap enough for today and tomorrow, you must also offer customers something else as well. With that in mind, let's look closely at the flips that are needed to create the value customers want in addition to fast, good, and cheap, beginning with the fact that you must *"Absolutely, Positively Sweat the Small Stuff"*!

FIVE THINGS TO DO NOW

1. Get five of your smartest people in a room and put the following scenario to them. You have to speed up your service delivery, or out-of-the-box performance, by 20 percent in the next three months. Ask them to figure out how you could do it.

2. While you have your best and brightest in the room, you should also ask them to give you two possible quality improvements that would leave your competitors for dead. Then give someone the project of figuring out how you could do them as cheaply as possible

3. Using some creative thinking tools, develop at least three potential Skype-like scenarios that, even though your industry dare not consider the possibility, would completely rip the margins out of your business.

4. Purchase your product or service, or at least pretend to, from your own business. Then ask yourself, was that fast, good, and cheap at the price?

5. Conduct a roundtable discussion with your team, the more senior the better, and discuss the potential changes in the demands of your customers and staff over the coming years. Remembering that what is good enough, fast enough, and cheap enough today won't be tomorrow. What challenges would these changes present your business?

ABSOLUTELY, POSITIVELY
SWEAT THE SMALL STUFF

Companies have to do more to win customers than offer a dependable, good-quality, reasonably priced product or service. That's what lots of ordinary companies do. Extraordinary companies do something else, and so do extraordinary career-minded professionals. These flipstars do the little things. They realize that in an oversupplied market, competitive advantage will increasingly be built on elements once considered superficial. You know, the small stuff. Flipstars sweat the small stuff. Big-time!

> FLIPSTARS SWEAT THE SMALL STUFF.

FAST, GOOD, CHEAP, X-FACTOR: PICK FOUR

Doing the little things right is all about figuring out how to fulfill customer wants, not just satisfy customer needs. Customer satisfaction is fleeting at best. You could invest your time and energy into meeting customer expectations only to have those expectations rise before you get there. Even if you met the expectations before they increased, it would not give you an advantage in the marketplace. The money will be in giving customers what they want, even when they don't yet know what that is.

People need fast, good, and cheap. Faced with the four forces of change, people want things to be simpler, easier, and more beautiful, among other qualities, and they want to feel good about the products, services, and experiences they consume.

It sounds a little ridiculous to say that fast, good, cheap is a need. It is! The ability of the market to turn features and standards that were once a luxury into a necessity in the competitive landscape is nothing short of fascinating. Remember the research discussed in chapter 1, which shows that 29 percent of Americans today consider fast broadband a necessity. And 5 percent said the same thing about flat-screen televisions.

In that same survey 4 percent of American consumers said an iPod is a necessity. This is their reality and over the coming years the percentage of people who feel that an iPod is a *need* not a *want* will increase markedly. Ask yourself, what feature or standard of service do you currently offer that was previously an added luxury that may have become commoditized without you even realizing? I can think of keyless entry for automobiles, as discussed in chapter 3. Or what about same-day service from a courier company? Internet banking for some. The ability to pay with a credit card. Or maybe you are a little spoiled and it is your home-delivery-and-pickup dry-cleaning service.

Purchasing decisions are rarely based on rational thought processes. Time and time again customer research has shown that emotions drive our decisions and behavior. According to the *Advertising Research Journal,* research has found that emotions are twice as important as any other consideration in customers' decisions about what to purchase. And not just purchasing decisions but all decisions we make in life. Only after making the decision emotionally do we call upon our cognitive processes to rationalize our behavior.

It is amazing how powerful our minds are at rationalizing some of the objectively insane decisions we make. In fact, we *automatically* rationalize everything we do. Psychologists and marketers refer to the cognitive bias we demonstrate after buying something as "postpurchase rationalization"—the willing self-delusion about the quality of a recent purchase. Even when we know we've made a terrible decision, we can convince ourselves it was worth it.

It will often be the smallest things that we grab on to to rationalize our emotionally driven behavior. This is why it is necessary to "Absolutely, Positively Sweat the Small Stuff," and why even with regard to utilitarian products and services:

- Style is substance.
- Fashion is function.
- Feelings are the most important facts.
- The soft stuff is the hardest stuff, and the hardest to get right.

Later in this chapter we'll look at these factors in terms of what I call the "total ownership experience." For now let's continue the discussion by looking at some of the most powerful X-factor positions you can take to market.

Fast, good, cheap + green

The X factor could be any number of things. Being green is popular right now, as environmental issues become ever more urgent. Innovating far ahead of the competition, Toyota has staked its new millennium play on being fast, good, cheap, and green. Starting with the Prius and then extending its hybrid technology through the rest of the Toyota and Lexus brands, Toyota has offered time-stressed, upwardly mobile, increasingly affluent customers the opportunity to control their personal contribution to the world's pollution. At the same time it has relieved them of guilt as conspicuous consumers, decreasing their sense of stress in that way as well. Toyota is not merely satisfying needs here. It is fulfilling wants.

Other companies that are defining themselves as fast, good, cheap, and green include Siemens, L'Oréal, and even what you might think of as old-industry Alcoa, which has staked out a position as the cleanest and most high-value-added company in the metals business, with large revenues coming from high-premium alloys and packaging and fastener expertise that extends from aluminum soda cans to plastics. In a highly publicized speech in May 2007, Rupert Murdoch, a real flipstar, declared that the News Corporation will go green.

One of the most widely reported examples is Wal-Mart. Undoubtedly one of the most successful companies of the last twenty years, Wal-Mart grew to enormous size by being the best combination of fast, good, cheap that customers in retail had ever seen. But having made fast, good, cheap the price of entry into their industry and spawned copycat behavior from a host of competitors, and with same-store sales down in 2006 for the first time in the company's history, Wal-Mart has had to look for something new to sustain growth.

At first they tried to be fast, good, cheap, and hip. But Wal-Mart has had trouble convincing customers that they should look to its stores not only for the cheapest deal, but for designer clothing and high-margin products like flat-screen televisions. In December 2006 Wal-Mart seemed to lose its nerve for this effort, firing the cutting-edge marketers that only a year earlier it had hired with fanfare. Where it has not lost its nerve is in a determined effort to be green.

Long criticized for low wages, inadequate health-care coverage, gender discrimination, and a devastating impact on small local businesses, Wal-Mart regularly faces a negative public relations picture. A number of American communities have lobbied successfully to keep Wal-Mart from opening a store in their vicinity. CEO Lee Scott admitted that when Wal-Mart began to explore an environmental sustainability agenda in 2004, it was simply "a defensive strategy."

Since then Wal-Mart has embraced sustainability with a passion, and Scott told *Fortune* magazine, "What I thought was going to be a defensive strategy is turning out to be precisely the opposite." Wal-Mart's environmental goals include a 25 percent increase in the efficiency of its truck fleet within three years, and a 100 percent increase within ten years; a 30 percent decrease in store energy use; and a 25 percent reduction in solid waste. Wal-Mart now sees sustainability not only as good public relations but as good for Wal-Mart, both in terms of millions of dollars saved in lower energy, packaging, and other costs and in terms of heightened morale and productivity on the part of employees, who have a new reason to be proud of where they work.

Wal-Mart is such a big company—fiscal-year 2007 revenues were almost $349 billion; there are 1.8 million direct Wal-Mart employees; and 176 million unique customers visit its more than 6,700 stores every week—that its decisions have a huge ripple effect. With a supply-chain network of sixty thousand suppliers around the world, Wal-Mart can shift many markets toward greater sustainability. For example, it has made a commitment to sell salmon only from sustainable fisheries, and Wal-Mart now has fourteen "sustainable value [supplier] networks" for everything from chemicals to food and paper products. One of its biggest recent successes with customers, organic cotton clothing, has helped to grow global organic cotton production by over 20 percent since 2001. Wal-Mart's flip into a green brand identity is now being emulated by North America's second largest retailer, the Home Depot, which is branding thousands of the products it sells with an "Eco Options" label.[1]

• • •

BP is an interesting case study in the pros and cons of offering a "green" alternative to the oil business as usual. In 1998 BP (at the time, the company's legal name was its original one, British Petroleum) acquired the Amoco Corporation, an American business. To make the deal palatable in the United States, it was presented as a merger and the company temporarily became BP Amoco. In 2001 Amoco was dropped from the name, and the company once more became BP. Only now BP no longer stood for British Petroleum, it was simply an "initialism" that company marketing presented in advertising as standing for "Beyond Petroleum." Along

PEOPLE WANT TO BE SEEN AS GREEN, AND THEY WANT TO WORK FOR AND BUY FROM COMPANIES THAT THEY SEE AS GREEN, TOO.

with this came a new logo, a green-and-gold disk representing Helios, the Greek God of the Sun.

No longer trading as "British Petroleum" was useful because it sidestepped long-standing criticism of corporate colonialism in BP's traditional market strongholds in ex-British colonies in Africa and elsewhere. But no doubt more important, especially in markets such as the UK, the United States, and Australia, was the desire to appeal to public concern about the environment.

In this regard, BP's former CEO John Browne showed remarkable prescience. In 1997 the company withdrew from the Global Climate Coalition, an oil-industry organization dedicated to promoting climate-change skepticism, with Browne commenting that "the time to consider [global warming] is not when the link between greenhouse gases and climate change has been conclusively proven, but when the possibility cannot be discounted and is taken seriously by the society of which we are a part. We in BP have reached that point." This was the first time an oil-industry executive had spoken out in support of doing something about the climate. Oil company or not, BP deserved some kudos for this.

In 2002 Browne gave a high-profile speech saying global warming was "real and required urgent action," and he was one of the most vocal industry advocates of signing the so-called

Kyoto Accords on international action to combat global warming. In 2004 BP started making low-sulfur diesel fuel, and they are creating a network of hydrogen fueling stations in California. In 2000 BP purchased Solarex and became a leading producer of solar panels. BP Solar, to be renamed BP Alternative Energy, accounts for 20 percent of world photovoltaic (solar panel) production.

All this was to the benefit of the company's image and its bottom line, and I am sure it is a viable strategy for the long term. But BP will face increasing pressure to make its operations as green in fact as its marketing is in spirit. This task will fall to Andy Hayward, the designated successor to John Browne, who resigned in 2007 in large part because of criticism of BP's environmental record.

BP's "Beyond Petroleum" play is a great strategy, if future reality matches present marketing. The marketing has been so successful that I believe it has won the company significant leeway with customers to get things right. In seminars on employment branding I flash a slide with the logos of the biggest oil companies in the world, and BP gets the most favorable response. It is mixed, of course, with some very cynical, but all in all the BP campaign has been successful. Almost unanimously audiences say that all other things being equal, they would accept a job at BP first. Some people even insist that "Beyond Petroleum" is actually BP's legal name.

One of BP's billboard and print ads reads, "BP: Solar, natural gas, hydrogen, wind. And oh yes, oil. It's a start." I agree. Now BP must execute and finish the job of "greening" itself, or a competitor will hijack the strategy and the customer goodwill that it has temporarily won.

An article titled "Green Is Good" in the *Bulletin* cited the following companies for doing good in the green space:

- Continental Airlines spent U.S.$16 billion to upgrade the efficiency of its aircraft, including fuel-saving winglets that have led to a 5 percent reduction in emissions.
- British Airways will sell customers offsets for their share of the carbon emissions generated by the flights they take. Sadly only one in two hundred consumers has stuck his hand in his own pockets for carbon offsets, but the trend is growing among affluent, socially conscious consumers. Even if the big impact from carbon offset trading will no doubt be on the part of multinational companies driven by regulatory pressure and general public sentiment, this offer effectively brands British Airways as green among a much wider customer demographic than the relatively small group who buy a per-flight carbon offset.
- Tesco, the UK supermarket giant, has bio-diesel delivery trucks and offers merchandise discounts to customers who bring in their own shopping bags.
- HP will take back any of its own machines from customers for recycling and has started to audit its suppliers for their recycling practices.

Companies are beginning to make serious money from investments in being "Fast, Good, Cheap + Green." Consider the following examples cited in the same edition of the *Bulletin* in April 2007:

- Goldman Sachs invested $1.5 billion into cellulosic ethanol, wind, and solar, a gamble that has more than paid off.
- Swiss RE, the Swiss insurance giant with revenues over $24 billion, pioneered derivative-based products to hedge against the risks of climate change.
- DuPont, once a poster child of environmental mismanagement, now receives $5 billion of its $29 billion in revenue from green end-use products such as chemical coatings for solar panels. It is developing a green replacement for nylon called bio-PDO, produces genetically engineered corn to make ethanol, and has partnered with BP on a new fuel called biobutanol.
- GE wind turbines are selling faster than they can produce them.

There is one final example that I like. I had an opportunity to work with Google in Silicon Valley, and was astonished to hear that if a Googler buys a Prius or any other hybrid-engine car, the company will contribute $5,000 toward the purchase, with some conditions around staying at the company and not selling the car the next day, of course. It is behavior in perfect alignment with Google's mission statement: "Don't be evil."

Fast, good, cheap + responsible

Not being evil is also the order of the day, and not a moment too soon. Corporate social responsibility, which includes both environmental and ethical issues, is proving to be important for both society and for the bottom line. It is not just how you treat the communities you operate in, but also how and where you source your raw materials and labor.

One of my favorite clients is the Commonwealth Bank Foundation. Born out of unclaimed savings accounts, the foundation is dedicated to helping Australians, especially the younger generations, improve their financial literacy skills. They run seminars, have Web sites, and basically invest millions of dollars in this social initiative. Commonwealth Bank refuses to allow any product information to be included in such activities, so it is not a hidden sales pitch. This does not mean, however, that the bank is not proud of its achievements in this area over quite a few decades, and as you would expect, it uses its foundation activity to help attract the best staff and the most profitable customers.

> IT IS NOT JUST HOW YOU TREAT THE COMMUNITIES YOU OPERATE IN, BUT ALSO HOW AND WHERE YOU SOURCE YOUR RAW MATERIALS AND LABOR.

Some companies have built their entire brands on being responsible. The Body Shop is an excellent example. In her brilliant book *Business as Unusual,* flipstar Anita Roddick told the story of how she built the Body Shop on a reputation of social and environmental activism. As far back as 1986, the Body Shop formed an alliance with Greenpeace on the "save the whales" campaign. The Body Shop actually promoted the fact that its products were banned in China because Chinese law required animal testing of cosmetics and skin-care products.

Roddick insisted that Body Shop marketing reflect a "values-based company." Shareholders even complained that maximum profits were not being achieved because profits were being funneled into social projects. But this is the drawing card of the company. Marketing along these lines, the company achieved phenomenal growth, expanding at a rate of 50 percent annually from its opening. When its stock was first floated on the Unlisted Securities Market in London in 1984, it was listed at 95p. Eighteen months later the stock was valued at 820p. After a patchy performance period in the early 2000s (not due to a failure of marketing, but due to a manufacturing outsourcing bungle and some internal turmoil), the company was valued at roughly $1 billion at the end of 2005.

In March 2006 the Body Shop's positive reputation for social and ethical responsibility was tested by its sale to cosmetics giant L'Oréal for 652 million pounds. Because of the sale, *Ethical Consumer* magazine dropped the Body Shop from 11 out of 20 on their "ethical rating" system to only 2.5 out of 20. Whoops! The sale (or sellout, as some have called it) has been seen as a bit of a betrayal of the company's ethical and unique roots. An index that tracks thousands of consumers in the UK (the daily BrandIndex UK) saw the perception of L'Oréal slump by almost *half*.

So important is no animal testing to the Body Shop's core customers that some of them actively promoted a Body Shop boycott, because L'Oréal has not banned animal testing of its products. Dame Anita Roddick, who died in 2007, swore to give away the 130 million pounds she made from the sale, but this did not stop the cry that she had "sold out."

On the flip side, the Body Shop boosts L'Oréal's image among socially conscious investors and customers. And despite the backlash of hard-core Body Shop fans, L'Oréal grew sales in key Body Shop lines 9.7 percent like-for-like in the year ending December 31, 2006, compared with 6.4 percent for all L'Oréal brands. Twenty-five new Body Shop stores opened in 2006, bringing the total to 2,290. The stores in Canada, Japan, and Russia performed particularly well, with U.S. stores lagging somewhat.

More broadly in the cosmetics industry, the Campaign for Safe Cosmetics lobbying group says that more than five hundred cosmetics and body-care producers have joined its campaign, pledging to eliminate toxic ingredients from their products. Interestingly, global giants L'Oréal, Revlon, Procter & Gamble, and Estée Lauder have not been quick to sign on, although L'Oréal has made substantial strides in reducing its energy and water use, waste products, and direct carbon-dioxide emissions through its SHE (Safety, Health, and the Environment) initiative.[2] It will be a while before the big cosmetics companies can make themselves green and socially responsible throughout their vast industrial operations. Until then the smaller players can effectively use their "responsible" position to differentiate themselves in the market.

In the fast-food industry, Burger King took an early lead in socially responsible positioning in March 2007. The company announced that in the near term it would source 2 percent of its eggs from providers that do not confine chickens in cages and 10 percent of its pork from providers that keep pigs in pens rather than in small crates. The numbers may seem small, but they will rise as more cage-free- and crate-free-produced eggs and pork become available, thanks to Burger King and to consumers' increasing concern about the ethical treatment of animals. I will bet dollars to Whoppers that Burger King's move will eventually be emulated by other fast-food companies.

. . .

Nike shows how customer preference is forcing companies to make their operations more environmentally friendly and socially responsible. Accusations about sweatshops and exploitation of workers in the third world impacted Nike's reputation badly. It has since made a concerted effort to be seen in a more responsible light.

In the 1970s Nike shoes were made primarily in Taiwan and South Korea. But as economic conditions in those countries improved and workers gained rights to organize, Nike found things more to its liking in Indonesia, China, and Vietnam. It chose these places because wage demands were low and because they had laws prohibiting workers banding together in unions.

The minimum wage in these countries would cover only 70 percent of the basic needs of one person, and Nike didn't even pay the minimum wage, petitioning governments for exemption citing "financial hardship." In Nike's first year of operation in Indonesia, factory officials were convicted of physically abusing workers and one even fled the country because of investigations into sexual abuse.

But in 2005 there was a massive turnaround. The company commissioned a 108-page independent audit of working conditions at 569 of its factories. It found that as many as 650,000 workers (mostly nineteen- to twenty-five-year-old women) in China, Vietnam, and even the United States and Australia are at risk from excessively long working weeks, being ripped off in wages, verbal abuse, and even horrific human rights abuses such as sexual exploitation and not being allowed toilet breaks.

Kudos should be paid because Nike executives have now publicly acknowledged the problem and promised to clean up their act, and they're seeing massive benefits. Nike appointed a staff of inspectors to go from facility to facility ensuring basic working standards are met, and the company has allowed random factory inspections by the Fair Labor Association. Nike will now also deal directly with organizations that make complaints, rather than just issue denials—which means the days of public street protests are pretty much over because those with legitimate grievances have direct recourse to the company. So it's actually doing real good for their public image.

Nike has also been very smart in sponsorship of Lance Armstrong, highlighting his philanthropic ventures as much as his extraordinary athletic achievement. In conjunction with Lance Armstrong's foundation, Nike has sold more than 50 million yellow wristbands to raise money for cancer research. This move has started a craze, with every charity or "movement" having its own wristband. One of the most visible is the white "make poverty history" band.

Fast, good, cheap + beautiful

Dell, another super performer over the last decade or so, has been suffering recently as its product lines, from desktops and laptops to servers, have become increasingly commoditized. Dell's cost basis has long been the envy of other computer manufacturers. What is truly amazing is that Dell has a better cost basis even than Lenovo and other Chinese manufacturers.

While it remains highly profitable, it knows it must freshen its appeal in customers' eyes. Dell's initial efforts to do so through a new emphasis on design in its XPS range have signally failed, however. Instead it is HP that is gaining in the design sweepstakes on the leader in that area, Apple.

Twenty years ago, Samsung was a commodity manufacturer for other consumer electronics companies and had a discount brand image for its own products. Not content with that, the company set its sights on design excellence, and diligently entered every industrial and consumer product-design contest it could. The result is that Samsung has become a recognized global design leader and a premium brand that does coventures in LCD panels and other areas with Sony as an equal, not a junior partner.

The power of Samsung's design story has an impact on customers far beyond the technical capabilities of any of their products. *BusinessWeek* calls the lead designers at Samsung "foot soldiers in Samsung's continuing assault on the world of the cool." Jong-Yong Yun, Samsung's chief executive, said he wanted to make Samsung the "Mercedes" of home electronics. Patrick Whitney, the director of the Institute of Design at Illinois Institute of Technology, said that Samsung is a "poster child for using design to increase brand value and market share."

SAMSUNG IS A "POSTER CHILD FOR USING DESIGN TO INCREASE BRAND VALUE AND MARKET SHARE."

The transformation has been happening since 1993, when the then chairman visited a technology show and was frustrated that Samsung products were lost in the crowd. Since then, they have revolutionized design practices at the company.

They shifted their design labs to a place that was closer to the best design schools, and started an in-house design school (Innovation Design Lab of Samsung, or IDS) where employees could study under the best in the business. They also started collaborating with design-focused partners, a strategy we will explore in chapter 7, "To Get Control, Give It Up," such as the U.S. design firm IDEO (their first collaboration was on a monitor in 1994).

Further, they have poured literally hundreds of millions of dollars into updating the look, feel, and function of every product from MP3 players to washing machines. Between 2003 and 2004 they upped their design budget from 20 percent to 30 percent, and more than doubled their design staff.

This has really paid off. Since 2000, they have received more than one hundred citations at the world's top design awards in the United States, Asia, and Europe. In 2004 they won five awards at IDEA (Industrial Design Excellence Awards). In 2005, completing more than a decade of reform, Samsung clocked up just over $10 billion in profits, making them the world's most profitable tech company. In his book *Change Begins with Me,* chairman Kun-Hee Lee said of their change from a me-too commodities producer into a leading design-based firm, that in a world where products were fast becoming commodities, Samsung would never thrive on scale and pricing power alone. They needed a creative, competitive edge.

A flipstar worth talking about!

Fast, good, cheap + easy

My favorite service in the world is Pronto Valet Parking at Sydney's domestic airport. I board a lot of flights. I also have young children, so I like to come home as much as possible in between commitments. It is not uncommon for me to park my car at the airport three times in a week. I have gotten to know the guys who park my car, and I can't tell you how much easier they have made my life.

You pull in less than a hundred yards from the terminal, leave your car running, give them your ticket, tell them roughly when you will be back, and off you go. I would have missed a dozen flights last year without them, so needless to say they are fast. This service costs an extra fifteen dollars for the first twenty-four hours and five dollars for every twenty-four hours after that, which considering how much time and stress it saves you would be a bargain at twice the price. So it's cheap, too. I have never had a problem with my car, so I would have to say it is also good. Most of all, though, it is easy. They go out of their way to make it so. The roster is perfectly designed to meet the high-demand spots. They open your trunk for you when they return your car. They take all forms of payment. And they are always well mannered. My mate Avril Henry, who leads a similar lifestyle, felt so indebted to the Pronto guys that she bought them a case of beer for Christmas last year.

Airline frequent-flyer programs, airline lounges, and hotel loyalty cards are all examples of ways in which a service provider can make a service easy. There is little doubt that we are feeling a lot of pressure around time. Not only that, though, I would suggest we are feeling even more pressure on our mental space. We feel like we just can't take anything more in, not because we don't have the time, but because we don't have the mental energy. Increasingly affluent customers who feel increasingly stressed about their mental energy being in short supply will pay a premium for *easy*.

On the topic of being a road warrior, consider the innovative new approach to servicing and selling cars from Mercedes-Benz called Airport Express. Mercedes set up both a service center and a dealership at major city airports. One compelling reason for doing this is that these are high-traffic areas where exposure is good. And the people who travel regularly are usually professionals and are likely to fit the Mercedes target market. But the real winner is that it makes it much easier for Mercedes owners who travel to have their cars serviced.

But it is not like they just service the car. They wash it, do the dry cleaning you left on the backseat, and run any other errands they can. Oh, and they drop you at the terminal and pick you up. Clients also get a goodies bag to take on their flight, with chocolate, bottled water, granola bars, and loose-leaf English breakfast tea. They have introduced flower delivery and gift-buying services, which Belinda Yabsley, my good friend and former manager of the Sydney Airport Express, describes as "things we have time to do but customers don't."

Clearly it's working. Within twelve weeks of opening, the Melbourne operation welcomed its five-hundredth service customer, and sales were running at *twice* the projected levels. But Mercedes-Benz Australia spokesman Toni Andreevski says it's not solely about revenue at each Airport Express location. It's about marketing. Says Andreevski, "It's a way for us to add value to the brand by giving customers a tangible benefit. It's a different form of advertising. We could have spent half a million dollars on a billboard, but instead we're giving people the Mercedes experience, rather than a photo of it." I would suggest that in time it will also prove to be very profitable. It is too good not to be.

Of course it is not just Mercedes that does this sort of thing. As discussed in chapter 3, "Fast, Good, Cheap: Pick Three—Then Add Something Extra," Lexus was the first luxury-car brand to offer a heightened level of service. Before that, luxury-car makers told the following story to their customers: "You're lucky we sell you our expensive cars, and we and our dealers will bleed you dry at frequent regular service intervals and even more frequent irregular repair visits." Lexus was the first brand to make everything easy for the luxury-car buyer, changing the game for every other manufacturer, as flipstar Toyota has done in every segment of the car business. Toyota has not yet set up a Lexus version of Airport Express, but if they decide to do so, they would be a good bet to change the game there, too.

As things stand, Lexus will pick up your car for service at your home or work. They leave a replacement for you to use, and when they picked up my wife's car recently, they happened to leave the new model, prompting her to express interest in upgrading. They, too, return the car clean, with Lindt chocolates on the front seat. However, Lexus goes a step further. As part of their brand story (something we will talk a lot more about in a few pages), they align themselves with "high society" establishments, but do so in an unusual way. In Sydney, for example, you get free parking at the opera house. It is not like Lexus owners can't afford parking, it's just *easy*.

Fast, good, cheap + fun

When you are selling a product or service that has substitutes, and your cost structure prevents you from doing much else, you could always just make it more fun. This is what Virgin Blue has done in the extremely competitive Australian airline industry.

Virgin Blue is not only about fun, of course. The airline is definitely fast, good, and cheap, and it is also green. The Virgin Blue Web site proudly announces that Virgin is "the first airline in Australia to have a comprehensive program for carbon offset."

But in keeping with the cheeky, iconoclastic brand identity of Richard Branson and all Virgin businesses, Virgin Blue pushes the envelope on the fun side of flying. It is not uncommon for the customer-service manager on a Virgin Blue flight to announce a fake destination. You see customers start freaking out, only to hear the customer-service manager say, "Just joking," and move on with the rest of the announcements. On one Virgin Blue flight, I was sitting with a chap who was carrying two boxes of Krispy Kreme doughnuts (which he certainly didn't need, mind you). When we touched down and people started collecting their carry-on luggage, the flight attendant told him that in the state of Victoria you couldn't bring any food or drinks into the terminal from the plane. He was very apologetic and asked where he should leave the doughnuts. She said she would take care of them. A full thirty seconds later, as the doors were opening, she told him she was joking and that he could have his doughnuts back. He was so embarrassed he almost refused them.

While Virgin Blue has not been a runaway, showstopping success, most suggested it would never last. Since its first flights in 2000 the airline has proven the doubters wrong. Even with the arrival of Jetstar, offering even lower fares, fun seems to be keeping Virgin competitive. In addition, one of the problems Virgin had was that its primary story—"We are the fun, low-cost airline"—didn't attract business-class passengers. Originally they thought it was just that the message wasn't getting through, so they poured money into advertising. But eventually they found that "fun" just doesn't cut it for business-class travelers, who really want to be treated with an extra level of care and to stay connected while in the air. Virgin introduced the Velocity club for frequent flyers and fully flexible fares. These options and the ability to pay an extra thirty dollars to get a front-row or extra-roomy exit-aisle seat make Virgin a viable option for businesspeople flying at times and to destinations that Qantas business-class service doesn't reach.

Now it is not that business-class travelers don't like fun. They do. They just need the travel experience to, above all, be easy. On its transatlantic flights, Virgin Atlantic has turned on the pampering style in their business-class offerings. Including an in-flight bar, flat-sleeper seats (like the Qantas SkyBed), and a masseuse, as well as a limousine transfer and a first-class waiting lounge, Virgin Atlantic offers the complete service experience—both at sea level and at forty thousand feet. Anything to help business travelers feel better prepared for that first day of meetings in London or New York.

In-flight extras include dining when you want and eating what you want (rather than the traditional "here it is, eat it now" style). And the amenities kit actually has beauty products as well as basic toiletries, complimentary noise-reducing head-phones, and extra-large in-chair video screens.

Fast, good, cheap + healthy

Healthy may be the strategy with the most competitive advan-tage over the next couple of decades. The only thing more im-portant to baby boomers than staying healthy in their retirement years is having the money to enjoy their youthful energy and appearance (bet on the plastic-surgery industry continuing to soar as part of that trend, too). And Generations X and Y are even more fitness-obsessed than their parents.

Let's have a look at some examples.

It has taken just twelve years for Fitness First to grow from a single club in the UK to a sixteen-country, five-hundred-plus gym and 1.2-million-member organization. Sydney was home to the five-hundredth Fitness First gym in 2006. In 2005 Fit-ness First was acquired by a private equity firm (BC Partners) for more than $2 billion. It is the fastest-growing health-club company in the world; in 2007 Fitness First registered an 18 percent growth in earnings and was on track to match its 2006 record of forty-nine new clubs worldwide. In the UK, the com-pany has enjoyed sixteen consecutive months of impressive like-for-like sales growth (even excluding the impact of new clubs).

Organic food accounts for only 1 to 2 percent of food sales worldwide, but the organic-food market is growing rapidly. The organic-food market in the United States has enjoyed 17 to 20 percent growth over the past few years. Meanwhile conventional-food sales grew only at 2 to 3 percent a year, or in other words about the same rate as the growth in population.

Multinationals are beginning to see the value of these products and to invest in organic produce, leading to increased competition, increasing economies of scale, and a subsequent decrease in price and increase in accessibility to fuel the market and force heavily polluting and industrialized agribusiness to clean up its act. Earlier in the chapter I talked about Wal-Mart's green push into practices and products that further environmental sustainability. That has included a substantial increase in organic-food offerings in the grocery sections of Wal-Mart stores.

HEALTHY MAY BE THE STRATEGY WITH THE MOST COMPETITIVE ADVANTAGE OVER THE NEXT COUPLE OF DECADES.

In the UK, the market has gone from just over 100 million pounds in 1994 to 1.2 billion pounds in 2006, and will break the 2-billion pound mark in 2010. Tesco's—the nation's largest supermarket chain—reported a 30 percent rise in organic sales in 2005–2006. In Australia the organic-food market is valued at almost $400 million a year and has enjoyed 25 to 30 percent growth per year over the last few years.

Vitamins and food supplements (the epitome of self-help and self-medication) are a booming market. For example, the European Molecular Biology Association puts the global food-supplements industry consumption at roughly 125 billion euros a year. In the European Union in 2001 sales of food supplements reached almost 2 billion euros, which was growth of almost 7.7 percent from the previous year.

In the UK in 2005 vitamin sales alone were valued at over 320 million pounds, which constituted over 15 percent of total over-the-counter drug sales. This was up from 280 million pounds in 2001 and included a massive 8 percent jump in 2004.

The UK Food Standards Agency places the consumption of high-dosage vitamins alone at between 30 and 40 million pounds a year, and says nearly 50 percent of the population takes vitamins.

· · ·

Obviously there are more X factors, and the list could probably go on forever. This is enough to get your mind clicking. What X factor can you offer to set you apart from your competitors?

When you come up with some possible X factors (the only limit on them is your imagination), you need a model to work from as you set about redesigning and repositioning your products and services. And you need to do it from the perspective of the customer. The model I suggest you work from is what I call the Total Ownership Experience.

THE TOTAL OWNERSHIP EXPERIENCE

A few years ago Joseph Pine and James Gilmore published a wonderful book called *The Experience Economy: Work Is Theatre & Every Business a Stage.* The book has been very successful—deservedly so—but sometimes I wonder at how little people seem to have learned from it. I meet real estate agents, retailers, and even bankers who say they've read the book and who talk urgently about the customer experience. When I ask them what they are doing about it, they say things like, "We brew coffee before we show a new house to stimulate the senses"; "We make sure there is lots of color in the stores"; or, "We have all our financial products online."

It is not that multisensory stimulation is not a key part of the total ownership experience. It is, but as you are about to find out, it is just one step in building a much-deeper connection to your customers. Just ask brand-building legend Howard Schultz, the founder of Starbucks. In a memo Schultz sent to senior management at Starbucks in February 2007 (later leaked to the media), he reflected on the importance of this relationship and criticized the company for losing sight of it:

> Over the past ten years, in order to achieve the growth, development, and scale necessary to go from less than 1,000 stores to 13,000 stores and beyond, we have had to make a series of decisions that, in retrospect, have led to the watering down of the Starbucks experience, and what some might call the commoditization of our brand. . . .
>
> For example, when we went to automatic espresso machines, we solved a major problem in terms of speed of service and efficiency. At the same time, we overlooked the fact that we would remove much of the romance and theater that was in play with the use of the La Marzocca machines.[3]

Schultz noted that the height of the new machines blocked customers' view: they were denied the "intimate experience" of watching the barista make their drink. He also criticized the new streamlined store design. Changes had been made to achieve greater efficiency and ensure that return on investment was satisfactory, but the result was that Starbucks coffee shops no longer had the "soul of the past." They felt like "a chain of stores," said Schultz, as opposed to conveying the "warm feeling of a neighborhood store."

Do you and your people strive to make your service delivery "romantic"? Are your call-center staff striving to have an "intimate experience" with your customers? When you see opportunities to gain efficiencies in your operations, do you first consider the impact they will have on how it "feels" to do business with you?

As you expand and grow your business you have to hold on to what made you you in the first place. It is exactly this kind of authenticity that makes brands attractive. And as you will learn in this chapter, it is these intangibles, like the "warm feeling of a neighborhood store" that will differentiate your brand, your product, your business from the competitors.

Schultz, however, does not stop there. Later in the memo he says: "Now that I have provided you with a list of some of the underlying issues that I believe we need to solve, let me say at the outset that we have all been part of these decisions. I take full responsibility myself . . ."

In a world of corporate executives who usually seem far more concerned about their stock options and golden parachutes than about customers or staff, you've got to love a leader who takes a full share of responsibility for moves that have put his company at risk, relishes the accountability of his position, and then uses it to rally the troops to join him in doing better:

Push for innovation and do the things necessary to
once again differentiate Starbucks from all others.
We source and buy the highest quality coffee. We

have built the most trusted brand in coffee in the world, and we have an enormous responsibility to both the people who have come before us and the 150,000 [Starbucks] partners and their families who are relying on our stewardship.

Let me stop a moment and say with unbridled admiration that these are the words of a true flipstar. Howard Schultz changed how the world drinks coffee and he obviously hasn't lost one iota of his imaginative passion for supplying customer wants as well as needs. And it is not just about the customer, it is as much about the staff, who by the way are not called staff—instead they are called "partners."

I once gave a presentation in Los Angeles about becoming a talent-centric organization, and at the session before me two Starbucks employees shared their experience of what it was like to work for Starbucks. A young man who still works in a store as a barista was in tears as he spoke about how proud he was to be part of Starbucks. In tears, and I was not the only one who noticed—the four-hundred-odd executives in the room did, too, and were clearly moved by the young man's story. And it was not as though the company had rescued him from the depths of despair. He just loved working there that much. But I digress—back to the point . . .

In regard to "Absolutely, Positively Sweat the Small Stuff," Schultz's memo makes clear that details like the smell of a Starbucks store can only have significance in the context of a total customer experience that is much bigger than using sensory stimulation to trigger emotional responses. In other words the customer experience goes beyond sensory stimulation or

THE TOTAL OWNERSHIP EXPERIENCE

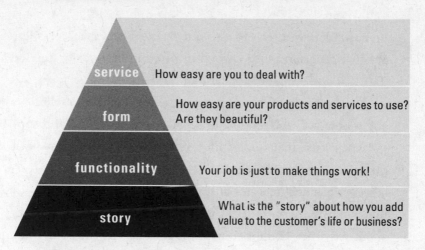

a jazzed-up transactional environment, whether bricks-and-mortar or online. Buying and selling coffee in a way that helps indigenous coffee growers is as much a part of the Starbucks brand story as good coffee. Being able to inject "romance and theater" into the making of a cup of coffee satisfies much more than customers' need for caffeine and a warm drink. It satisfies their desire to be taken to another place. The "third place" besides home and work or school, as Starbucks calls it. The customer's total ownership experience includes four things:

- service
- form
- functionality
- story

Service is essentially how it feels to *buy* your product or service. It is about store design and layout, the behavior of staff, the simplicity and ease of navigation on a Web site, and so on.

Form is how it feels to *use* your product or service. It is less about the performance capability of your product or service than it is about design, appearance, and ergonomics.

Function is how it feels to *own* the product or service. It is about how well the product or service integrates into and supports the customer's daily life and lifestyle. It is about both quality and integration.

Story is how it feels for customers to *say* they own the product or use the service. It is the story customers tell themselves about why they bought your product or service and why they have an affinity with your brand. And it is the most powerful of the four elements of the total ownership experience.

Notice the use of the word *feel* in the above descriptions. This is deliberate, considering that emotions drive our decisions, not our rational thought processes. As I said a few paragraphs ago, we tend to seize on relatively superficial qualities to justify our emotional imperatives.

Consider an experience I had with my wife. A few years ago we were shopping on the Gold Coast in Surfer's Paradise, Australia. We were on a street of boutiques for Gucci, Prada, Bally, and other high-end fashion brands. In the Gucci shop, my wife, Sharon, admired a handbag that cost something near $1,500. I could tell she really wanted it and I also thought it looked nice, as it should for the money.

But something about the bag bugged me. And no, it wasn't just the price tag. It looked strangely familiar. And then it hit me. I had recently seen the same bag in a street vendor's stall in Southeast Asia.

Wise spouses will know that I should have kept my mouth shut. But I couldn't help myself. I said, "I saw the exact same bag last week in Malaysia at a fraction of the price. I will be there again next month and I will bring you one home." (Interesting how a developed-world luxury item generates economic activity in the developing world, too, isn't it?)

"What? I don't want some bag from a street stall in Malaysia. That bag would be a fake!" Sharon said.

"Nobody will ever be able to tell the difference," I said.

"Of course they will. You may not be able to see it, but people who really know about these things will."

"Bollocks. I bet if you put the bags side by side, ninety-nine out of one hundred women and assorted experts who think they know all about Gucci will pick the wrong one as fake. I will bring you back two of those bags in different colors. We will still save a bundle, and I guarantee you, no one will ever know they're fake. I am pretty sure they have them in Chinatown in Sydney as well. Maybe we could look there."

"No way. I want a real one. First, most people *will* know the difference, and even if they didn't, I will. And every time I wear the bag, I will know it's a fake. So quit making a fuss and buy me the damn bag."

So we bought the bag. Typically, however, I could not let it rest. "So tell me, darling"—a poor choice of words—"how will people know this is a real Gucci bag?"

"The stitching!" she yelled as we left the store.

Stitching was her justification. Damn stitching.

"So, $1,450 worth of stitching!" I yelled back as I realized I had made a similar decision some years before based on exactly the same thing—stitching. More on that soon.

Reflecting on that experience now, I have decided we were both right. The bag in Malaysia was quite a convincing knock-off. But buying and owning a fake, and pretending it was real, were not part of my wife's story about herself. She wasn't just buying the bag because of what it said about her to the world, but because of what it said to her about her. This is what I call Aspirational Inside.

Howard Schultz's memo is all about matching aspirational inside with aspirational outside. If Starbucks does something that helps the bottom line in the short term but over the long term undermines the way customers feel about the experience of buying coffee in a Starbucks store, drinking it, or being seen with a Starbucks coffee cup, the extraordinary success of the Starbucks brand will come undone. This includes not only the smell of the coffee and the look of the stores, but also the coffee having been harvested, packaged, and distributed in an environmentally sound, socially responsible way.

Starbucks' ability to offer customers fast, good, cheap at the premium price, plus an X factor of feeling good for, in, and about themselves, has recently been extended to sales of specially chosen books and music CDs. Starbucks has achieved great success with exclusive retailing windows for CDs by artists such as Bob Dylan and Alanis Morissette, including outselling all other retail outlets on the late Ray Charles's final CD, *Genius Loves Company*. In 2007 Starbucks formed its own music label, Hear Music, in collaboration with the Concord Music Group, and signed Paul McCartney as the label's first artist. The same year it gave a huge boost to number one on the *New York Times* hardcover bestseller list to Ishmael Beah's memoir *A Long Way Gone: Memoirs of a Boy Soldier,* about being conscripted as a child soldier in Africa.[4]

Note that the crucial thing is not the retailing power of Starbucks. It is how the company uses its hard-earned power to achieve true synergy, choosing musicians and books that fit key customer demographics and resonate with socially responsible themes such as economic justice for coffee growers in the developing world.

I've already referred to what I call aspirational outside, the conspicuous consumption of something flashy and expensive—such as a ridiculously expensive cup of Starbucks coffee—so other people know you have money, good taste, or good values.

You may have said the following things yourself at times:

"A car is just a car. As long as it gets you from A to B."

"You can't tell the difference between a real Gucci bag and a fake one. You are getting conned."

"You would have to be stupid to spend four dollars on a bottle of water when you can get the same stuff for free from the tap."

However, as we gain a deeper understanding of the human mind and how we make decisions, it actually becomes inaccurate to say that such decisions are illogical or irrational. They are not. They are in fact very "normal" ways to behave. The story we tell ourselves about the products we buy and other such decisions we make is so deep to the human experience of the world that "because it feels good" is actually the only rational explanation for why we do

THE STORY WE TELL OURSELVES ABOUT THE PRODUCTS WE BUY AND OTHER SUCH DECISIONS WE MAKE IS SO DEEP TO THE HUMAN EXPERIENCE OF THE WORLD THAT "BECAUSE IT FEELS GOOD" IS ACTUALLY THE ONLY RATIONAL EXPLANATION FOR WHY WE DO ANYTHING.

anything. If it is aspirational inside, the story only has to appeal to me, and who cares what anyone else thinks. If it is aspirational outside, then what is important is the story being presented to other people.

The point is that what we think is rarely what drives our behaviors. It is what we *feel* that matters. I started this conversation by introducing you to the four fundamentals of the total ownership experience, before explaining in some detail why I deliberately used the word *feel* when describing what they were. Let's go deeper into each of these elements now. Please note that even though I have separated them here for ease of understanding, the four fundamentals are inextricably linked—especially the story. The story of a product and why someone would buy it is intertwined through each of the other three fundamentals, as well as being the basis for all PR and marketing conducted by a company (remembering, of course, that this includes you).

THE STORY

No one, least of all my wife, wants to admit that we all shop for things that help tell a good story about us. Through the things we buy and the experience of owning and using them, we tell that story to ourselves and others. This is the foundation of the ownership experience and the most powerful driver of value and premium pricing.

Nothing makes us more loyal and less price conscious than a product or service that integrates into our favored story about ourselves. You know what I mean. The story that we are popular, hip, savvy, enlightened, successful, and socially and environmentally conscious. You fill in the blank, and that's why we're wearing the labels we wear, driving the cars we drive, and so on. The products and services we buy say that we are exactly who we want to be. That we are in with the in crowd. That we are worthy of respect, admiration, or envy.

Your job is to find what your story is. The story of your brand, the story of your company, the story of your product, and scream it from the rooftops—or whisper it through the streets, which is usually the more powerful of the two. Notice that I did not say to find out what story the market wants and to yell that story from the rooftops. A story needs to be real, authentic, and original.

Let me give you some examples of the stories we tell.

I have a friend who works for a company with an extremely powerful reputation in the market. She hates it. When asked to describe the workplace, she uses words like *toxic, disorganized, small-minded,* and *the least friendly place she has ever worked.* So I asked her why she stays there. Her response was: because of how it will look on her résumé. She knows that if she wants to move on to another organization, her next prospective employer will think, *Wow, you worked at ABC company. You must be good.*

Another very close friend recently purchased a BMW, despite saying he never would. He used to say, "Everyone has a BMW," which is why he never would. After his purchase, he told me a story that sounded like he had already related it a few times before. Probably most of all to himself.

"I went to buy a new car and my next-door neighbor made me promise not to buy anything until I had at least test-driven a BMW. I agreed. After test driving a Mercedes, Audi, Lexus, and Porsche, I begrudgingly headed to the BMW dealer and took the new 335i coupé for a drive. Omigosh! Dude, it left the rest seriously for dead. I now know why the BMW is such a popular car. Because it is better than all the others."

Now he is not an expert judge of a good car, so don't take this as the hard truth, but this experience had made it okay for him to buy a BMW. If you probe a little deeper, as I did, you learn why he could justify spending the money he did.

"So what makes this car so good?" I asked.

He began to describe the car's many fine features in surprising detail. He capped the story off by saying, "This is my kind of car. And you know what? I work so hard, I deserve it."

Whether you realize it or not, you do exactly the same thing when you make any kind of decision. It may not be as elaborate as this example, but the story is there somewhere.

So the question becomes, *How do we construct a better story so we can build more powerful brands? What makes up the story?*

All good stories have three things:

- detail
- characters
- language

Using the BMW example, consider the following elements of a good story and remember that this is what you need to do for your brand, too.

Detail: My friend knew the BMW 335i went from zero to one hundred kilometers per hour in 5.5 seconds. He referred specifically to its competitors: Audi, Mercedes, Porsche, and so on. Plus ATC (Automatic Traction Control) and ABS (Antilock Braking System).

Character: Statements like "no one else has one" are evidence of the identity and character he associates with the car. "My kind of car" and "I deserve it" personalize the purchase, a process that is clinched by calling the piece of metal a "baby."

Language: This bit cracks me up. If you knew this guy, you would think it was ridiculous for him to talk about "twin turbos" and "turbo lag." Less obvious, but more powerful perhaps, was when he described the car as "perfectly understated." I would just bet that the salesperson used that exact language. This is an example of how the story is not separate from, say, the service experience.

Let's look at some more flipstars as we go into a little more depth on these three elements.

Detail

Detail is anything specific you can reference to say why a product is better, more suited, etc. Companies use statistics, competitive benchmarking, surveys, information about the manufacturing process, or the history of the company to distinguish their offerings from the competition.

For a recent trip to Las Vegas, I asked a travel agent about the best place to stay. I had thought the Bellagio (based on the movie *Ocean's 11,* which is an extreme example of characterization). However, in an attempt to convince me to stay at Wynn's she said that since MGM Grand had taken over the Bellagio, staff-to-room ratios had dropped to 1.5 staff members per room. And that Wynn's, which is owned by Steve Wynn, who previously developed both the Bellagio and the Mirage casino hotels on the Las Vegas Strip before selling them to MGM Grand, Inc., has a ratio of three staff per room. This detail is a powerful piece in the story that will build Wynn's competitive advantage on the Strip.

Or what about American Airlines, whose December 2006 *American Way* in-flight magazine began with a letter from the editor noting the following specific details:

- 120 million passengers
- 1.3 billion miles
- 1.5 million flights
- 100 million bags
- Every twenty-two seconds an American Airlines or American Eagle flight takes off somewhere around the world
- 250 cities in forty countries
- 95 million calls received
- 450 million visits to AA.com
- 100 aircraft

I am not sure this is the kind of detail I would use to build my story. Sure American Airlines is building a story of a global, regular, experienced airline. As a very frequent flyer I think, *Too many people, just a seat number, and no one really caring about me.* But hey, that's just me.

One more example of detail: Sirius satellite radio tells its story as made up from the following elements:

- over 130 channels
- 69 channels of commercial-free music
- Available two hundred miles offshore
- 60 channels of sports, news, talk, and entertainment

Characterization

Characterization is what brings a story to life. It does not actually have to be about a specific person, but it must convey human attributes. This is difficult to do, often expensive, but extremely powerful.

In the past the most popular form of characterization was celebrity endorsement. Although it is still very powerful, many would suggest that this technique is losing its effectiveness. Let me share with you some examples of characterization that I like.

Skins Compression Garments are highly styled athletic clothing made with "engineered gradient compression" to improve circulation, minimize soreness from exertion and injury, and speed recovery. Skins are marketed with the constant statement that "We don't pay sports stars to wear our products." This small piece of information is extremely powerful because star athletes and well-known teams in Australia and elsewhere—including all the major Australian Rugby League teams and players from Arsenal, Chelsea, and Manchester United in British soccer—have been seen wearing Skins clothing while training and competing. This is a great story, but it will need to keep evolving.

In the United States, the Under Armour sports clothing line, which includes but is not limited to compression garments, has become a huge success with paid endorsements from major sports teams. If Under Armour matches Skins in technical features, style, and price, and both brands are fast, good, and cheap, celebrity endorsement could well make the difference in the competition between the two.

There is much debate about whether organizations need a

charismatic leader who is in the public eye. I am decided. *Yes!* It does not need to be the CEO. It could be the founder. The founder could even be dead but have an identity or an image that people can associate with.

I would like to avoid using another fashion example here, and am feeling that you may be bored with my shopping preferences, but it is only fair that I include this example. I have criticized my wife for spending too much money on a handbag; I should at least fess up to doing so on suits.

Many years ago I was looking at the suits in a Hugo Boss boutique. It is important to note that this was in my first year in business and I could hardly afford a suit, let alone one from Hugo Boss.

After looking at a few suits in their entry-level range which were about $800 a pop, I had almost built a powerful enough story in my mind that I needed to buy one of these suits. My story was that if I wanted to be a player, I had to look like a player.

Then the sales assistant said, "But you really shouldn't buy any suit until you have seen this range here." It was the over $1,500 range. The sales assistant pulled a $1,700 suit from the rack that, except for being a different shade of black, looked pretty much like the $800 suit I was trying on for size. Neither of them actually looked that good on me. They were a bad fit for my body shape, but I convinced myself, with a lot of assistance from the man serving me, that tailoring could fix this.

I took a good hard look at the two jackets side by side, like a nonmechanical person looking under the hood of a car. I couldn't see or feel any difference, and I said so.

"What's the difference?" I asked. "Show me."

The sales assistant said, "This suit has higher-quality fabric and stitching, and it will 'sit' on you better. It will feel like a part of you." (Are you picking up the use of both characterization and language?)

"What does that mean? You can't be serious. It is more than double the price."

The sales assistant repeated himself, more or less word for word, and I said, "Doesn't anyone in the store have a better explanation than that?" At that point, the sales assistant went to get help from the manager. He needed reinforcements, because I had the bit between my teeth.

The store manager walked over, and before he could say a word I asked him, "In addition to the fabric, which doesn't look any better to me, why is this suit worth nine hundred dollars more than the one I'm trying on?"

"Well, if I could just say a word about the fabric, sir, it's what we in the business know as a super 120 and it weighs two hundred and eighty grams per square meter." (Notice the use of detail.) Looking at me in a way that made me want to punch him in the nose, he continued, "In layman's terms, this simply means that it is a better-quality fabric and will 'sit' better on you."

"Really?" I wasn't about to give in easily.

"Yes, sir. And then as to the stitching"—remember my wife in the Gucci boutique—"much of this garment is hand-sewn for effect, while the fabric is laser-cut, ensuring perfect shape."

"Uh-huh." It was a good, practiced spiel, but I wasn't ready to throw in the towel. "That may be, but is that really going to make a difference to me over the life of the suit? I still can't see any of the differences you're talking about."

"Why don't you try it on?" the store manager said. He got

me to put on the jacket, and repeated some of what he'd just told me about the fabric and the stitching.

I switched jackets a couple of times, looking for a difference in the mirror.

"I still can't tell the difference!" I exclaimed.

"Ah, sir," the store manager said in a mocking tone, "I know you can't tell the difference, but I assure you the people you want to be doing business with can."

I bought the $1,700 suit, not to save face in front of the all-knowing store manager, but because I actually bought into the story that the suit would help me become more successful and respected in my business. I believed it would give me a seat at the right tables, so to speak, in front of the right kind of people. Guided deftly by the store manager, I built a story of myself in my head as a powerful player in the business community.

Language

Finally, on language. We looked at the importance of language in the sections on *detail* and *characterization,* so let me just share two dramatic examples of how language influences our experiences as customers.

One of Australia's favorite sons, Steve Irwin, epitomized a stereotypical view of what Australia is like. For the record, it is nothing like what the overwhelming majority of Australians are like, but the world loved him anyway. The best thing about Steve Irwin was that he wasn't putting on an act. He really was that enthusiastic and he really did use words like *crikey.* This image of Australians is rampant around the world. While I was eating alone at a counter in San Francisco, a road warrior from

New York asked me if kangaroos really hop up and down the main street of Sydney. "Of course not," I said. I would be willing to bet that more than half of Sydney's residents have never seen a kangaroo outside of a zoo. And we don't all talk like Steve Irwin either, but this hasn't stopped people making serious money from the outback "Aussie" story.

The best example of the use of language is when your brand name becomes part of the language:

- "Just Google it," for a search online.
- "I need a Kleenex," when you want a tissue.
- "Get me a Coke," when you couldn't distinguish one brand of cola from another.

And so on.

You may be aware that the founder of Illy coffee invented the first steam coffee machine and coined the term *espresso*. When his grandson recently tried to trademark the word, the powers that be said it was too late, but he had the right idea.

I have spent a lot of time on *the story*. I make no apologies for this. This is fundamental. The story, as superficial as it seems, is anything but. It is the reason why people do what they do. It is also the reason why it forms the basis for the pyramid. The model in this case is also the metaphor.

FUNCTION

Design without functionality will fade. That is, of course, unless the core purpose of the product is simply to look good, as is the case with, say, a piece of art or a collectible for your shelf. I think the best way to market your product is to build a good one. However, when talking about the total ownership experience, I am assuming that you have fast, good, and cheap already figured out. What I mean by functionality here is how well the product or service integrates into a customer's life.

Pronto valet parking is a functional service. So is Mercedes-Benz Airport Express. And here are a couple more that were designed purely with functionality in mind.

First Luggage offers on their Web site that they will "arrange for your luggage to be collected anywhere in the world and delivered to your chosen destination without stress or hassle." Yeah, that's a cool idea.

They note in their FAQs that many companies will ship your luggage for you if you ask, but First Luggage has it as their sole business—so there is no complex paperwork, no weighing of luggage, no boxing of your own goods, nothing like that. You just tell them where you are, where you're going, and when you need to be there. Your bags will be there first. They also text you to let you know your bags have arrived ahead of you (ah, peace of mind).

First Luggage has also partnered with British Airways, and offers a 5 percent discount to passengers who book through BA. The timing for First Luggage could not have been better, as they cash in on increased baggage restrictions caused by terrorism in the UK.

Progressive Insurance has pioneered what they call the Immediate Response System—it's a high-tech network of GPS systems and emergency response vehicles equipped with laptops, printers, and wireless net connections. It makes sure claims assessors are on the spot ASAP—sometimes even before the police—and can assess the claim, organize towing if it's required, write out a claim check *on the spot* in some situations, and generally provide one hell of a customer experience.

Progressive's Peter Lewis says of this system and his philosophy, "We're not in the car-insurance business, we're in the business of reducing human trauma and the economic cost of auto accidents." Or as I call it, they are in the business of functionality—integrating, in this case during very traumatic experiences, as simply and powerfully into people's lives as possible.

The truth is the Progressive example is driven by cost savings, too, because it cuts out a lot of costly back-office paperwork and administration while vastly enhancing customer loyalty. What a win-win. Save costs, and improve the customer's experience.

Online grocery shopping is another example of an offering based on functionality. Online grocery shopping has soared in popularity—although along its route to popularity there have been a few bumps. In 1999 the U.S. online grocery-shopping market was $200 million. By 2004 Forrester Research reported that the market had grown 1,200 percent to $2.4 billion. The Food Institute predicts that online grocery sales in the United States will reach around $5.4 billion by the end of 2007, and *Key Note* predicts the online grocery-shopping market will continue to enjoy 20 percent year-on-year growth until at least 2010.

In mid-2006 Amazon.com quietly launched an online grocery-shopping service that could deliver to anywhere in the United States. Almost all major supermarket chains in Australia, the UK, and the United States now offer online ordering and delivery. For instance, Tesco captured almost 70 percent of the online grocery market in the UK in 2006 and their sales hit almost one billion pounds.

If you don't order your groceries online, you should—at least, in my opinion. It is so much easier than going to the store, it is almost a joke. Given that 25 percent of our grocery purchases are convenience purchases, it is fair to say that most of the 75 percent remaining are well-thought-out, planned purchases. In fact, if you are like most shoppers, you will do one big shop (weekly, monthly, or whatever) and smaller fresh-produce or convenience shops in between. The big shop rarely changes, in terms of the types of items you need and the brands you would normally buy.

> ASK YOURSELF, HOW MANY OF YOUR PROCESSES OR POLICIES ARE SET UP TO MAKE LIFE EASIER FOR YOU OR TO MAKE LIFE EASIER FOR THE CUSTOMER? CHANGE THE ONES THAT ARE ABOUT YOU.

Online shopping remembers what you buy, has it all ready in a convenient list, knows your credit-card details and where you live. You then select your delivery time slot, usually in the next twenty-four hours. They deliver outside of normal business hours, which is important for the modern, busy person.

On a different note, the emergence of home services and outsourcing of traditional chores has grown exponentially over the last decade. According to IBISWorld, the average Australian household now spends $14,000 per annum outsourcing things they used to do themselves. Cooking, cleaning, washing, ironing, child care, gardening—you name it, there is someone who can do it for you. This whole industry has emerged to service the needs of people's time-pressured lifestyles. That is, the service integrates into the customers' lives by removing the need for them to do it themselves.

Ask yourself, how many of your processes or policies are set up to make life easier for you, and to make life easier for the customer? Change the ones that are about you.

FORM

Beautiful can be a form of competitive advantage. Apple is beautiful. Well-presented food is beautiful. The choreography of Cirque du Soleil is beautiful. Google's home page is beautiful.

Bang & Olufsen is beautiful, and this Danish consumer electronics manufacturer also understands the often counter-intuitive dynamic of consumer behavior. Bang & Olufsen products are extremely expensive, but they are sneered at by audiophiles and home-theater enthusiasts. I would agree with them on the sound, but I think the B&O panels are stunning. But hey, that is just my opinion. It sounds crazy, but whether it is the best audio or video is actually beside the point. Bang & Olufsen's success is built on the superficial areas of design and simplicity.

People don't just buy B&O stereo equipment and flat-panel televisions to listen to music and watch DVDs, they buy them to look good in their homes. The essential functionality of B&O products in customers' lives is as stunningly beautiful interior-design objects. Oh, and so their friends know they can afford to buy B&O products (part of their story).

The company understands this very well. At one point Bang & Olufsen CEO Torben Ballegaard Sorensen wanted designers of B&O's Beomax 8000 television set to include a microchip that would be compatible with future HDTV standards. That would conceivably extend the Beomax 8000's selling life and its annual $10 million contribution to cash flow.

The B&O designers refused to consider the CEO's suggestion, saying it was an "affront" to the design of the television as a sculptural object in its own right. When the new HDTV standards were operational, they would design a new television as a sculptural object from scratch.

I am not criticizing the CEO. He is a flipstar in its purist form, sacrificing what would be potentially more revenue and profits in the short term to stay true to B&O's source of competitive advantage. Form not function. I also love that he was willing to be directed by the design team. He was willing to give up some control, the topic of chapter 7, "To Get Control, Give It Up."

The very affluent people who spend as much on a Bang & Olufsen stereo or home-theater setup as the cost of a well-respected motor vehicle aren't doing it to rock out with killer sounds and images, but so that the products themselves will look fabulous in their homes. Likewise, the equally affluent customers who are focused on audio and home-theater performance will never buy Bang & Olufsen.

As I said, you can't actually separate form or functionality or service from the story. As Rich Teerlink, the former CEO of Harley-Davidson Motorcycles, once said, "We don't sell motorcycles. What we sell is the ability for a forty-three-year-old accountant to drive through a small town dressed in black leather and have people be afraid of him."

The functional mechanical performance and the distinctive appearance of Harley-Davidson's products obviously combine to make that customer experience possible for communities of Harley-Davidson fans all over the world. But it is equally obvious that for Harleys, appearance counts for more than performance. It is the design that enhances the story.

In 2006 Harley-Davidson's net income climbed to $312.7 million from $265 million a year earlier. Revenue rose 14 percent to $1.64 billion, and shipments jumped 11 percent, including a substantial 22 percent jump for their high-end models. Predictions are that earnings per share will rise 11 percent to 17 percent by the end of 2009.

There are lots of good motorcycles that cost a lot less to buy than a Harley-Davidson. In every category of product and service, the winning providers, the ones who have the highest profit margins in their fees and prices, are the ones who sweat the small stuff to create a distinctive customer experience, as Harley-Davidson has done with the look and feel of their motorcycles. They go beyond the product or service itself to the total ownership experience that customers have, and they do everything in their power to make that experience truly unique.

Apple, an obvious flipstar, has understood this for a long time. As my Apple mates say to me when comparing PCs to Apple computers, "Apples are beautiful." And it seems as though other computer makers are getting on board, too.

Dell's new cases look much more modern and slick. Asus has teamed up with Lamborghini (an instance of characterization by association with a well-established brand) and produced a leather-bound laptop—not the carrying case, but the actual casing of the laptop itself is leather. Hewlett-Packard has reclaimed the number one PC spot over Dell in large part because they have been faster to jump on the trend that design is a powerful differentiator.

SERVICE

Service usually gets the most attention in discussions of the customer experience. But in many cases, product form, functionality, and story can trump bad service. That was my experience with an Italian sofa that I bought recently. Actually it only *arrived* recently; in fact I bought it some time ago.

I want to mention the establishment where I bought it by name because the service was appalling, and they deserve to be called on the carpet, but my publisher's lawyers prevent me from doing so. It was a struggle to get assistance from the staff from the first moment my wife and I walked in the door. And after we had ordered the sofa, they sprang the good news that delivery would take three months. I was thinking more like three days, so we'd be able to start enjoying the new sofa that weekend.

Four months later, *nothing*. No sofa, no phone call, no e-mail, no letter from the store explaining what was happening. I rang and asked where in blazes my sofa was, and they told me there were always shipping delays from the Italian manufacturer. They assumed I was an idiot and didn't realize the true situation: that they were a small importer of specialty furniture with no inventory except what was on display, and that they minimized costs by shipping only full containers. They had obviously had a slow year, and as the rare customer who had actually bought something, I was paying the price.

Half a year after I had initiated the purchase, the store called to say that my sofa would arrive the next day. I called back to say that tomorrow was not convenient, but they would have none of that. I asked if they could at least tell me when it would be arriving so I could arrange for someone to be home to receive it. No information there either.

To top it all off, the delivery people refused to take away the cardboard and plastic the sofa arrived in. "Sorry, sir," they said, "that is your responsibility."

I could not imagine a worse service experience. But we absolutely love the sofa. It is both gorgeous and amazingly comfortable. It is a perfect fit for our space. In other words, the service was appalling but the total ownership experience is still awesome. The trials and tribulations we went through to get the sofa only make the story better in the end.

I'm not discounting the importance of the buying experience in general. If a product or service is highly commoditized, then it is the buying experience that may be the only thing that differentiates one offer from another. Such is often the case in banking. In the case of, say, a hotel room, the customer service is the form and function and the primary reason for the transaction. One hotel room is very much like another, and it is the service wrapped around the room that makes the difference.

Like all frequent business travelers, I stay in lots of hotels. Some I visit regularly, perhaps a few times a month. I belong to every loyalty program, but none of the hotels seems able to remember what my preferences are. Every time I check in, I still have to fill in my details on the form and say I like a nonsmoking room with a down comforter and a firm pillow. I am sure any corporate road warriors reading this have similar stories.

Four Seasons, in the words of founder Isadore Sharp, set out to "redefine luxury as service." For instance, they implemented what they call "curbside check-in" for frequent guests. When you arrive, you are literally handed your room keys as you get out of the car. Oh, and how many visits makes a "frequent" guest? As few as *five*.

The "no luggage required" policy is a feature of the hotel to help guests who lose luggage in transit. Lose or forget a tie? No worries—tell them what your suit looks like, they'll fix you up. Wait, what was that? You lost your *whole suit*? No worries— we'll fit you for a new one you can borrow.

Consider the following stories that appeared in a *U.S. News & World Report* article about Four Seasons hotels:

- A concierge donned fins and a snorkel to find a wedding band lost in a lagoon.
- A hotel telephone operator spent forty-five minutes on the phone directing a lost guest all the way to the hotel's entrance.
- A man asked room service for a martini shaker, only to find a tuxedoed server standing at his door—accessories in hand—ready to do the shaking.

The results speak for themselves. A tiny 2 percent of guests in 2006 reported problems or registered complaints about service (down from the lofty heights of 4 percent in 2005 . . . oh, the shame!) at the hotel. RevPAR (revenue per available room) was up 11.8 percent in 2006, on the back of a 10.6 percent rise in room rates, at the same time as a seventy-basis-point rise in occupancy. Gross revenues were up 10.5 percent and operating profits 18.2 percent. Four Seasons is committing to 20 percent earnings growth for each of the next five years. Only an unstinting commitment to the customer experience makes this possible.

DESIGNING THE TOTAL OWNERSHIP EXPERIENCE

The Apple iPod is an excellent example of how all four fundamentals of the total ownership experience can come together to create profitable global hits.

When the iPod first came along it didn't sound better than existing digital music players and it cost more. Worse, it had lousy batteries that often failed long before they were supposed to.

But thanks to Steve Jobs, whose brainchild it was, the iPod had a great marketing slogan, "A thousand songs in your pocket," that instantly told the story of what it would do for customers in their daily lives. It had a distinctive look, a pure white rectangle with white bud earphones where other digital music players were all shiny metal and awkward, complicated shapes. And it set a new standard in ease of use and ergonomics on its own and in tandem with Apple's iTunes software. When Apple later launched the iTunes Music Store, customers finally had a compelling, comprehensive answer to their digital-music-player needs and wants. The battery problems, serious though they were and are, have never held the iPod back because of the way the product and iTunes work in customers' lives.

As important to the rise and rise of the iPod is how people think they look when they're carrying their iPods and listening to them. They feel cool and believe they look cool. They are part of a hip community formed around and experienced through a commodity, just like the middle-aged businessmen who buy Harley-Davidson motorcycles and fantasize about riding them with Marlon Brando in the movie *The Wild One* or with Peter Fonda, Dennis Hopper, and Jack Nicholson in *Easy Rider.*

Part of the story of the iPod community even came to include the fact that the white bud earphones were a signal to muggers and pickpockets: this person has an iPod. In places like New York City, it became a weird badge of honor that the pickpockets on the subway or the bus would focus on you because of your iPod earphones. It meant you were cool enough to have something worth stealing, something hot that everyone wanted. The New York Police Department and the Metropolitan Transit Authority even put up posters specifically warning iPod owners to be wary of pickpockets and other thieves.

Popular retail chains can also engender loyal communities of customers. In the last chapter I spoke about the Uniqlo chain in Japan and elsewhere. In the United States the discount superstore Target has carved out a distinctive brand identity as the favorite discount store of design-conscious, hip consumers in all demographics. Company executives are proud of the fact that the millions who are in on the secret often refer to Target as "Tar-zhay," as if it were a fancy French boutique. This brand identity, summed up in the slogan "Expect More, Pay Less," allows Target to enjoy a price premium over rival retailers such as Wal-Mart and Kmart.

Target's formula for flipping the discount-store brand DNA around and making itself known for high-quality, well-designed products that it can charge a little extra for has two main prongs. One is a product mix that is on the cutting edge of customer trends, thanks to in-house trend spotting and product lines from style gurus such as the fashion designers Isaac Mizrahi and Todd Oldham and the architect Michael Graves. The other is to have superb customer service that would be more expected from a luxury department store than a discounter. Both support the hip design- and value-conscious story that Target customers experience in shopping there.

Target has not yet succeeded in establishing the same brand identity in Australia, but their recent deal with Stella McCartney, Paul McCartney's designer daughter, was a good step in the right direction. In March of 2007 Target launched an exclusive Stella McCartney clothing line at its Australian stores and triggered scenes of chaos. Racks in Target stores in Melbourne (where there were lines before opening—unheard of for Target in Australia) were cleared of the Stella garments in as little as forty-five seconds, with security having to be called in to keep the commotion under control. The scenes of women fighting with one another to "steal a bargain" was the stuff of reality-TV programming. One hundred Target stores across Australia opened at 8:30 A.M., only to be emptied of midsize Stella-range dresses by twenty past nine. It was absolute chaos. People were actually stripping clothes off mannequins as they cleared the stores of stock.

Target paid McCartney $1.27 million to design the one-of-a-kind range of clothes exclusively for their stores. This is part of their move to position themselves at the "high end" of the discount market. Target U.S. last year teamed up with style guru Isaac Mizrahi for the same reason, and H&M has Madonna designing clothes for them in a similar push.

However, a few weeks after these scenes of mass hysteria, a few articles surfaced that mentioned that the glow had worn off rather quickly. After the initial run on the products, they actually sold really slowly after the hype wore off, with Target only marginally (and sometimes not at all) meeting its sales targets (no pun intended). So while it was a great marketing coup initially, there wasn't really a sufficient follow-through.

Nevertheless, this is a sign of things to come as Target attempts to carry its "cheap chic" story to other countries. In the United States the company is going strong, and in 2006 total revenues rose to almost $60 billion.

Cirque du Soleil is another great example of bringing the four elements together. From its origins in the work of two Montreal street performers, Guy Laliberté and Daniel Gauthier, Cirque du Soleil has become a worldwide phenomenon with multiple touring companies and permanent residencies at Walt Disney World in Orlando, Florida, and no fewer than four Las Vegas casino hotels.

The secret of Cirque du Soleil is nothing other than "Absolutely, Positively Sweat the Small Stuff." Cirque du Soleil takes the traditional elements of circus performance and stitches them together into story concepts such as *Love,* an interpretive stage production done to a musical score of Beatles songs at a specially built theater in the Mirage Hotel and Casino in Las Vegas; *Kà,* a martial-arts fantasy performed at the MGM Grand in Las Vegas; or *Mystère* at the Treasure Island Resort in Las Vegas, which has as its theme the origins of life in the universe. Cirque du Soleil's other resident companies and touring productions have similar narrative themes that make the familiar circus formula of acrobats, animals, and clowns new again.

EVERY DECISION YOU MAKE MUST BUILD THE STORY

In the last chapter I mentioned how Toyota has responded to the fact that all the major car manufacturers in North America,

Europe, Japan, and Korea now build good-quality cars. Toyota not only leverages its manufacturing and service expertise to maintain a lead in being fast, good, and cheap in customers' eyes, it also adds an X factor with innovations like its pioneering hybrid-engine technology, and thus appeals to customers as "Fast, Good, Cheap+Green."

Toyota has developed another X factor in its Scion range of vehicles for Generation-Y consumers, offering the attributes of "Fast, Good, Cheap+Hip." The Scion is a perfect fit for the customized car culture that has sprung up within the hip-hop generation of consumers worldwide, as evidenced by television shows

DESIGNING THE TOTAL OWNERSHIP EXPERIENCE CAN HAVE AS MUCH TO DO WITH PROCESS AS PRODUCT.

such as MTV's *Pimp My Ride.* Toyota acted on this trend much sooner than the competition, just as it did with hybrid engines.

Launched at the end of 2003, the Scion has been a smash hit. Dealers sell Scions as fast as they get them, and they reap significant marginal profit from customers' desire to customize their cars with side-panel graphics, roof racks for snowboards and mountain bikes, and so on. (More on this in chapter 7, "To Get Control, Give It Up.") These things have absolutely nothing to do with the vehicles' performance on the road and absolutely everything to do with their performance in customers' lives.

At launch, Toyota set a 2006 target for U.S. sales of 150,000 cars. In the early autumn of 2006, Toyota saw that it was on track to sell 175,000 Scions in the United States by the end of the year. Industry analysts said that Toyota could quickly ramp up U.S. sales to 250,000 Scions a year. But instead, Toyota did another flip. It looked at the prospect of those easy sales and huge profits and decided that ramping up sales volume would be foolish in the long run, because it would undermine the Scion's brand identity.

Toyota announced that it would limit Scion production and distribution to ensure that it sold no more than 150,000 vehicles in the United States in 2007. Further, it restricted Scion television advertising, which was never very extensive anyway, to a few late-night television shows that are popular with Generation-Y consumers. It also suggested that it might eliminate Scion television advertising entirely, and concentrate instead on event marketing and branded entertainment. The Scion play already includes a music label for emerging artists and the Scion Release clothing line. Instead of using commonly known sports stars or celebrities to sell their cars, they use DJs.

The way Toyota USA vice president Mark Templin puts it is, "Because we no longer have to focus on brand awareness, we can be even more edgy and more risky." The way I put it is, to sustain Scion's brand awareness as a vehicle that helps customers make their lives into a hip and exciting story, they have to become even more edgy and more risky. They have to do an even better job on "Absolutely, Postively Sweat the Small Stuff," and as the moves I've sketched out indicate, they're well on the way to doing so.[5]

As this and previous examples show, designing the total ownership experience can have as much to do with process as product. In the case of Scion marketing, the medium is the message. DJs instead of celebrities. Edgy shows instead of mainstream. The differentiator in any particular instance depends on the story a product or service tells about customers' lives, their aspirations, and their sense of community. The story could emphasize simplicity, ease of use, beautiful design, community, or some combination of these things.

So the big question is:

WHAT'S *YOUR* STORY?

Bang & Olufsen, Toyota, and other flipstars understand not just what their customers' desired story is, but what their individual brands stand for. They find the common ground between their own reality and the wants of their customer, and design everything they do to build and tell a story that reflects that common ground. As you consider the competitive position of your company or your hopes for your own career, what compelling story can you offer to your customers, your staff, or your employers?

Finally, as you take the flip that you must "Absolutely, Positively Sweat the Small Stuff" to heart, don't lose sight of the previous flip. "Fast, Good, Cheap: Pick Three—Then Add Something Extra" is still the price of entry, and the standard of what constitutes fast, good, and cheap is continually on the rise. You must have a solid product or service that matches your competitors' offerings, but to make your own offerings stand out from the pack, you must also make magic from the small stuff.

The story changes slightly from customer to customer, and it also changes for each individual customer over time. The point is that if you really want to distinguish your products or services from the competition, you have to give thought to the experiences that they engender or support. The products and services we love the most become part of the story of our lives.

FIVE THINGS TO DO NOW

1. Make a short list of your potential X factors.

2. Think about the sort of activities you could engage in for the social good. How could you leverage these to impress customers and attract more staff, *in an authentic way*?

3. Pick one of your products and develop a story about why someone should buy that product, including detail, characterization, and language. Role-play it, pretending you are a customer telling his or her best friend why they spent a premium to buy your product instead of your competitor's.

4. Do an "easy audit." That is, review your "buying" and "ongoing service" experience and decide if they are designed to make life easier for you or for your customer. If they are designed to help you, change them so that they help the customer.

5. Create two compelling lists. One for how your brand appeals to aspirational inside, and another for how your brand appeals to aspirational outside.

BUSINESS IS PERSONAL

Ultimately, everything is personal.
—Jonathan Schwartz, President/COO, Sun Microsystems

FROM THE INFORMATION AGE TO
THE RELATIONSHIP ERA

In *The Godfather: Part III,* the murderous Don Licio Lucchesi (played by Enzo Robutti) tries to smooth over a difficult moment by telling Michael Corleone (Al Pacino), "It's not personal, it's only business."

A client recently said the same thing to me in justifying why his bank was offshoring its call-center operations to Bangalore, India.

"What do you mean, it is not personal?" I said. "Forgetting that you are my client for a second, I am a customer of this bank, and I have to tell you it is very personal to me. I have choices, you know. So do the rest of your customers. Other banks ask for our business every day with credit-card offers in the mail and in their advertising. The interest you pay or charge us is no different from what it is at other banks, and you tell me that banking is not a personal relationship. It is now, it was yesterday, and it will be tomorrow. Besides, all business is personal. It is about trust, and when customers call the bank they want to feel comfortable with the people who answer and they want to feel respected by them. Outsourcing the customer service part of your business is insane in my view."

WE ARE FAST EXITING THE KNOWLEDGE AGE AND ENTERING THE RELATIONSHIP ERA.

My protest was in vain. The decision had been made. I bet the decision will be reversed at some point in the future, but by then there is no telling how much damage the bank will have done to its customer relationships and its bottom line.

I'll be coming back to my bank's decision a little later in this chapter, but I want to register the general point that more than anything else in life, relationships rule. Most pundits in and out of the business world consider this the information age, or its recent fashionable variant, the knowledge age. We hear a lot about knowledge workers being the most important part of the economy, the ones who will reap the greatest benefits from the dizzying pace of technological, social, and cultural change.

Sure—but lots of people in developed countries are knowledge workers in some way, shape, or form. Being a knowledge worker is not going to offer competitive advantage in and of itself. It certainly won't in a truly global economy. The pundits are living in the past. We are fast exiting the knowledge age and entering the relationship era. The flip in this case is that "Business Is Personal."

I am not talking about the traditional cynical adage "It's not what you know, it's who you know that counts." It is certainly a part of human nature that people look out for their family and friends and scratch one another's backs to get ahead. But the ideas that what you know and who you know determine your success are both flawed, because they are static rather than dynamic.

In the years ahead, two things will count the most. The first is your ability to unlearn the things that are losing relevance, to flip yourself free of old scripts, and to learn the things that are gaining relevance. The second is whether people come to know and trust you as they struggle to bring their own learning forward. That is, do you really care about and respect them? Sounds soft I know, but especially in Western economies, it will be the hardest of business imperatives.

Ben Stein wrote an article in the *New York Times* that focused on a fear that capitalism's success over the last fifty years may not be carried on into the future.[1] Why? Because of a lack of trust. Capitalism is built on trust. Trust that the goods you send to your neighbor will be paid for. Or that the check you send to your supplier will be honored by the bank where you have previously deposited your hard-earned cash, and that your supplier will in turn send you the goods. Goods you bought, by the way, trusting that they will live up to the quality and performance that your supplier said they would. Capitalism is built on trust, and building trust with your clients, customers, and staff will future-proof your organization.

In fact, building trust is having a well-deserved renaissance in contemporary business thinking. A comprehensive 2004 study by leading market research company Yankelovich, Inc.—they coined the term *baby boomers* and have the longest continually running database on American consumer attitudes, lifestyles, and buying behaviors—found that "trust increases retention, boosts spending, enables premium pricing, and provides lasting competitive advantage."[2]

Globalization has made knowledge and expertise pure commodities in the same way that goods and services are. In a world as fast changing as ours, no body of expertise, no matter how valuable it may seem, lasts five years, much less an entire career. And no proprietary system is immune to competitors who can cheaply acquire the information necessary to copycat you on quality and beat you on price. Trust cannot be commoditized, and neither can the ability to engage others to believe in your vision, and to inspire them to behave in a certain way.

Any product or service your company offers, any knowledge and expertise you possess, however, can be outsourced and commoditized. And it is not just about people. It is about technology as well. Legal contracts that were once the domain of lawyers billing hundreds of dollars an hour can be downloaded from the Internet for twenty dollars a piece. Health care is one of India's fastest-growing industries, and middle-class Americans are going there for their heart operations.

Let me share a personal example of how trust can ensure that your product, service, and knowledge remain the favored choice in the eyes of the marketplace. I recently had laser eye surgery. Two friends had both had it done and raved about the results. One had paid about $6,000 for his surgery at a Sydney clinic. The other had his done in Kuala Lumpur at one-fifth of the price. Both were successful, each having 20/20 vision as a result.

I decided to do a little more research. As it turns out, both places use the same equipment to perform the operation. And as someone who's had it done, let me tell you this is significant. With the exception of administering some Valium (much needed, by the way), putting in some eyedrops, and placing a clamp under my eyelid (I am getting freaked out just thinking about it), the doctor basically did nothing. Oh, he did input the appropriate data into his computer based on the presurgery tests that I had done, which were also conducted almost entirely by machine.

Furthermore, both surgeons had trained in exactly the same place: Australia. I was headed to KL in a couple of weeks anyway, so it was a simple choice. Did I want to pay $1,200 or $6,000?

I paid the $6,000, and would do it again in a heartbeat. Why? I met the Sydney-area surgeon before deciding to have him do the operation. I even quizzed him on the Malaysia option.

"Pete, I care," he replied, "because my livelihood depends on it. If something happens to you, in addition to my feeling awful about that, my business would suffer a blow that it would struggle to recover from. If something went wrong, the media would be all over it. Do you think the surgeon in Malaysia has that much on the line when he operates on you? Who would you trust with your sight? Me, who lives and works in the same neighborhood as you and your wife, or someone who lives many thousands of miles away?"

Sold! Why? Because like everyone else in the world, I want to do business with people I know, like, and trust, and whom I can rely on for help in the future.

Recall my friend the Plexiglas manufacturer who was struggling with Chinese imports. He eventually sidestepped those imports, and avoided having his business torpedoed by them, through capitalizing on the relationship of trust and confidence that he had built up with his customers. They stood by him as he transitioned his business to concentrate on doing the most profitable special orders and one-of-a-kind displays, rather than fight a losing battle to keep a sustainable share of bulk-order displays. Special orders and one-of-a-kind displays require a deep understanding of the client's business and customer base. You can't outsource that! This does not mean that "Business Is Personal" negates the need to be fast, good, and cheap (at the cost). As a partner in a law firm said to me recently, his long-term clients were increasingly demanding that he meet the price of his competitors.

At the end of the day, it is people—customers and staff—who keep you in business. What applies to your relationship with customers applies equally to your relationship with staff. Both groups are looking for experiences that give their lives meaning and enhance their sense of purpose in the world. Both want to be treated with respect, and be stimulated to grow and learn. In short, people not only want to do business with, they also want to work for and with, people they know, like, and trust.

STAFF AND CUSTOMERS: CO-NUMBER ONES

If there is a disconnect between how you treat staff and how you treat customers, it is going to hurt your business. I don't mean that you can't expect high performance from staff. In fact, treating staff with care and respect supports higher and higher staff performance, especially with regard to meeting customer needs and wants, a subject I'll return to at the end of the chapter.

For now, let me give you an example of how this applies in the legal fraternity, which is definitely dealing with the four forces of change. The Australian market for legal services is a mature one, with most national firms looking to Southeast Asia for new market opportunities. Building market share domestically means stealing that share from the competition. This is as opposed to the organic growth that increased the size of the market for all until recent times. A rising tide lifts all boats. But when that tide turns, it's easy to get stuck in the mud.

Australia's big national law firms are now in a nasty fight for market share and profitability. They are all throwing tens of thousands of dollars at the graduate market to woo the best young lawyers, and then spending hundreds of thousands of dollars to develop them. The net result after three years, however, has been a disappointing 50 percent employee retention rate for many of these firms.

In this regard, Australian law firms are experiencing a common global dynamic. In every category of goods and services, there is global oversupply and global underdemand. At the same time the opposite is true of skilled labor. The world-class talent required to out-innovate and out-market the competition is scarce, and the most talented and best-credentialed individuals are accordingly making demands of organizations that are rocking the foundations of the way business gets done. These circumstances put firms into a squeeze between downward pressure on fees and prices and upward pressure on wages and perks for highly skilled staff.

Recently I conducted a consulting project on recruitment and retention for one of the largest law firms in Australia. This firm has enjoyed an extremely successful commercial practice in Australia, New Zealand, and Southeast Asia over the last several years. But upon receiving a prestigious award for market-leading client service, the firm's former chief executive partner said that "the client comes first," that staff "did not have a right to a personal life," and that they must all sacrifice whatever is important to them in order to meet the expectations of clients. The mainstream business media caught on to the story and wouldn't let it go.

The former chief executive partner was not trying to attack his staff. He was trying to say that the law firm's clients were number one. But his comments were not reported that way, nor were they viewed in that light by those the firm needs to recruit, the top Generation-Y graduates who, like all of their peers, are renowned for placing the balance between work and personal life at a premium in deciding which job to accept.

The firm had to start sucking up, and hard. They had to reposition themselves in the talent market, or watch their competitors get the cream of the crop of each year's law graduates. For a firm so used to getting the best of the best, nothing short of a mind-set flip would be required to keep them at the cutting edge of their market.

I advised the firm to go soft. That is, to give their potential new recruits the same world-class service they gave their clients. Or put another way, to make the attraction and recruitment experience more "personal."

Making staff co–number one with customers is a flip that many organizations resist. But if you don't have talented, creative, dedicated staff on board with your thinking and mission, you can't reach clients and customers effectively. A competitor who gives staff higher priority will get there before you.

In conjunction with the partner responsible for staff issues, I presented these ideas to a group of key partners in the firm. As you can imagine, there was some protest from the hardened senior lawyers in the room, most of them ambitious baby boomers and Gen X-ers. All of them had played by the old rules and sacrificed everything outside the job to get where they were. They expected Generation Y to do the same. "Young lawyers should be grateful to work in such a prestigious firm," was the cry. But to their credit they were able to come to terms with the new dynamic in the labor market and *flip!*

Through a series of educational sessions for the interviewing partners and changes to the way recruitment was handled, they made the process much friendlier and more personal. Although I am not at liberty to share exactly what we did, you can use your own team to come up with a strategy to personalize your recruitment process in a way that is aligned with your company's employer brand.

DO NOT TREAT STAFF IN A WAY THAT IS AT ODDS WITH THE WAY YOU TREAT CUSTOMERS.

The result for this firm was a 50 percent increase in job-offer acceptance. This is an example of a company recognizing that business is personal for both customers and staff, and successfully managing a PR disaster at the same time as attracting more of the best and brightest job candidates. It also illustrates that in today's business world you must flip and "Think AND, Not OR." Conventional business thinking holds that it is an "either/or" world in which customers or staff can be number one, but not both. In fact, as my clients discovered, making both customers and staff number one creates a win-win situation. There's no better evidence of that than the fact that the firm won the same client-service award the following year as well.

Making sure that you do not treat staff in a way that is at odds with the way you treat customers is essential, even if you are a sole proprietor and you are the only staff your business has. The way you treat yourself will inevitably be reflected in the way you treat customers, for good or ill.

With that in mind, let's take a close look at the decisions on outsourcing and offshoring by my bank and numerous other companies throughout the world.

GETTING HIT IN THE HEAD WITH A *BRIC*, OR, GETTING OUTSOURCING RIGHT

Over the next forty years, according to projections by Goldman Sachs, the BRIC economies (Brazil, Russia, India, and China) will likely become larger as a group than those of the so-called G6 (the United States, the UK, Japan, Germany, France, and Italy). Most of that growth will come in the next thirty years, when China will probably surpass the United States as the world's largest economy. Of the current G6, only the U.S. and Japanese economies will remain among the six biggest.

Many people find this prospect frightening. A study by one of Europe's leading business schools, IMD (the International Institute for Management Development in Lausanne, Switzerland), found that of 1,962 U.S. businesses surveyed, 49 percent believe they have lost their competitive edge to China, and a further 47 percent believe they have lost their edge to India as well. These fears are fueled by the undeniable growth of outsourcing from the developed to the developing world. China and India are now consolidating positions as the world's manufacturing and service-provider destinations, and former Soviet-bloc countries such as the Czech Republic are increasingly doing the back-office work for big companies in France and Germany. For example, General Electric has 48 percent of its software development done in India, where the engineers earn on average $12,000 a year compared to $72,000 in the United States. It is estimated that salaries rose 17 percent in India in such professions in 2006, which could quickly erode this price advantage.

Much of the anxiety in developed economies about outsourcing to developing economies betrays a poor understanding of the value chain. It is one thing to make products or do the paperwork for the world cheaply, and it is another entirely to innovate, design, and sell goods and services in the world's advanced consumer markets. Own the manufacturing of a product or the back-office part of a business, and you own the links in the supply chain that are farthest away from the consumer and most easily commoditized. They are the least valuable. Own the links that are closest to the customer, however, and you own the links that represent the highest value and profitability. Thus, as Goldman Sachs also projects, per capita incomes in the United States and other currently developed economies will remain higher than those in Brazil, Russia, India, and China, even after these economies have grown bigger than the G6.

Doomsayers on the threat of the BRIC economies to the developed world often cite such figures as the numbers of engineering graduates per year in different countries. In 2006, according to multiple media sources, China graduated 650,000 engineers and India graduated 500,000. By contrast, the United States graduated 70,000, Australia, 4,600, and New Zealand 1,500. Even given the vastly larger populations of China and India, this seems like a huge disparity.

But these figures, though often quoted in respected media such as *Fortune* magazine, are in fact misleading. In December 2006 researchers at Duke University reported that in the case of India only 112,000 engineers graduated with a bachelor's degree or higher, and that the higher figure publicized by India's National Association of Software and Service Companies (NASSCOM) was not only wrong but included those with less than degree-level qualifications. In the case of China the correct figure for graduating engineers was more like 350,000 than 650,000. Using the definitions that generate these numbers for India and China, the United States actually graduates as many as 137,000 engineers per year. This places it at 470 engineers per million people, compared to 270 per million in China and just 100 per million in India.

The Chinese and Indian numbers are still impressive, of course, and are growing every day. Consider that according to the UN, 8.5 million regional Chinese citizens move to urban areas every year. So, should freshly minted engineers in the United States, Australia, and New Zealand be freaking out? That depends. If you're a new engineer who is betting on applying the rote engineering knowledge you learned in engineering school for the indefinite future, yes, you should be worried. If you're betting instead on a lifetime of learning and unlearning, and of leveraging relationships with valued customers and clients, you should be confident of your ability to make your way.

You might even find it very profitable to go to work in India for a time. India graduates 112,000 engineers a year, but the average quality of these engineers is so poor that Indian business and government officials fear it could be a limiting factor on India's economic growth. To find engineers who can think creatively, Indian firms are beginning to recruit aggressively in the United States and other developed countries.

It is the same if you are an accountant. The big-four accounting firms are beginning to adapt their existing "get them in young, work them hard, pay them as little as possible, and charge them out at as much as possible" to the developing world. In 2006 alone, more than 360,000 U.S. tax returns were completed in India at a value of over $40 million. That figure is projected to rise to anywhere between 1.6 million and 22 million returns by 2011. Instead of getting some young graduates in developed countries to crunch the numbers and do the basic work, they found they could get it done cheaper and to a better standard in India.

Again, should you be worried if you are an accountant? Yes, if you believe your value is the left-brain number-crunching skills that are rapidly being replaced not just by some graduate in India for one-fifth of the wage but by software that those same graduates are developing. No, if you realize your value is making your clients feel secure by minimizing their exposure to risk, and remembering the little things about their businesses. And definitely no, if you realize the value of aligning yourself strategically with those clients, building deep relationships that enable you to help them grow and develop their businesses. Skills are becoming commoditized. Relationships are not!

The trend to outsourcing the production of goods and services will continue, as well it should. There are huge cost savings to be had, and companies that do not achieve these savings will be at a disadvantage to companies that do. Successful projects orchestrated by firms such as Accenture are evidence of how powerful such outsourcing activities can be for businesses.

The biggest, most profitable companies in the world are among the most aggressive outsourcers. In the early 1990s, Jack Welch, then CEO of GE, mandated a 70/70/70 rule: 70 percent of business processes to be outsourced, with 70 percent going offshore and 70 percent of that (around 30 percent of the total) going to India. That policy has been continued by current GE CEO Jeffrey Immelt, and the company has around thirteen thousand employees in Delhi alone. It's no wonder that according to *CIO* magazine, 73 percent of Fortune 500 companies see outsourcing and offshoring as an important part of their strat-

OUTSOURCE KNOWLEDGE AND EXPERTISE BY ALL MEANS. BUT OWN, NURTURE, AND LEVERAGE THE FINAL LINK IN THE VALUE CHAIN, THE RELATIONSHIP WITH YOUR CUSTOMERS.

egy, and that Gartner estimates the global offshoring market in 2007 as worth around $50 billion.

But if you can achieve short-term savings from outsourcing almost anything, you can also suffer serious long-term costs from outsourcing the wrong things. The right things to outsource, depending on your business, could be manufacturing, process engineering, or back-office services. The wrong things to outsource, no matter what your business, are those that touch the customer. Outsource knowledge and expertise by all means. But it is vital to own, nurture, and leverage the final link in the value chain, the relationship with your customers.

Let's go back for a minute to my bank's decision to outsource their customer-service call centers to Bangalore, India. My client at the bank told me, "Don't worry, Pete, your call is unlikely to be answered by someone in India."

"Why not?"

"Well, if you are a low-net-worth customer, then in all honesty you are not that valuable to the bank and your calls will be answered in Bangalore. If you are a medium-net-worth customer, we will have your calls answered locally but most likely by an outsourced provider. But if you are a high-net-worth customer, your call will not only be answered by a bank employee here as a priority matter, but you will also be assigned a personal representative and given their cell-phone number."

Every business needs to target its most profitable customers (although as we will discuss in chapter 7, "To Get Control, Give It Up," the old 80/20 rule may not apply as readily today), but the risk is that some high-net-worth customers were once low-net-worth customers. If my business had not taken off before my bank moved its low-net-worth customer service overseas, I might have moved to another bank before I became a valued high-net-worth customer. Or a low-net-worth customer may actually have plenty of assets elsewhere and be giving the bank a trial run to see how they're treated.

It reminds me of the sequence in the movie *Pretty Woman* where the Julia Roberts character goes shopping with credit cards given to her by Richard Gere's character. Julia's character goes into a fancy Rodeo Drive store wearing her own cheap (maybe almost nonexistent would be more accurate) clothes and is treated so disdainfully by a sales assistant that she retreats in embarrassment. After getting her courage up and venturing into another store, where she is treated better, Julia returns to the first store, wearing beautiful new clothes and carrying bags with thousands of dollars of other purchases in them. Julia asks the sales assistant if she is paid on commission. When the sales assistant says yes, Julia holds up her overflowing shopping bags and says, "Big mistake. Huge."

That moment drew cheers from audiences when the movie was first released. And it carries a powerful lesson: All customers have potential, and you'd better be very sure what you're doing before you deliberately treat some customers worse than others.

Just ask Dell Computer, which has spent millions of dollars bringing call-center operations for North American customers back to North America. Yes, it is more expensive to operate your customer call center in the United States than it is in Bangalore. But not as expensive as losing thousands of clients who, rightly or wrongly, do not want to deal with a staff member twenty-five thousand kilometers away.

JPMorgan Chase, CapitalOne, and IBM have done the same thing. The reason is not necessarily that the outsourced customer service was bad, although in Dell's case it certainly seems to have been, judging from the stories on Web sites such as crmlowdown.com. In cases where the outsourced-to-India customer service was actually quite good, North American customers could still feel ill served and condescended to by call-center staffers speaking English with British-inflected Indian accents. The fact is that North American customers of a company with North American operations want to speak to someone who sounds like them, like a neighbor.

The same thing is true in the UK and the rest of Europe. Citing consumer preference, the Irish arm of the Swedish telecom firm Tele2AG recently switched its call-center operations out of India and back to Ireland. British bank Lloyds TSB closed all of its call centers in India in response to a petition signed by four hundred thousand customers.[3] One in three respondents in a British survey said they would stop doing business with a bank that relocated its call centers offshore. Another study reported that only 5 percent of British customers are satisfied with offshore call centers.

If my bank genuinely had customers at the center of all they did, they would have considered outsourcing something besides their call centers. They would have outsourced the payroll division, the risk assessment division, or another back-office function that does not face and touch customers. There is no question that these, or potentially some other functions, can be adequately handled in parts of the world where labor is cheaper.

The bank could then use these cost savings to create better working conditions, increased remuneration, and improved training for the staff on the front line. This would improve the bank's ability to attract higher-quality customer-service staff to their call centers (one of the major motivations for offshoring customer service in the first place), which in turn would deliver a better experience to customers. All of which would give the bank a genuine competitive boost against the other Australian banks. After all, there is no competitive advantage to speak of in financial products themselves, which are largely the same in every bank.

That's not mere assertion. Australia's ANZ Bank has already proven it true. ANZ Bank has made customer service a prime focus and has announced a strict policy of not outsourcing or offshoring this function. ANZ does outsource, but nothing that is customer facing. They have about a thousand employees in Bangalore, but all work in back-office and ICT roles. Customer-facing roles stay in-country.

The bank has also mounted an advertising campaign focused on service rather than product. I do not think it is coincidence that from 2002 to 2006, ANZ saw a 77 percent rise in share price, beating their nearest rival by 15 percent and the large-bank average by 25 percent.

OUTSOURCE EXPERTISE, NOT RELATIONSHIPS. OR TO PUT IT DIFFERENTLY, OUTSOURCE COMPETITIVE NECESSITY, NOT COMPETITIVE ADVANTAGE.

Again, outsource expertise, not relationships. Or to put it differently, outsource competitive necessity, not competitive advantage. It is essential for large organizations to have efficient payroll operations. It is essential to have people who can crunch the numbers. These things are the price of entry, and from a customer point of view offer no real advantage over the competition. Handle them locally, outsource them, or offshore them to the moon, because they have little impact on the customer experience. But don't outsource and send beyond your daily control the part of your business that connects and interacts with the customer. The cost benefits in the short term may be good, but the impact on your brand will be far more expensive in the medium to long term.

A 2005 study by Gartner found that 60 percent of companies that outsource customer-facing operations risk losing customers. Perhaps that is why 80 percent of companies that outsource customer service fail to reach their cost-saving targets. The lost customers cost far more than is saved in lower operational expenditures.

In light of all these factors, let me share with you the outsourcing model for—wait for it . . . an outsourcing company. Syntel, which specializes in advising companies on business-process outsourcing, moved 6,000 of 6,500 U.S. jobs to Pune, India. But it kept the sales staff—the employees with the deepest and closest customer relationships—in the United States. I think they might be onto something here.

Making the most of your relationships with customers is hard work. Fortunately there are some new technologies that can help.

HIGH-TECH CAN BE HIGHLY PERSONAL: MINING THE POTENTIAL OF NEW COMMUNICATIONS TECHNOLOGIES

Recently I was presenting a master class for clients of CB Richard Ellis, the commercial property giant. During a break, a very senior manager of a listed property trust pulled me aside and said, "Pete, I have a problem with the whole Gen-Y thing."

"What is it?" I asked. "I thought we were making real progress with your managers."

"No. This time it is not a work thing. It is my daughter."

I was not sure I wanted to continue the conversation. But I figured I had better try. "What's the problem?"

"Last night she said she was ready for a boyfriend."

"How old is she?" I inquired.

"Fifteen."

"Were you supportive, or did you remind her that you had a gun collection?" I got a bit of a laugh.

"Of course I was supportive," he said. "I asked if she had met someone. She said no. I asked if there was someone she liked. Again she said no. Finally, I asked if there was someone she thought she would like if she got a chance to meet them. Again she said no. I was like, what's going on?

"Then she told me, 'Dad! I don't need to meet them. I am just going to do a search.' Some site called Facebook or something like that."

Now I understood the problem. This baby-boomer dad did not know about social-networking Web sites, and that Facebook is one of the most popular of them, along with MySpace. Even after I described them, he still had difficulty understanding that his daughter could take seriously a relationship she might form on some "stupid Web site," as he called it.

This same executive used Internet technology every day to connect and stay in touch with colleagues, suppliers, clients, friends, and family via e-mail, and he regularly pulled information off Web sites. In fact the very same sorts of technology had completely changed the way his industry, real estate, advertises, too, and stays in touch with its clients. Despite this, he could not fathom how his daughter could use this technology to start and develop a relationship.

To him that did not seem like a real relationship. But it did to his daughter, and he needed to recognize that not only can such relationships feel real, they can have real world consequences. In the summer of 2006 a sixteen-year-old American girl tricked her parents into getting her a passport, and then flew to the Middle East to be with a twenty-year-old guy she met on MySpace. The authorities at the airport where her plane landed convinced her to return home to her parents. In December 2006 MySpace announced that it was taking measures to protect MySpace's youngest users from sexual predators who troll its pages looking for victims whose trust and confidence they can gain online before luring them into a real-world meeting.

FOR GENERATION Y, SOCIAL-NETWORKING SITES ARE NOT A SUBSTITUTE FOR GENUINE PERSONAL CONNECTIONS. THEY ARE INSTEAD AMONG THE BEST CURRENT MEANS OF INITIATING AND SUSTAINING GENUINE PERSONAL CONNECTIONS, EQUALLY VALID AND MEANINGFUL AS AN ENCOUNTER AT A SCHOOL DISCO OR OTHER "REAL WORLD" SOCIAL EVENT.

Facebook and MySpace are part of what has been dubbed Web 2.0. In terms of the evolution of the Internet, Web 2.0 signifies the mass adoption of online collaborations and user-generated content sharing, things that have been part of the Internet from its inception but that have now moved from being specialist activities to being commonplace ones, at least among Generation Y. For Generation Y, social-networking sites are not a substitute for genuine personal connections. They are instead among the best current means of initiating and sustaining genuine personal connections, equally valid and meaningful as an encounter at a school disco or other "real world" social event.

Also important in the Web 2.0 space are massively multiplayer online games such as EverQuest, World of Warcraft, and Second Life. Owned and produced by Sony, EverQuest is a sword-and-sorcery fantasy world in which groups of gamers as large as fifty to seventy individuals join to complete various quests in nearly four hundred three-dimensional zones, including all sorts of geographical terrain and cityscapes. The game is so addictive that it has been called EverCrack, as in crack cocaine, and an online support group, EverQuest Widows, exists for people whose real-world relationships have been affected because of their partners' obsession with it. Lately, World of Warcraft, a similar game with some eight million subscribers worldwide, has overtaken EverQuest in popularity, although the latter remains a hugely important product for Sony.

Although its current user population is so far only a fraction of that of EverQuest or World of Warcraft, Second Life is an even more interesting phenomenon long term. Owned and operated by Linden Lab, Second Life offers a virtual second world in which all sorts of fantasy activity can take place, but which is designed to look much more like the real world than the EverQuest-type realm. Users of Second Life create an avatar through which they can lead their second lives online, including buying and selling property, building houses, and carrying on other economic activity with Linden Dollars, which can actually be exchanged for U.S. dollars in the real world. In early 2007 $655,000 was traded on Second Life every month, and many people earn their real-world livings by designing virtual products and selling them to users who want to spruce up their avatars or their avatars' homes and businesses. The top-ten Second Life entrepreneurs have an annual average income of around $200,000. That's real money.

Multiplayer online games and social-networking sites have become arenas that stimulate creativity and offer opportunities for connection between people and organizations, including businesses and governments, in a way that is becoming an almost seamless extension of the real world. And they are doing so with no regard for geographical boundaries.

The Chicago- and later Los Angeles–based band OK Go were relatively unknown before launching a hilarious video clip on YouTube, a Web site where people can upload and download video for free. A single continuous shot showing the four band members dancing on eight treadmills, the "Here It Goes Again" video attracted more than 15 million viewings on YouTube and carried the band to stardom. The video was named the most creative video of 2006 on YouTube and won the 2007 Grammy Award for best short-form video.

MULTIPLAYER ONLINE GAMES AND SOCIAL-NETWORKING SITES HAVE BECOME ARENAS THAT STIMULATE CREATIVITY AND OFFER OPPORTUNITIES FOR CONNECTION BETWEEN PEOPLE AND ORGANIZATIONS, INCLUDING BUSINESSES AND GOVERNMENTS, IN A WAY THAT IS BECOMING AN ALMOST SEAMLESS EXTENSION OF THE REAL WORLD. AND THEY ARE DOING SO WITHOUT REGARD FOR GEOGRAPHICAL BOUNDARIES.

Musicians are particularly active users of MySpace, which allows them to upload up to four songs in MP3 format. Music companies in turn pay close attention to MySpace in deciding which new artists to sign. And both music and film companies use MySpace to spread the word about new releases. Given that MySpace has more than 100 million users and is currently the third-most-popular Web site in the United States and the sixth-most-popular Web site in the world, it is clear that social networking has ceased to be a fringe activity and has become part of the mainstream of popular culture.

In recognition of this fact, politicians are putting their avatars on Second Life. The first woman Speaker of the House, Nancy Pelosi, made her debut in Second Life in the same week that she first brought the gavel down to open a new session of Congress. And Second Life's news bureau filed regular reports from the 2007 meeting of the World Economic Forum in Davos, Switzerland, including interviews with the avatars of attendees such as the American political commentator and blogger Arianna Huffington.

THE KEY IS NOT THAT YOU SHOULD BE IN SECOND LIFE, BUT THAT THE INTERNET AND ITS RELATED TECHNOLOGIES ARE PROVIDING NEW AND POWERFUL WAYS FOR YOU TO CONNECT WITH YOUR CUSTOMERS. BUSINESS IS PERSONAL, AND THE NEXT GENERATION OF COMMUNICATION IS A BIG PART OF BUILDING RELATIONSHIPS THAT CUSTOMERS WILL VALUE.

Some of the most intriguing uses of Second Life and other online communities are in education. Peter Yellowlees, a psychologist at the University of California, has built a practice and hospital environment on Second Life where he takes his students to help them understand what it is like to be a schizophrenic. He has created a chilling experience filled with the types of hallucinations that schizophrenics suffer. People walk through a hospital and mirrors suddenly flash "shitface" at them, or they watch a televised address by a political figure who suddenly starts screaming, "Go kill yourself, you wretch!"

After Yellowlees opened up the virtual hospital to the public on Second Life, 73 percent of visitors said it improved their understanding of schizophrenia. Many other educational initiatives are popping up in Web 2.0, seeking to harness the immersive experience of virtual worlds in order to accelerate learning.

Harcourts, the New Zealand–founded real estate agency, with offices in Australia and Southeast Asia, has set up an office in Second Life. Initially it was to display real-world property, but in time they may find they are able to list and sell Second Life property and make a commission in Linden Dollars that they can convert into real-world dollars. I also have a client in the charity sector looking at how they might be able to have avatars donate Linden Dollars to their charity, in exchange for something that shows they have engaged in philanthropic activity, which again can be converted to real dollars. One of my major banking clients is looking at how they might offer money management, investment, and other informational seminars using Second Life as the venue. In June 2007 five companies—Alstom, Areva, Capgemini, L'Oréal, and Unilog—held a "neojobs meeting" on Second Life, telling job seekers, "In the fantastic Second Life universe, join us for your first 100% virtual interview and discover 100% real positions."[4] The list could go on.

The key is not that you should be in Second Life, but that the Internet and its related technologies are providing new and powerful ways for you to connect with your customers. Business is personal, and the next generation of communication is a big part of building relationships that customers will value.

In case you are still not convinced

In November 2006 Toyota launched a virtual Scion City on Second Life, where all of its Scion range can be bought and customized with Linden Dollars. Other companies that have a presence on Second Life include Adidas, Sears, and Sun Microsystems. Dutch banking giant ABN Amro has set up a financial consultancy on Second Life. The value of this activity for the companies is twofold. On the one hand it helps stimulate some business in the real world. But more important is that it puts the companies in touch with the leading edge of consumer trends.

Likewise the cleverly crafted MySpace profiles of Wendy's and Burger King each have over 100,000 registered friends, who post suggestions for new meals and funny stories about their experiences at the fast-food chains. They also sometimes post hateful vitriol, which the companies leave untouched. Toyota, Microsoft, Ernst & Young, Apple, and EA Games are among the large brands to get actively involved with similar personal/business branding exercises in Facebook, the largest student community site in the world. The CIA has even begun to recruit prospects on Facebook!

Companies, organizations, and governments are paying serious attention to developments in online communities because of hard demographic facts:

- According to the Pew Research Center, 87 percent of U.S. teens are now regular Internet users, and *more than 50 percent* are daily users.
- Over 81 percent play games online and almost half make purchases.
- With a combined purchasing power of around $139 billion a year (according to market research firm Harris Interactive) and even greater influence over others' purchasing decisions, they're not just playing Pac-Man anymore!

This kind of technology use is clearly not just a fad. It is also no longer a small enough niche to ignore. In fact, it's large and powerful enough to warrant serious attention and to be utilized as an extremely powerful new forum for building relationships and brands. But it doesn't have to be massively serious relationships, nor does the use of technology have to be cutting edge. What this section is talking about is using the lessons from high-tech-can-be-highly-personal, and having the courage to apply them to your world.

A car service center could use it (and some do) very effectively. When a customer's car is due for a service, they can send a gentle reminder and enable the customer to book any available time slot via text message, and after the service work is done, they can send another text message to let the customer know the car is ready. This is such a simple, unobtrusive, and useful application of very cheap technology. It simultaneously enhances the customer experience and streamlines the company's operations, making them more profitable.

Or look at other uses of technology to streamline a highly personalized life. The LG Internet refrigerator is designed so it can order your groceries through an online grocery gateway. It is also a music player and PDA. The DIVA (among other) air conditioners can be activated via an "I'm on my way home" text message.

First Direct, the online banking subsidiary of HSBC in the UK, sends a text message to its customers when they are approaching their credit limits to save them embarrassment when they try to make a purchase they can't afford, and also to spare them unwanted overdraft fees. In early 2007 the bank estimated that the text-message program had saved customers around 32 million UK pounds in overdraft fees.[5] The program has proven so popular that other banks in the UK, such as Nationwide, have begun to follow suit.

Or you could get slightly more sophisticated in your use of technology in building relationships in the way Amazon.com does. It remembers not only what you have purchased, but what other people who have bought things that you have purchased have purchased, and then makes recommendations that are often very useful of other things you may find interesting. There is no human involved, yet you feel valued and as though the site understands you. At the end of the day, so long as people feel it is relevant and personal, they are happy to listen to your message. This is what Seth Godin discussed in *Permission Marketing,* and also may help explain why a survey of five hundred cell-phone users by London PR firm Rainier PR found that 74 percent of users were happy to receive short-range, location-specific promotional information via technologies such as Bluetooth.

The list of companies that have employed technology in really cool ways to stay in close relationship is getting longer every day, but is still nowhere near as long as it should be. Wireless technology is now allowing personalization based on where you are and what you are doing like never before.

Companies also are now employing high-tech means to empower traditional relationship-building exercises. That is, rather than completely change their marketing to adopt high-tech measures, they are using some new technology in conjunction with their old campaigns. For instance, McDonald's has introduced Web access to some of its larger urban restaurants where some clientele (often travelers) want to eat and check their e-mail. Their restaurants and advertising message remain unchanged, but the technology is placed on top. Proctor & Gamble tied an extensive print and broadcast campaign for Dove soap to an interactive campaign in Times Square where people could vote in real time for the images they found most attractive.

An Australian example of using technology to connect is United International Pictures' partnering with a number of interactive marketing and technology companies to launch *Mission: Impossible III*. Nearly ten thousand people registered online to use their cell phones to participate in a huge, citywide

WIRELESS TECHNOLOGY IS NOW ALLOWING PERSONALIZATION BASED ON WHERE YOU ARE AND WHAT YOU ARE DOING LIKE NEVER BEFORE.

puzzle game with clues delivered by text message and "hypertag" technology embedded in bus shelters and signs. This was a far better way than thirty-second television ads to attract a large number of committed fans who would swell opening weekend ticket sales for *MI3* and create an all-important buzz about the movie.[6]

As these examples show, relatively simple interactivity can increase customer participation massively and hence the feeling of a reciprocated relationship with your brand. Not delivering the kind of customer service people expect today is all the more shortsighted when you consider how simple and cost-effective it can be to interact with customers through new technology. Reconnect with one of the governing flips of this book. "Think AND, Not OR," and recognize that high tech can indeed be highly personal.

And for Pete's sake (pardon the pun), start engaging in a dialogue with your marketplace, whether through technology-enabled means or not, instead of talking at them all the time. Monologues are soooooo yesterday!

Finally, if you think social networking is just for people with too much time on their hands, consider that IBM has named a vice president for social software. On January 24, 2007, the holder of that new position, Jeff Schick, announced that IBM's Lotus Connections would use text message, blogging, personal profiles, and so on to empower employees to establish virtual worlds for mutual brainstorming. This new software—exactly the same thing as teenagers, college stu-

DON'T MAKE TECHNOLOGY YOUR OBSESSION. MAKE CONNECTING WITH CUSTOMERS AND STAFF YOUR OBSESSION. IF TECHNOLOGY HELPS YOU DO THIS, AND IT CAN, THEN USE IT.

dents, and others use on MySpace, Facebook, and Second Life—will be used within IBM and sold to other companies. IBM's prototype for Lotus Connections contains 450,000 profiles. As Schick put it, IBM believes Lotus Connections will "unlock the latent expertise in an organization." It will do so by making that expertise relational.[7]

(I'll have more to say about business uses of social-neworking software in chapter 7, "To Get Control, Give It Up.")

In closing this section on high-tech-is-highly-personal, let me say emphatically that *it is not about the technology.* It is about connecting with the customers. Despite what you read about younger consumers, it is not really about the technology for them either. They, and increasingly the broader market, just use technology as their means to connect. *Don't* make technology your obsession—make connecting with your customers and staff your obsession. If technology helps you do this, then use it.

THE REBIRTH OF THE MIDDLEMAN

This chapter's tour of the new relationship technologies would not be complete without a brief look at one of the most interesting business developments in recent years, the rebirth of the middleman. Way back in the late 1990s—doesn't that seem like a long time ago already?—many people predicted that the Internet would bring the end of the middleman. No longer would consumers need any intermediaries between them and product or service providers, the argument went. The Internet would provide direct connections for everything from A to Z.

And so it did. But in multiplying the connections between customers and a whole world of suppliers, it also reestablished the value of intermediaries that could sift through all the offerings and sort them in a variety of ways. I say "that" advisedly, rather than "who," because the new middleman may just as well be a technology as a person or a firm.

The most successful sites in the world are in fact middlemen. With the exception perhaps of Amazon.com, the most visited sites are those that connect the supplier to the customer, such as eBay, or people to people in the case of MySpace, or people to content as in the case of Wikipedia and YouTube. Google is no different. As the old saying goes: "When everyone is panning for gold, sell pans."

It is not just about the Internet, though. It is about services—be they personal or technologically driven—that help us to sift through the clutter. The phenomenal growth of mortgage brokers around the world is a very good example of what I am talking about.

The home-loan market is becoming increasingly competitive with consumers spoiled for choice, at least ostensibly. However, the variety of sources of home loans, the complexity of comparing them, and also the fact that it has to happen on your time have led to a reintroduction of the importance of the middleman. Mortgage brokers are as a result becoming increasingly popular. In the United States, for instance, mortgage brokers handle more than 80 percent of home loans.

In fact, the more these places distance themselves from the people who are providing the loans, the more popular they become. In Australia, the biggest mortgage broker is Mortgage Choice, which, in the last financial year, boasted a record after-tax profit of $13 million and generated $9.3 billion in housing-loan approvals.

Key to the Mortgage Choice brand story is that they have no stake in which lender provides the mortgage to the home buyer. Their brokers receive the same commission no matter which mortgage you choose, so that effectively they work for you, not the banks. The customer appeal of this brand story is manifest in the steep positive growth of Mortgage Choice's revenues and bottom-line profit.

Companies like Mortgage Choice and eBay demonstrate that there is a business model in just owning the relationship, and not actually selling anything. In the example of Skype in chapter 3, "Fast, Good, Cheap: Pick Three—Then Add Something Extra," I talked about eBay buying Skype for $2.6 billion. Might I be so bold as to suggest that what made Skype so valuable was in fact their relationships, not their business model, which could easily be duplicated. Although Skype did not yet have profits, it did have a large and growing base of loyal users, and as I mentioned earlier, eBay was making significant progress in monetizing the value of that customer base within a year after buying the company.

> AS THE KNOWLEDGE AND INFORMATION AGE EVOLVES INTO THE RELATIONSHIP ERA, THE ULTIMATE MIDDLEMEN BECOME THOSE WHO CONNECT US TO VALUABLE INFORMATION.

Product review sites are another increasingly popular form of middleman. Sites such as zdnet.com and cnet.com in information technology and consumer electronics—there are similar sites in other product and service categories—have become trusted partners in customers' purchasing. In the same space there are pure price-shopping assistants such as bizrate.com, pricegrabber.com, nextag.com, and Google's Froogle. Once you have found exactly which product you want to buy with the help of a review site, you can instantly find out who has it cheapest. Not only that, you can also often see whether a vendor has been found reliable by other shoppers. These sites continue to grow in popularity as people look for advice from people they feel they can trust, rather than people they think are simply trying to sell them something. Again, people want to do business with and take advice from people they know, like, and trust.

As the knowledge and information age evolves into the relationship era, the ultimate middlemen become those who connect us to valuable information and people. The Internet is full of so much garbage, the middleman has become essential. This is why some of the most valuable property on the net is Google. With a market capitalization of more than $155 billion and with 380 *million* unique users every month, Google is the middleman to end all middlemen. The market has become so oversupplied that being a "trusted adviser," even if it is only in the form of an algorithm, is a very powerful proposition.

Although I have used behemoth brands such as Google to make the point, there is nothing to disqualify local businesses from servicing their markets better than global ones can. In countries like South Korea, local search engines have been more popular than Google because customers consider the results to be more relevant. Relevance is key in a world where time and mental energy are scarce resources.

The point here is simple: from the customer's perspective, increasing complexity and the proliferation of choice (partially fueled by companies mistaking "choice" for "customer service"), in an environment suffering rapid compression of time, have made middlemen attractive again. From the corporate perspective, this could be an excellent opportunity to see if your services can be construed as making you a "middleman." Do you solve complex problems? Do you get people the information or service they need? Are you *saving* them time, or *costing* them time?

Reframing questions about the services and products you provide in this way can help you clarify your position and role, and also may give you great ideas for changes you could make.

TALENT WANTS TO BE "PERSONALLY" CONNECTED TO THE COMPANIES THEY WORK FOR AND THE PEOPLE THEY WORK WITH

You need great people if you want your business to be truly competitive, not just in terms of building relationships with clients, but in terms of being the most innovative, telling the best stories, and having the courage to take the most action. In the talent-scarce environment that we are operating in, this requires a well-thought-out and coordinated campaign to promote your employment brand to attract the best talent. Although building the employer brand is not the subject of this book, I am interested more broadly in what makes a workplace attractive to great talent and how it develops and retains the talent it already employs.

I believe there are two key things that both attract and retain great talent:

1. The work you do
2. The relationships you build with your people

This may seem a little simplified, but it is not. I deliberately leave out things like money and benefits because, as with

"Fast, Good, Cheap: Pick Three—Then Add Something Extra," these are the price of entry. At the end of the day, all other things being equal, people stay at jobs because they like doing quality work and because they like the people they work with.

The quality of the work includes several things. It is about whether people find the work interesting, but it is also about whether the work challenges them and forces them to grow. My attention is obviously going to be on number two, again making my point that business is personal. But before I go on, let me just say that if your organization was a flipstar business, the work would rock. How could it not?

To the point: your ability to acquire, develop, and retain the best talent is directly proportional to the quality of the relationship you build with that talent.

Every client I have tells me, "Our people are our greatest asset." They mean it, but most of the time they don't know how to show it. All of my work in companies to date on generational change and workforce trends suggests that this will be one of the most important strategic issues for companies over the coming years.

RELATIONSHIPS ARE SIMPLE, BUT NOT EASY

To finish the chapter, here are five keys to building great relationships so you can profit from a world where people think business is very personal.

Shift your mind-set

Relationships are only as good as you think they are. An insightful study published in the *Journal of Experimental Social*

Psychology found that the single-most-important factor in determining whether or not a marriage was happy was *not* understanding, sex, love, kids, or anything like that; it was whether spouses rated their partners as better or worse than they rated themselves. Couples where each partner rated the other person higher on a list of qualities were happiest.

Relationships work only if you think they can work, and if you think they are worth investing in, because only then will you put in the appropriate effort to ensure they are a success. In other words, relationships are a self-fulfilling prophecy.

Ensure competence

Competence is fundamental to striking up a quality relationship. Simply put, if you are no good at what you do, no one will want to be your friend. To understand what is required to meet basic expectations of competence, think back to chapter 3, "Fast, Good, Cheap: Pick Three—Then Add Something Extra."

Deliver on what you promise

Once you are comfortable with what you offer, make sure you don't overstretch on what you promise. The only thing worse than a bad job is a bad job that you thought was going to be fantastic. If you can't meet an expectation, level with the customer. Don't be like the handyman who came to our house, gave my wife a quote to fix our stairs, and told us when he would be back to do the work, but then never showed up.

Build trust

In many ways, building trust in a relationship is contingent on the three steps above (having the right mind-set, doing a good job, and delivering on what you promise), but trust is also established independently of the above. Medical researcher Wendy Levinson found that doctors with good patient relationships were less likely to be sued, *even when they did things wrong.*[8] Patients who trusted their doctors believed they had their best interests at heart. The way patients judged whether their doctors had their best interests at heart were very simple things like:

- Did they "sound" interested
- How "long" they spent in consultation
- How intently they "listened"
- The "tone" of voice

I would like to make the point here that not getting sued and getting return business are different prospects altogether. I may not sue a doctor who misdiagnosed me, but I doubt I will ever go back. So again, competence is still a *huge* part of this equation.

Have good manners

People might think this sounds simple. If that's the case, why is it that Web sites like crmlowdown.com are *filled* with examples of straight-out rude customer service? And why is it that I can issue you an airtight, money-back guarantee that every single person reading this book can give an example of when they have been treated appallingly by a company? Ensuring that people on your front line (and yes, this includes *you*) treat customers with respect and courtesy (and actually go the extra mile in helping them solve their problems) might seem obvious, but I will talk about it until it actually starts happening!

Consider the following. After buying an expensive phone from a Vodafone store (a brand to which I have been loyal for ten years), I was thrilled to learn that their generous warranty period covered full replacement of my new handset after failure for—wait for it—*two weeks*.

Well, my phone did break (*three* weeks after I bought it), and given that I was about to head out on a five-city business trip to the United States, I really needed a functioning phone. So I took it back to the store hoping they could help me out. After I spent an hour dealing with three different and equally unhelpful people, not only did it become clear that they could not repair the phone in the two weeks before I left for the United States, but I was told that if I wanted a temporary replacement phone I was going to have to pay $100.

They certainly acted as though they didn't value my business and didn't want to help. If this is how you feel, then get out of business! Fast!

The unfortunate thing for Vodafone is that they do not own this store. It is a franchise, but it still bears their logo. This is outsourcing the relationship you have with the customer. And to make it worse they did not make the phone either. But hey, my relationship was not with the phone manufacturer, it was with Vodafone.

FIVE THINGS TO DO NOW

1. Get a group of key customers in a room and ask them whether they believe your company goes out of its way to build quality relationships with them. Do the same with a group of suppliers, too.

2. Run a focus group with employees and find out whether they think you are a staff-focused organization. Be prepared to do something with the feedback you get.

3. List three new technologies you are going to explore that could be valuable tools to help you stay in touch with existing customers and staff. Might I suggest you start with text messaging, blogs, and podcasts. Then, just for fun, start a personal MySpace profile and visit Second Life and start playing around in these new spaces. Who knows, you might enjoy it.

4. Start one initiative that centers around reconnecting with customers you have lost touch with.

5. Do an audit of your systems and processes and ask if they have positive, neutral, or negative effects on your relationships with customers. Ask yourself for each policy, is it helping you or the customer?

MASS-MARKET SUCCESS:
FIND IT ON THE FRINGE

"Generation Y is too small, Pete. No one is asking for it."

This was the response I got from the managing director of one of Australia's leading speakers' bureaus when I introduced her to my area of expertise. I didn't think she was right. So I ignored the advice, and thereby laid the foundations of a profitable consulting business.

"This is every organization's number one problem!" I exclaimed. "They just don't have a label for it yet."

If you do what everyone else is doing and you draw your wisdom from the crowd, you will find yourself just another supplier of the same product, service, or skill set. Go your own way. Invent a new wheel. Cirque du Soleil did. The Apple iPod did. Burton Snowboards did.

Let's go back for a minute to one of the dominant parameters in today's business world: global oversupply, and global underdemand, in every category of goods and services. In these circumstances, you can putter along doing what everyone else is doing, and if you're more or less industry standard in your offerings and practices, you'll likely do okay. But if you want to stand out from the crowd, you have to differentiate yourself in one of two ways. You can either differentiate your approach to an existing market, or you can differentiate the market you're operating in by discovering or creating a new market. If you want to be a flipstar, you've got to zig when all your competitors zag (kill me if I use another cliché).

"WHEN YOU FIND YOURSELF ON THE SIDE OF THE MAJORITY, IT'S TIME TO REFORM."
—MARK TWAIN

This is not to deny that there is a kind of wisdom in crowds. "The wisdom of crowds" has become a popular notion lately, as studies have shown that patterns of behavior among large groups of people can be self-organizing in an optimal way. For example, pedestrian traffic on a crowded street will naturally flow in the most efficient way possible, and the stock-market picks of millions of investors determine very accurate share values over time. Likewise, large groups of people responding instinctively to various options can reveal the best product among competing alternatives and give important clues to emerging trends. This is the kind of risk-averse, largely unconscious wisdom that James Surowiecki, the "Financial Page" columnist of *The New Yorker* magazine, celebrates in his recent book called *The Wisdom of Crowds*. (I'll be talking about some of the same phenomena from my own perspective a little later in this book in chapter 7, "To Get Control, Give It Up.")

Notice, however, that crowds do not make something new

on their own. Crowds don't create innovations, they validate them. In a global marketplace, the crowd will recognize and celebrate the best innovations. But those innovations don't come from the center of the crowd. They come from the fringe, from bold companies and individuals who are willing to risk doing something different from what

YOU CAN EITHER DIFFERENTI-ATE YOUR APPROACH TO AN EXISTING MARKET, OR YOU CAN DIFFERENTIATE THE MARKET YOU'RE OPERATING IN BY DISCOVERING OR CREAT-ING A NEW MARKET.

competitors are doing, and to offer something different from what the crowds are currently embracing.

Let me put it another way. Crowds are where the big profits are, no question. You don't make money with unpopular goods and services. But market-changing and market-making products don't come from a crowd. They cannot be designed and developed by committee to the crowd's existing

CROWDS DON'T CREATE INNOVATIONS, THEY VALIDATE THEM.

specifications. Market-changing and market-making products like the iPod, ING Direct, Progressive Car Insurance, Cirque du Soleil, Starbucks, Virgin Blue, or Callaway Golf's oversize Big Bertha drivers spring from the minds of maverick companies and individuals who have the guts to gamble on attracting the crowds that will eventually validate them.

One thing crowds are definitely wise about is in telling whether something is a true innovation or a tarted-up version of the same old thing. Crowds have a nose for things that are really new and exciting. The best way to differentiate your offerings from the competition is therefore always to take the route least traveled. *Flips* are the way to locate those routes, and speed from the fringe to the mainstream. Or, to adopt some language from Kim and Mauborgne's insightful book *Blue Ocean Strategy,* flips are the course to adopt to discover entirely new markets—blue oceans—and leave behind mature, over-crowded markets—red oceans.

FROM THE FRINGE TO THE CENTER

Say you're looking at the iPod and thinking, *If we come up with a better MP3 player with a better link to the home computer, the Internet, and the home-entertainment system, we can take the iPod's place*. Even if you could do it right away you are probably still too late. With sayings like "iPod therefore I am" being bandied about, you are likely to get slaughtered. In reality it will take any organization a while (Samsung, Sandisk, and others are trying), which may actually mean that by the time you have built a better MP3 player, the world may well have left MP3 players behind in favor of a totally new way of getting, carrying, and enjoying music, pictures, and other things.

When Apple introduced the iPod in 2001, the MP3 market was still a blue ocean, despite the presence of a number of MP3-player manufacturers, because no one had yet created a total customer solution for managing digital entertainment. The market is a red ocean now—red with the blood of Apple's shark-bitten rivals, you might say—thanks to the iPod–iTunes combination. It will take an entirely new blue-ocean strategy to supplant the iPod with something else.

The market-changing power of the iPod is manifest not only in its huge sales, but in Apple's decision to change its name and start trading as Apple, Inc., rather than as Apple Computer, Inc. In 2007 Apple extended the iPod line with the iPhone, a combination of a wide-screen iPod with a mobile phone, a camera, and a wireless touch-screen internet communicator for Web-browsing, e-mail, and text messaging.

The last feature led a friend of mine to say, "Hey, now might be a good time to short shares in RIM," the makers of the BlackBerry. I don't think so. As initial sales showed, the iPhone is unlikely to be a BlackBerry killer, so long as the only way to input text is via a clumsy virtual keyboard on the iPhone's touch screen.

But that doesn't make me any less excited about Apple's phone. The iPhone is not a short-term threat to the BlackBerry as a business device, because that's not what the iPhone is trying to be. Instead the iPhone is the culmination to date of Apple's efforts to make the iPod the first "infotainment" device, a multipurpose gadget for talking to friends, listening to music, looking at video and still pictures, and surfing the Web, all with the characteristic ease and simplicity of the Apple interface. It marks a journey from the onetime fringe of MP3 players to the center of consumer culture.

The iPhone's initial price showed that Apple does not worry about hitting the center of the mass market when it launches a product. Instead it uses what I call market gravity, which Apple has steadily acquired over its innovative history, to pull the market to its new products and their unusual mix of appealing design and seamless functionality. At $499 or $599, for four or eight gigabytes of storage respectively, the iPhone was priced well above the mass market in cell phones in the United States,

ACTING ON THE FRINGE AND
TAKING A RISK WILL ALWAYS
FEEL UNCOMFORTABLE. IN
FACT ALL OF THE FLIPS WILL.
THEY ARE COUNTERINTUI-
TIVE AND CAN CREATE A
SENSE OF FEAR (I WOULD
RATHER THINK OF IT AS
EXCITEMENT).

just as the original iPod was priced well above other MP3 players. Obviously this shows Apple's love of fat profit margins, but it also shows its confidence that launching the iPhone on the fringe of affordability, so to speak, would help fuel the product's long-term popularity. And again, just as it did with the iPod, Apple lowered the price of the iPhone in due course, but still maintained a price premium over competitors.

Acting on the fringe and taking a risk like this will always feel uncomfortable. In fact all of the flips will. They are counterintuitive and can create a sense of fear (I would rather think of it as excitement). It takes guts to take action without a fully formed plan or a sense of certainty about your success. It takes guts to invest heavily in becoming fast, good, and cheap. It takes more guts to get to the heart of who you are and to yell that story from the rooftops as you build a total ownership experience. It takes even more guts to know who you are and not yell it from the rooftops, allowing people to discover you. And it takes guts to venture from the known, from the crowd. This is why it is a *flip*. But even if the iPhone had not taken off, Apple would have learned from the process. That is why Apple is, well, Apple!

The benefits of shuttling between the fringe and the mainstream show up plainly in Apple's bottom line. In the final quarter of 2006 Apple reported a 78 percent increase in profit to $1 billion on sales of $7.1 billion. During those three months the company sold 21 million iPods, an increase of 50 percent over the same period a year before. It also sold 1.6 million computers, an increase of 28 percent over the same period a year before, showing how the iPod cast a halo of enhanced desirability over the entire Apple range. As Apple CFO Peter Oppenheimer said, "This one was for the record books."

Apple would never have boosted its bottom line so much without making big, bold bets. In a study of new product launches by 108 companies, 86 percent of the "new" products were actually line extensions and they accounted for only 39 percent of total profits. The 14 percent of remaining launches that involved genuinely new products accounted for 61 percent of total profits.[1]

Toyota took the route less traveled by making an early bet on bringing a mass-production hybrid-engine vehicle to market with the Prius. Now it's leveraging the success of the Prius throughout the Toyota and Lexus product range. Making the part-electric and part-internal-combustion-powered Prius hatchback a viable mass-market product is one of Toyota's most notable achievements. All the major manufacturers had developed some form of hybrid technology. But General Motors and Ford, despite losing dominance to Toyota in almost all categories, sat on that technology and did nothing. Toyota, already winning the game, took action to change the game in a way that could have backfired.

Some may argue they didn't need to do this, as they were already dominating the market. Well, that's why they were dominating. They were prepared to innovate. To try new things. To challenge the conventional paradigm that "if it's not broken, don't

TODAY'S BEST PRACTICE IS TOMORROW'S BAD PRACTICE.

fix it." Toyota has not only positioned itself favorably in an increasingly "green" developed market, but sales of the Prius have been so successful that the company cannot meet demand in many countries, with waiting lists up to nine months in some.

The Prius's sales are limited to some extent by the greater up-front cost to the consumer of hybrid technology. Toyota has accordingly extended hybrid technology from the volume-car business to their luxury line, Lexus, whose affluent purchasers will find the additional cost less significant. The strategy doesn't depend primarily on the performance of the car, however, but on its appeal to a segment of affluent purchasers who want to be seen as "green."

In March 2007 the Prius sold almost twenty thousand units in the United States, a 133 percent increase from the previous year, and the competition is feeling the pinch. Toyota experienced an 8 percent rise in demand for its hybrids last year, at the same time as GM felt an 8 percent drop in demand for their cars. On Wall Street, Toyota's stock price rose 38 percent and GM's fell 20 percent.

Many attribute Toyota's pioneering foray into hybrids to rising gas prices and growing concern over global warming. The truth is, however, that Toyota had moved far ahead of the crowd long before these issues took center stage as they have in the last couple of years.

Perhaps the biggest hybrid-related flip on Toyota's part is that instead of trying to keep a proprietary lock on their technology, they are licensing it to all comers, creating a new income stream for technology that is already ten years old. The companies that are buying this technology will likely achieve Toyota's 2007 "best practice" only in 2010 or 2017. Current business advice is to benchmark your efforts against winning companies' best practices. But the flip is that today's best practice is tomorrow's bad practice. In ten years' time there will be much better ways to build a "green" automobile, and Toyota's competitors should be racing it to that new benchmark, not plodding along in Toyota's footsteps. (More on this in chapter 7, "To Get Control, Give It Up.")

What is your company's or industry's version of the Prius? Are you doing enough to develop this potential new market opportunity?

Again, Toyota was the first to identify and act on the mass-market potential in the fringe custom-car culture. As I discussed earlier, Toyota scored an instant hit with the Scion and then demonstrated its ability to flip perspective by deciding to limit production, selling many fewer Scions over the short term than it could, in order to protect its hip, edgy brand story over the long term. Toyota knows exactly how to sell to the masses without losing the fringe characteristics that spell innovation in the marketplace.

When Toyota was developing the Scion, the other car companies' marketing research—they all spend millions of dollars a year on trend spotting and consumer surveys—undoubtedly told them that customizing small cars was a growing phenomenon among young urban customers. But only Toyota had the guts to bet that what was still a fringe activity could generate a high enough volume of sales to support an entire new range of vehicles like the Scion.

After the Scion scored a huge hit with young customers, Honda and Nissan rushed out copycat vehicles, but they had missed the chance to establish themselves as cutting-edge brands for Generation Y. The Detroit three were even further behind the curve. Toyota performed a similar market-changing flip from the fringe into the mainstream with the Prius.

More recently, other carmakers are trailing Toyota down the electricity–gas hybrid path. Not BMW. The most successful of the European car brands over the last decade, BMW is going against the grain of the hybrid-engine technology Toyota has popularized and is instead developing hydrogen-powered cars.

Like Toyota, BMW took serious alternative-energy action long before it was cool to do so. First demonstrated at Expo 2000, the BMW 750hL is the culmination of three decades of research into hydrogen-powered vehicles. BMW sees the use of hydrogen as the answer to many environmental problems, since there are no harmful emissions, no depleting of resources, and no danger to the atmosphere.

The heart of the 750hL is a hybrid, twelve-cylinder combustion engine with two independent electronically controlled fuel-induction systems. These systems allow the 750hL to run on either gas or hydrogen. Now, before you get too cynical about this high-priced twelve-cylinder beast, BMW has also developed a Mini Cooper using the same BMW clean energy system.

Working with Shell Oil Company, BMW has developed a technology for dispensing hydrogen from a filling station's pumps into a car's fuel tanks. The world's first fully automatic hydrogen filling station was opened in May 1999 at the Munich Airport.[2]

My hat is off to both Toyota and BMW—and Honda has been aggressively developing similar technologies, too—for investing long before the market demanded that they do. Markets may be efficient, but in this case they would not have driven innovation soon enough.

Or consider the snowboard. In 1966 an American entre-preneur named Sherman Poppen bolted two pairs of chil-dren's skis together to make a stand-up sled for his daughter. As soon as they saw it, all the neighborhood kids wanted one. Mrs. Poppen combined "snow" with "surfer" and dubbed the contraption the Snurfer. The Poppens licensed it to the Bruns-wick Company, a sporting-goods manufacturer, which sold more than a million Snurfers through sporting-goods, toy, and department stores.

The Snurfer was a toy that couldn't be used on more than a sledding hill. But in the mid to late 1970s, a handful of entre-preneurs who had been exposed to it as teenagers began trying to develop what became the snowboard. One of the handful was Jake Burton, who in 1977 began shaping snowboards in a barn in Vermont and selling them out of the back of his Volvo.

The proto-market seeded by the Snurfer kept Burton's busi-ness alive, but snowboarding remained an underground sport confined to sledding hills. It took five years of lobbying on Burton's part before the first ski resort, Suicide Six Resort in Vermont, opened its slopes to snowboards. Throughout the 1980s and early 1990s, snowboarders struggled to reverse bans against them at major skiing areas.

By the mid-1990s, ski-resort operators were finally waking up to the flip: snowboarding didn't threaten their businesses, it brought a populous new, young demographic to their facili-ties, an infusion of fresh blood that they desperately needed.

IS YOUR COMPANY AS ACTIVE A MEMBER AND PROPONENT FOR ITS MARKETPLACE AS JAKE BURTON IS FOR SNOW-BOARDING? IF NOT, HOW COULD YOU BECOME AN AUTHENTIC PLAYER IN THAT COMMUNITY?

Snowboards appealed to young people who loved surfing and skateboarding, cost a lot less than skis, and were much easier to learn to use. As the snowboard express gathered speed, ESPN mounted the first Winter X Games in 1997, scoring big ratings, and in 1998 snowboarding competition came to the Winter Olympics, where it has become as high profile as alpine skiing and figure skating.

Today snowboards and related products are a $3-billion-a-year industry that continues to grow by leaps and bounds. And at the center of it all, reaping a substantial percentage of the business, stands Jake Burton, whose entrepreneurial portfolio includes both snowboard equipment and clothing companies.

Is your company as active a member and proponent for its marketplace as Jake Burton is for snowboarding? If not, how could you start to be an authentic player in that community?

For a digital example of the success of a product that started on the fringe and went mainstream, consider Stephen Cakebread, who designs games for the Xbox. While the rest of the market was looking for the next Doom, Halo, or Project Gotham Racing, he pioneered the online download of small, retro-styled games for consoles. Ironically, his most successful release is Geometry Wars: Retro Evolved, which began while he was working for Bizarre Creations on the hit game Project Gotham Racing. With more than two hundred thousand trial versions downloaded, and forty-five thousand paid downloads, this is impressive for a game that took one person less than three months to write as a hobby.

Although these numbers are dwarfed by the sales of "on disk" games, including the disk version of Project Gotham Racing, what Cakebread has started with downloadable games for consoles seems to be the next big trend. Downloadable gaming as an idea is not new. Publishers small and large have been developing and distributing small games over the PC for years via services from Yahoo and Real Networks with great success. But offering downloadable games on a console through things like Xbox Live Arcade is new.

Wired magazine has reported that from 2004 to 2005, console disk sales in the United States dropped by $700 million, according to market research firm NPD Group. Meanwhile, game companies earned $143 million from online console gaming in 2005, a figure JupiterResearch predicts will grow to $2 billion domestically by 2011.

Take this old-is-new-again style of gaming, couple it with the power of new technology to create new results (think back to complexity as a force of change), and you have a new trend starting on the fringe and heading to the mainstream.[3]

On the topic of console gaming, a really cool innovation from Nintendo has taken the market by storm. While Sony and Microsoft battle it out for supremacy with the hard-core gaming community, Nintendo's Wii is proving to be a huge hit with a target demographic of families and females. If that is not a departure from the gaming crowd, I don't know what is. Most notable is the use of the one-handed controller, about the size of a TV remote, with motion sensors that allow the gamer to use body movements rather than a joystick and half a dozen buttons.

It is so simple that grandparents can do it. In early 2007 many of my clients were talking about how fun it was to play Wii over Christmas with their kids and actually be competitive. I think some of them are even a little addicted. It helps that the Wii landed at half the price of the PlayStation 3, which was released around the same time (a good example of "Fast, Good, Cheap: Pick Three—Then Add Something Extra").

This "simple and easy to use" strategy has also fueled the phenomenal success of Nintendo's handheld gaming device, the Nintendo DS. The DS has a touch-screen function that allows users to navigate menus with a stylus rather than a kepad. Sales of the DS in 2007 are likely to hit 20 million.

The bottom-line impact is compelling. In the nine months following the release of the Nintendo DS in January 2006 and in anticipation of the release of the Wii, Nintendo's share price jumped 71 percent. And these investors' hopes have been more than met. The Wii became the fastest-selling console in history, selling more than six hundred thousand in the first eight days following its release. As a comparison, Sony's PS3 sold around two hundred thousand units.

In the year ending March 31, 2007, Nintendo increased its revenue and profit forecasts three times. In advance of final results for the year, Nintendo said it expected an operating profit of ¥185 billion and net profit of ¥120 billion. Nintendo also said it expected a foreign-exchange-related profit of ¥20 billion, rather than its previous forecast of a ¥10 billion loss. Finally, the stock nearly doubled over Nintendo's 2006–2007 fiscal year, while the Nikkei remained virtually flat.

In the words of Nintendo itself, the success of the Wii is that it has broad appeal as a "family-oriented game." Its success globally was reflected in Australia with sales of 32,901 units in the first four days, beating the Xbox 360 record of 30,421. In North America, Nintendo expected to sell four million Wii units by the end of 2007, compared with one million for Sony's PS3.

I have said little of the Xbox 360, also about twice the price of the Wii when it was released, which was sold at a loss for Microsoft in an attempt to "win the war for the living room." Nintendo's flip—ignoring the conventional wisdom of the gaming crowd—catapulted it to a market-leading position in less than twelve months, and in August 2007 it exceeded the lifetime sales of Microsoft's Xbox 360. To think that before the Wii's release, pundits were suggesting it was all over for Nintendo.[4]

> ASK YOURSELF, ARE YOU SO BUSY COMPETING ON FAMILIAR STANDARDS AND ASSUMPTIONS THAT YOU ARE MISSING A BRAND-NEW MARKET SEGMENT AS A RESULT?

Ask yourself, are you so busy competing on familiar standards and assumptions that you are missing a brand-new market segment as a result? What would happen if you stopped competing in the way you always have and went in a whole new direction? What direction might that be?

OUT ON A LIMB

To be a flipstar, you've got to venture out onto a limb. You can make okay money doing what everyone else is doing, but that's not the way to put a dent in the universe, as Steve Jobs puts it. Although the odds against immediate success may be daunting, the more you are willing to keep trying new things the more the odds change in your favor.

Let's briefly look again at Progressive Car Insurance, which I discussed in chapter 4, "Absolutely, Positively Sweat the Small Stuff." In 1996 Progressive Car Insurance was the thirteenth biggest insurer in the U.S. market on the basis of serving a limited population of high-risk drivers. A change in California state law that limited the insurance premiums Progressive could charge threatened its entire business model. So it found a new one.

TO BE A FLIPSTAR, YOU'VE GOT TO VENTURE OUT ONTO A LIMB.

Progressive figured out that not all high-risk drivers were the same. The telltale factor was customers' credit ratings. High-risk drivers with good credit ratings were much less likely to have expensive accidents than high-risk drivers with poor credit ratings. Pushing the analysis further, Progressive found that the relationship between a customer's credit rating and the cost of serving that customer held for drivers at all risk levels.

With this customer understanding, Progressive saw that it could shift from offering insurance only to high-risk drivers at high premiums to offering insurance to all drivers at widely varying premiums. The company could become a low-cost provider and undercut every other insurer in the marketplace.

The brilliance of Progressive's play was that it didn't stop its analysis there. Progressive understood that if it gained enough market share, the biggest automotive insurance providers in the U.S. market—State Farm, Allstate and Geico—could easily lower their own rates. In a price war alone, these three had deeper pockets and could beat Progressive at this game.

But remember the need to "Think AND, Not OR." Progressive did. Along with offering the lowest prices, Progressive ramped up its claims-adjustment service to match the capabilities of the big players and it staked its brand story on being the easiest insurer to do business with as well as being cheap and offering good service. That led to innovations such as offering prospective customers rate quotes from competitors (they became the middleman), even when the competition was offering a lower quote than Progressive itself was, and—in states that allowed it—enabling customers to register their vehicles with the state motor-vehicles department from the Progressive Web site.

Progressive gained immediate traction in the marketplace. Although State Farm, Allstate, and Geico all emulated Progressive's innovations, Progressive had successfully established its brand story in customers' minds as the insurer that would save them time, money, and mental energy. The result is that Progressive has jumped from thirteenth place to become number three in the U.S. market behind State Farm and Allstate, largely on the strength of its "easy to use" brand identity, including sending claims adjusters to the accident site or the customer's home or office and writing claims checks on the spot. From the fringe, Progressive has moved right into the center of the mass market.[5]

While on the topic of insurers, how about AAMI in Australia? While their competition is pushing toward "efficiency of operation" with push-button this and voice-recognition that on all of their customer "help" lines, AAMI refuses to let a machine do the talking. They have real people answering your call, and from recent experience provide one of the most friendly and helpful services I have ever received anywhere. It was almost a pleasure to have crashed my wife's car so I could make a claim. Well, that's an overstatement, but my experiences with AAMI and their customer service have been outstanding. They not only understand that business is personal, but they are prepared to stake their competitive advantage in this area even though no one else is. Not only that, but they have been the most competitively priced insurer for the three cars I have insured in the last couple of years.

ING Direct is following a similar path in financial services. Instead of trying to compete with the traditional banks in the traditional way, ING has created a new banking brand story by offering savings accounts online at higher rates than those available at your local brick-and-mortar bank. Appealing especially to younger customers who are comfortable with the Internet and love the ease of use of online banking, these accounts give ING a chance to achieve impressive long-term growth by offering new services to these young customers as they progress through the life cycle, develop in their careers, and become more affluent.

The traditional banks eventually wised up and began making similar offerings available. But the advantage in online banking lies with ING Direct, the company that was willing to go out on a limb and be the first to offer the customer a simpler, easier, more rewarding way of doing business. It helped that their product was also fast, good, cheap, and *easy*!

At the same time ING never forgot the fundamentals of the consumer financial-services market, and made sure that customers associated the company with security and reliability as well as innovation. In this regard it's interesting to see the stumble that

Virgin, one of my all-time flipstars, made recently with the Virgin Superannuation [retirement] Fund. As in most Virgin-branded businesses except the airlines, Virgin holds a relatively small stake in the enterprise, which is offered by Virgin Money but is operated by Macquarie Bank's Macquarie Fund Manager.

On the face of it, a Virgin Superannuation Fund makes great sense as a logical extension of the Virgin credit-card and Virgin mortgage services. But there was an embarrassing disconnect between Virgin's hip, edgy, risk-taking brand story, which reflects the persona of Richard Branson himself, and the overwhelmingly "old" advertising and marketing. Under the Virgin logo, known worldwide for being youthful and cutting edge, there were the same images of sixtyish couples walking on the beach as in every other retirement-fund marketing campaign. It was a disconnect both for the young customers using Virgin Money's credit-card and mortgage services on their first important purchases, and for baby boomers approaching the end of their working lives and wanting to make sure they had enough money to stroll worry-free on the beach.

Virgin may well ultimately succeed with its retirement fund. The company continues to be a flipstar precisely because of its willingness to keep on taking chances to enter and create new markets, and its successes far outweigh its failures. Richard Branson's R&D process is reportedly to listen when someone brings in an interesting idea and then put that person in charge of developing it. But here is a case that shows that your innovations have to be in line with the fundamentals of your

> YOUR INNOVATIONS HAVE TO BE IN LINE WITH THE FUNDAMENTALS OF YOUR OWN BRAND IDENTITY.

own brand identity. So far, that is not true of the Virgin Super-annuation Fund, at least not in my opinion.

Given the mention of Macquarie Bank as we talked about Virgin, I thought it useful to include an example of going out on a limb in your employment proposition. *Everyone* is talking about work–life balance, so much so that companies traditionally known for punishing hours are starting to talk about work–life balance (even when it is a lie) when trying to recruit good talent. Not Macquarie Bank. Sure, there may be some discussion of this at the HR level, but when you are being interviewed by a director at Macquarie, he or she will tell you in no uncertain terms, "Say good-bye to your friends, say good-bye to your family. You are not going to see them for ten years. But when you do, you will be rich! Interested?"

Having worked with Macquarie and knowing some senior people there, I can tell you that this is not entirely true. The people I know don't think they have sacrificed anything to be at Macquarie. They feel as though their work is an integrated part of their life, and they love how Macquarie celebrates success, develops its people, and promotes from within. Most of all they love that they are given "freedom within boundaries," to steal an old Macquarie phrase. Macquarie Bank gets more unsolicited applications than any company I have worked with in Australia (not as many as Google in the United States, but a lot by Australian standards) and they have very low attrition rates. Phenomenal when you think about how "off trend" their positioning is.

It is not surprising really that Macquarie would carve its own path in the employment landscape because they have a worldwide reputation for innovation in deal-making models and banking in general.

In the consumer financial-services market, Prudential Financial has shown how even a giant company can be a flipstar. When the technology bubble that burst in March 2000 was still expanding, Prudential's competitors such as Fidelity were all running advertising that emphasized dramatic growth in wealth. But Prudential was sending a different message: grow and protect your wealth. This was in keeping with Prudential's "rock solid" brand identity and its longtime logo featuring the Rock of Gibraltar.

Refusing to follow the herd of competitors, Prudential kept its eye squarely on affluent baby boomers who were middle-aged or nearing retirement. When the tech bubble finally burst, the company's "grow and protect your wealth" message resonated even louder with the most desirable customers. It also resonated with the competition, who all migrated sooner or later to a copycat positioning with some variation of the "grow and protect your wealth" promise to customers. Prudential was a flipstar because it took a divergent path from the competition. Then the market flipped in the direction Prudential was already pointing, and the competition had to flip their messages. But the copycats' messages don't resonate with customers like the original flipstar's do.

KEEP EXPLORING

Going out on a limb is not something you do once. Suppose you fall and never take another chance? Then you'll never get anywhere exciting. Flipstars go out on a limb many times in the course of their careers or their existence as companies. Richard Branson and

FLIPSTARS GO OUT ON A LIMB MANY TIMES IN THE COURSE OF THEIR CAREERS OR THEIR EXISTENCE AS COMPANIES.

Steve Jobs are the guiding risk takers at Virgin and Apple respectively, and all four are flipstars.

Chapter 2, "Action Creates Clarity," speaks about the importance of risk taking in terms of an action orientation in general. A key element in a successful action orientation is contact with fringe areas within society where new ideas percolate. Many, if not most, of these ideas will turn out to be fizzles, or just a case of old wine in new bottles, but a few of them will set an agenda for the center. They represent the future of the mass market in embryonic form.

This is why Nike assiduously tracks trends within minority urban communities, striving to identify what suburban consumers will later buy in even greater quantities. Retailers such as Uniqlo in Japan, Sportsgirl in Australia, and Target in the United States do the same to maintain a hip, cutting-edge identity in customers' eyes.

Nike has become such a big company, with 2006 revenues of $15 billion, that it is easy to forget that it started out on the fringe. In 1962 University of Oregon track-and-field coach Bill Bowerman visited a fellow coach, Arthur Lydiard, in New Zealand. Today movie fans across the world know New Zealand as the place where Peter Jackson filmed the *Lord of the Rings* trilogy, but in 1962 the country was definitely on the fringe of awareness for most North Americans and Europeans.

Bowerman was fascinated to observe that many people in New Zealand regularly engaged in an activity they called "jogging" for fun and fitness. Even though Bowerman was an ex–competitive runner and coached runners, he had never seen ordinary people running for pleasure. He joined Arthur Lydiard's jogging club for outings, kept jogging on his return home, and legend has it that it took four inches off his waistline. More important, he had witnessed a fringe activity that he thought had enormous potential for an increasingly

health- and fitness-conscious population in the United States and elsewhere.

Back home, Bowerman teamed up with cardiologist Waldo Harris to write a twenty-page pamphlet titled "Jogging." The pamphlet morphed into a book that sold more than one million copies. It also gave impetus to a handshake deal between Bowerman and an ex-athlete he had once coached named Phil Knight. Their enterprise, Blue Ribbon Sports, was eventually renamed Nike, and the rest is history, as they say.

Let me highlight one milestone in that history, the 1978 signing of tennis bad boy John McEnroe to a Nike endorsement contract. Those with long tennis memories might say that Nike would have gone after Björn Borg if he hadn't already been signed to Fila. But I doubt it. Picking a renegade from the mainstream like McEnroe has typified all of Nike's marketing, right up to the present, including its controversial 1987 use of the Beatles song "Revolution" in an ad (Nike apparently had the permission of EMI/Capitol Records and John Lennon's widow, Yoko Ono, but not that of Apple Records and the other Beatles), its 2004 Chinese martial-arts-themed advertising featuring NBA superstar LeBron James (some thought it was racist), and the 2005 use of a slightly modified but easily identifiable Minor Threat album cover in a promotion for Nike skateboarding shoes (easily identifiable by Gen Y, that is).

Consistently innovative companies like 3M recognize the need for regular visits to the fringe. There is a long-standing mandate at 3M that research staff should spend 15 percent of their time on whatever interests them. Most of what results from this are harebrained failures, but it also gave 3M the Post-it note, structured abrasive belts, new lenses for computer-monitor manufacturing, and other products. Following much the same practice, although it credits Stanford University's Ph.D. program in computer science as inspiration, Google got its Froogle shopping service, Google Earth, and Gmail from this practice. 3M goes so far as to arbitrarily demand that a certain percentage of its revenues in a set number of years should come from products or services that currently do not exist at 3M.

UNIQUE

Learning to innovate on the fringe and then carry the fruit of your innovations to the mass market makes you unique. Though all of the companies we've looked at in this chapter are not without competition, they each have distinctive brand stories that can never be mistaken for those of competitors.

It is possible for several companies in a single industry to be unique in customers' eyes. Think of the luxury-automobile market, where BMW, Ferrari, Lexus, Mercedes-Benz, and Porsche have unique brand identities and tell unique stories to customers.

I mentioned the example of Generation Y and my work at the start of this chapter. A few people were talking about

Generation Y when I decided to publish my research in this area. But more than one person told me I would need to position myself more broadly than Generation Y, and even

LEARNING TO INNOVATE ON THE FRINGE AND THEN CARRY THE FRUITS OF YOUR INNOVATIONS TO THE MASS MARKET MAKES YOU UNIQUE.

though I had some expertise in other areas, I knew this would be bad advice. To be perceived as unique you really have to go out on a limb.

I took action. By action I mean I spent three years (seriously) writing a book, even before I had a publisher. I knew this generational shift was a cause of many challenges facing businesses today, both from a staff and a customer point of view, and I knew over time businesses would come to recognize that.

They did, and the result for me was a very successful (meaning fun and profitable) consulting business that takes me all over the world, and has allowed me to work with some of the most powerful businesses and people on the planet. Through my work in this well-defined area I have been able to build my expertise in other areas.

Of course I am not the only person in my category anymore, and there are plenty of me-toos in the market. The truth is I was not the first into the market either, but I was the first to be focused primarily on Gen Y and I was able to tell a story that people remembered and passed on, and this rising tide has raised all of our ships.

Peter Switzer, a well-known finance commentator in Australia, is changing the way financial planning is done. Switzer Financial Services offers investment advice but makes a point of taking no commission and having their consultants paid by the hour to ensure absolute impartiality. Any commissions paid by the institutions that receive investment at their advice are actually *paid back to the client.* This is a Virgin-like assault on the margins of the financial planning sector.

Even though Switzer is not the only advisory business with this fee-for-service model, it is certainly a fringe model in the industry. Gaining in popularity, it goes against the product-centric, commission-driven model that has built this industry to date. Imagine it was you. Think of how your adviser buddies would have felt about you "undercutting" their pricing model as you went out on your limb. Might I suggest that not only were they not agreeing with you, but there was no love out on that fringe either.

Interestingly enough my own advisers, Ark Financial, have gone in the other direction. They not only charge a fee for service, they also charge commission plus a yearly retainer. They are staking their positioning on what they call "Beyond Wealth." They regularly coach and educate their clients on how to enjoy their wealth, and get more out of life while at the same time helping to grow their finances. It is more like going to see a life coach than just a financial planner. The point is both approaches are unique and both are being endorsed by growing numbers of customers.

Two other Aussie examples of companies that bucked industry wisdom and started whole new trends are Boost Juice and McGrath real estate.

While accompanying her husband, Jeff, on a trip to the United States, Boost founder Janine Allis saw a hole in the Australian market for a healthy fast-food alternative. Upon returning to Australia, she developed a business plan and raised $250,000 start-up money by recruiting her friends as investors.

The first store opened in Adelaide in 2000, and since then more than 170 stores have opened throughout Australia, with the first international franchises in Chile, Indonesia, Kuwait, and Singapore. With revenues expected to exceed $100 million in 2007, this is a great achievement in just six years. So much of what Janine has done at Boost has been against the crowd. The staff are the most obvious. Unlike so many retailers, the staff at Boost more often than not are having fun. Boost deliberately hires some of the wackiest and weirdest people who apply. They want that kind of edginess in their stores.

Their PR activities have included giving away a franchise at Sydney's Manly Beach through popular local radio station TodayFM. Not to mention Janine being the charismatic leader and voice behind so much of this PR activity. Instead of running a typical radio ad, Janine would instead give motivational advice to listeners, building the Boost story of a healthy lifestyle through proper nutrition. Although there has been some debate as to how healthy Boost juices really are, few could argue with their phenomenal success. It is cool just because it is different.

DIFFERENT IS COOL JUST BECAUSE IT IS DIFFERENT.

McGrath real estate is another of my favorite examples. During the property boom of the late 1990s and early 2000s John McGrath turned McGrath Partners into a force in the Sydney real estate market. They were among the very first agencies—along with Ray White Double Bay, Di Jones, and others—to raise the bar in the industry. Their advertising was clean, modern, and of the very best quality. Their signs used color and images and matched the high standard of their print advertising. They moved away from private treaty sales and helped spur the auction culture that drove Sydney property prices to dizzying heights. They motivated vendors to make significant improvements on their homes before listing them for sale, including renovations and renting furniture to make a better presentation.

This market leadership enabled McGrath Partners to charge a commission 1 percent higher than the industry average, which on sales exceeding $1 billion is a handy, full-profit premium. Today, many agencies have raised their standards, and with a flat market McGrath will need to become faster, better, and cheaper, as well as easier and more inspiring to deal with if they want to continue to be a leader in their industry. Innovation never ends!

On the flip side in the real estate business is longtime friend of mine and flipstar Pete Gilchrist. He recently established a real estate agency in New Zealand called The Joneses. It sets out to totally reinvent the way real estate is sold. Their biggest changes from the usual model? A flat fee rather than commission, and team members with dedicated tasks (rather than a single agent who does the whole thing end-to-end).

The absence of commission has been calculated to save up to $6,000 on a $300,000 sale, $10,000 on a $500,000 house sale, and over $21,000 in a million-dollar sale. The team structure means that instead of having a "jack of all trades" (who has to manage the whole thing because her commission rides on it) doing everything from photographs to booking

INNOVATION NEVER ENDS!

advertising space, a dedicated professional is assigned to each task.

This turns the real estate model on its head, and it's being met with very positive reactions from consumers who have long felt that the traditional model wasn't giving them very good value for their money.

For companies and individuals the flip that mass-market success is found on the fringe has two implications. Either differentiate yourself from competitors within an existing market, or differentiate the market you operate in by creating a new market. Or for best results, do both; but in any case, start exploring.

Ten things to do when you finish this chapter

I know I have been finishing each chapter with five things to do, but in the spirit of being a little different, here are ten.

1. Start an innovators' club. Pull together a group (or multiple groups) of high performers in your company and host a meeting with them once a month where you look at fringe activities, new market opportunities, and your current competitive strategy.

2. Keep a journal of fringe stuff. Not just fringe stuff in your market but any market. If your bank does something new and unique, write it down, give it some thought, and then present it at your innovators' club.

3. Do something random at least four times per year. Visit an art gallery. Buy a different magazine than the ones you normally read. Or maybe even take a new course in something that does not relate to your job.

4. Surf the net. If a new activity, product, or service is becoming popular, the first signs of it will appear on the Internet. Spend some time regularly looking at MySpace, Facebook, YouTube, and other popular sites to see what's becoming hot—or not.

5. Check out what teenagers are doing at the local shopping mall. What are they wearing? How are they talking? What are they talking about?

6. Block out three hours at least four times per year and ask yourself: If I was going to build this company from the ground up, what would I do? Or if I was going to redesign this product, or develop a different product for this same need, what would it look like? What else could this product or service be used for?

7. Hire someone a little wacky. If you are big enough and have the resources, employ someone in marketing with an industrial-design background. Or as Google does, hire an actuary to work in HR.

8. Don't attend your industry conference this year. Attend some other industry's conference instead. Or at least attend both.

9. Apply the 3M test. For example, demand that 20 percent of your revenue in three years comes from a product or service (or many of them) you don't yet have.

10. Stop being such a wimp! Do something new, different, and cool with your product or service. Something that has never been done before.

TO GET CONTROL,
GIVE IT UP

You don't own your brand, you don't own your customers, and you don't own your staff. These days you have less control and influence over them than ever before. If you are smart, you will flip this negative into a positive, and use it to tell a better story, add more value, and turn your business into an awesome place to work. This is the only way you will get significant control. And even then it won't really be control, just influence.

This is the hardest flip to master. Human nature makes us want to expand the territory we control, whether as companies or as individuals, and at the very least hold on to the power we already have. But some of today's biggest success stories are being written by companies and entrepreneurs who have learned to give up control.

This chapter will describe some companies that are struggling to let go of the business models that have made them successful. This is understandable. Take for example the music industry, which—it is no secret—is struggling with the onset of digital music downloads, peer-to-peer file sharing, and a host of technologies that are completely changing the way music is made, distributed, and listened to. You can hardly blame the music executives who have made their careers, and their companies' profits, with a tightly controlled selection and distribution model for clinging to the only way of doing business they have known.

I don't say this because I think a music executive reading this chapter will be surprised to hear that his business model is under threat. Nor will a chief marketing officer be shocked to learn that the customer owns his company's brand. What I do hope is that these executives, and you, feel inspired by the possibilities

YOU DON'T OWN YOUR BRAND, YOU DON'T OWN YOUR CUSTOMERS, AND YOU DON'T OWN YOUR STAFF.

these changes present rather than crippled by them. Specifically, this chapter will suggest you do the following things:

- Stop resisting and learn to embrace the changes forced upon your business model and distribution networks.
- Connect and interact with your customers, not just so you can build valuable relationships with them, but so you can benefit from their expertise for better product development.
- Let go of your desire to control the research-and-development process, and tap into the pockets of individual brilliance that can be found when you open your doors to a wider community.
- Resist the urge to own and hold on to everything. Explore new business models where money can be made

in relationship with suppliers, and even competitors, that your organization cannot or will not exploit in isolation.

- Empower your staff to do their job.

THE POWER SHIFT

The four forces of change have shifted power from the organization to the individual. Technology like the Internet, increasingly affluent customers, oversupply in customer markets, and extremely tight labor markets are putting the individual in the driver's seat. This is forcing businesses to be more accountable for what they do, and it is also undermining some of the most successful business models on the planet. Read on to learn what flipstars are doing about this new center of power.

Winning through losing

Change is like a wave. You either ride it, get pummeled by it, or you sit on the beach and watch it run its course. For a while in the early 1990s Microsoft decided to watch the Internet wave from the beach. That was until 1994, when two twentysomethings who were still wet behind the ears—J. Allard, a programmer, and Steven Sinofsky, a technical adviser to Bill Gates—sent separate memos to senior managers at Microsoft. Their memos said: get into the water, the Internet is huge, and it is here to stay.

In effect they were saying that the Internet was no ordinary wave, it was a tsunami that would consume even those sitting on the beach. These memos were enough to get Microsoft off the sand and into the water. A little more than twelve months later Bill Gates himself issued a memo aptly titled "The Internet Tidal Wave." He exclaimed, rightly, that the net was the "most important single development" in the computer industry since the IBM PC. "I have gone through several stages of increasing my views of its importance. Now, I assign the internet the highest level," he wrote.[1]

The battle began to develop the ubiquitous browser for the Internet. Microsoft, among a few others, wanted to be the gateway to the World Wide Web. Or more technically, the facilitator of people's Web experience. The battle, mostly with Netscape in the late 1990s, was won by Microsoft—first, because they built a better browser, and second, because they embedded Microsoft Explorer into the Windows operating system, which created problems with antitrust regulators in the United States and Europe, but not before it had won the company market dominance in Internet browsers.

The cool thing is that the Internet wave is still coming in, and some would argue it has not even gotten started yet. Are you riding the wave, getting pummeled by it, or standing on the beach? No one is immune. The Internet is redefining distribution networks and consumer power, at the same time compromising the "trust" and property rights that the capitalist system is built on. And it is not just the Internet either. For Microsoft the next tidal wave, apart from perhaps Google and online applications, is probably China. The battle there is twofold, with the software pirates and with Linux, a free opensource operating system.

In the United States, Europe, and Australia, the battle for the computer desktop was won by Microsoft with more than 90 percent of all PCs using Windows as their operating system. While piracy is still alive and well in these markets, it is nothing compared to what happens in China. According to the

Business Software Alliance, piracy in the United States hovers around 22 percent, the UK 29 percent, and Australia surprisingly a little higher at 31 percent. This sounds high, but it is estimated that 90 percent of all software in China is pirated.

In China, piracy seems to be a way of life. It is in everything from pharmaceuticals (making Viagra rip-offs using ancient Chinese herbal ingredients) to airplane and automobile parts.[2] There was around $2.2 billion worth of software, movies, and music piracy in China in 2006. China accounts for roughly two-thirds of the world's pirated goods and is the point of origin of around 80 percent of counterfeit goods seized at U.S. borders.[3]

Add to this equation that Linux is basically a free operating system that can be customized to your company's needs. Despite being a model of "open source," which I will discuss later in this chapter, Linux has done little to dent the phenomenal market share that Microsoft has in the developed markets mentioned above. In China, it is a different story. Although it is believed that Microsoft is ahead, Linux is not far behind in what is obviously a more immature and rapidly growing market, including deals with the Chinese government and education system.

It is likely there are more illegal copies of Windows than there are legitimate ones. It may seem obvious to someone in Australia, the United States, or the UK to put a stop to this piracy, and install controls on piracy in China like those imposed in Western countries. But it could be argued that this is the opposite of what Microsoft should try to accomplish. To win this war, Microsoft needs to lose the battle. For now anyway, some industry experts have suggested that Microsoft turn a blind eye to the piracy that is rampant in China for two main reasons. Once you build your IT infrastructure around Windows, it is highly likely you will want a number of add-on applications, many of which are made and sold by Microsoft. And it will be easier to enforce antipiracy measures when China's own nascent computer and software industries begin to value a more "respectful" relationship with intellectual-property-rights advocates, and when Linux no longer poses a significant strategic threat.

This may seem like a very audacious thing to suggest for a company that has profited hugely from exercising strict control of their product, but winning by losing is actually a strategy Microsoft knows well. The Xbox 360 is built on this strategy. Microsoft and others see the next big market opportunity in becoming the operating system and hardware vendor of choice in the living room, as people increasingly receive and store their home-entertainment content digitally. In an attempt to gain the dominant position over its main rival, Sony, and more recently Nintendo, Microsoft sells the Xbox 360, which ironically is still very expensive, at a loss.

IT MIGHT SOUND AUDACIOUS, BUT WINNING BY LOSING IS A STRATEGY MICROSOFT KNOWS WELL. THE XBOX 360 IS BUILT ON IT.

It seems that Microsoft has been doing this in China for longer than we may think. In a 1998 presentation at the University of Washington, Bill Gates remarked, "Although about three million computers get sold every year in China, people don't pay for the software. Someday they will, though. And as long as they are going to steal it, we want them to steal ours. They'll get sort of addicted and then we'll somehow figure out how to collect sometime in the next decade." That decade is basically up, and Microsoft has yet to collect. However, any attempt to control piracy in China is likely to be ineffective. In this case Microsoft will get some control only if it is willing to give it up for a little longer.[4]

If the Internet is the main power shifter, it would be remiss not to examine the company that has probably best worked out how to profit from the democratization the net has created: eBay. This book's second chapter is about the flip that action creates clarity, and eBay is the ultimate "action creates clarity" business. Pierre Omidyar began eBay as a "thought experiment." He was interested in finding out if people could trust one another enough to buy and sell items without being controlled by some third party. Obviously it worked. In terms of the preceding chapter, Omidyar brought his innovation in from the fringe of his own mind and had it validated by the wisdom of crowds. Not only that, but even the name was not part of some grand master plan. Prior to starting eBay in 1995, Omidyar owned a business consultancy called Echo Bay Technology Group and tried to register EchoBay.com, but it was owned by Echo Bay Mines. So he abbreviated it to eBay.

To grasp the power of eBay, consider the following numbers:

- Since its launch in 1995, eBay has grown from nothing to a $6.35-billion-a year business. Not bad for a company barely a decade old.
- More than $52 billion worth of goods were sold on eBay in 2006.
- Approximately 1.3 million people around the world used eBay as their primary or secondary source of income in 2006, including almost 13,000 in India.

- There are more than 222 million members of eBay who trade in over 50,000 categories of goods. More than $729 is spent on eBay every second.
- There are more than 100 million items on the site at any given second, with 6.6 million added per day.
- Last year, more than 2 billion British pounds' worth of goods were sold on eBay in the UK alone (a car was sold every two minutes), and eBay claims that as many as 70,000 Britons now make their living from eBay.

The cool thing about eBay is that it is not a middleman in the same way Mortgage Choice is a middleman. In my mind it is not a middleman at all. It is simply a marketplace, a platform with very limited controls, policed mainly by the users (the wisdom of the crowd), where buyers and sellers meet and transact.

Pierre and eBay still kick Yahoo Auction's butt because they moved fast, had a simple and fast procedure, and gathered a critical mass of users that makes it seem crazy to buy or sell anywhere else. As a result they are able to charge a premium for listings and sales that is three times that of other competitive online auction sites. And it is not just eBay that wins. By removing the layers of control (all the people trying to skim dollars off the top), the eBay buyer gets a better deal and quite often so does the seller because of the excitement of the auction and the absence of high-rent retail space.

EBAY OWNS, STORES, AND SHIPS *NOTHING*! YOU DON'T HAVE TO OWN EVERYTHING!

Best of all eBay owns, stores, and ships *nothing*! You don't have to own everything!

The next big frontier for eBay

is in the Chinese market, where they have had little success to date. After pouring tens of millions of dollars into China, eBay recently announced a partnership with Chinese company Tom Online Inc.

FREE MUSIC ANYONE?

Until recently the recorded music business was very resistant to the changing nature of their marketplace. The entire industry is being dragged kicking and screaming into letting customers purchase and download individual songs online rather than buy an entire CD. The record companies are still wasting time and money suing illegal downloaders in court. But that war is already lost to the overwhelming force of consumer action. Some day an economics student will make a nice Ph.D. thesis calculating how much money the record companies lost in refusing to face reality.

One of the hall-of-shame examples in that thesis will be Sony's misguided attempt to copy-protect its CDs with a computer virus. Sony secretly implanted two types of viruses on around one hundred different CD titles that it sold to consumers.

When CD owners played their CDs on their computers, the viruses installed themselves on the computer operating systems and started making changes to the media player to ensure the CD could be played only on that one computer. The program automatically installed itself, and the user was never informed about its presence. The changes it made to the computer also rendered it susceptible to other viruses and hackers. And the program communicated personal information to Sony about the computer on which it was installed, including the user's IP address (enabling Sony to identify each user).

This was seen by many, justifiably, as a massive breach of

privacy and trust. Eventually, Sony had to recall the CDs from distribution. They also faced legal action over the copy-protection programs, due to breaches of privacy and damage to personal computer property.

Sony Pictures Entertainment senior vice president Steve Heckler said this in 2001: "The industry will take whatever steps it needs to protect itself and protect its revenue streams. . . . Sony is going to take aggressive steps to stop this. We will develop technology that transcends the individual user. We will firewall Napster at source—we will block it at your cable company, we will block it at your phone company, we will block it at your ISP. We will firewall it at your PC. . . . These strategies are being aggressively pursued because there is simply too much at stake."[5]

As music-CD sales continue to fall because of downloading, music-industry executives are understandably worried, but the position outlined by Mr. Heckler is not a winning one, as Sony's copy-protected-CD fiasco shows. Illegal downloading is wrong, but when asked, many who engage in this activity have said that if it was easier and less restrictive to download legally they would. Obviously iTunes and a host of other such sites have now made it easier to download, but still make it restrictive to use.

Digital music sales from legitimate downloading have not yet offset the decrease in CD sales. But they are getting closer to doing so. Legitimate downloads increased 89 percent in 2006 over 2005, and global digital music sales over some five hundred legitimate downloading sites reached $2 billion. Total music sales were down 4 percent from 2005 to 2006, but the trend is clear: music customers want to pick and choose the songs they buy, not be locked into the music company's, or the artist's, album offerings.

THE MUSIC INDUSTRY IS BEGINNING TO RECOGNIZE THAT IT IS BETTER TO ADAPT TO, RATHER THAN FIGHT, GLOBAL CUSTOMER BEHAVIOR.

Not only do customers value control over the songs they buy, but buying online is faster and easier. The simplicity builds on the power of ideas presented in chapter 4, "Absolutely, Positively Sweat the Small Stuff."

There are signs of the music industry beginning to recognize that it is better to adapt to, rather than fight, global customer behavior. The *New York Times* reported from the January 2007 meeting of Midem, the music industry's annual trade fair, that the major recorded-music companies are feeling their way toward joining independent labels in selling digital downloads in MP3 format without copy-protection, based on the realization that hackers will quickly figure a way around any copy protection software and on the expectation that these unrestricted file formats will serve as a form of advertising.

Customers' demands to access music however they want will also put pressure on Apple's digital-rights management model, which makes iTunes available only on the iPod. Fueled by customer frustration, legislative proposals to mandate interoperability are being made in France and other European countries. Throughout its history, Apple (like Sony) has made things that work only on its proprietary platforms. The first important signs of moving away from this were making iTunes compatible with Windows and using Intel chips in the latest Macs.

The introduction of an Intel chip into Macs and Apple laptops was a major reason for the strong growth in Apple sales in very recent years. iTunes will need to follow a similar path. Sooner rather than later, I hope. Considering the rate of downloads, Apple will likely find more money and better margins selling digital music than they did selling digital music players.

Meanwhile the musicians who have embraced free downloading have vastly broadened their potential markets. This is not surprising considering that until such technology came along, their ability to access the market was very low without a record label, and in effect they had no control.

The Arctic Monkeys are one of several bands to use MySpace to attract the interest of record labels. But some now argue that in a downloadable world, bands no longer need a record label at all. January 2007 was the first time a band hit a major music chart without being signed to any record label. The "unsigned" rock band Koopa, from Essex, England, hit the UK Top 40 with a single called "Blag, Steal & Borrow" that was available only as a downloaded MP3 file.

Unfortunately the PR frenzy that surrounded Koopa being the first "unsigned" band on the UK charts was a little misleading. You technically can't be on the UK charts without being signed, because you need what is called a Catco ID for royalties to be paid. So in truth Koopa was actually "signed" with a company called Ditto Music, owned by brothers Lee and Matt Parsons. The key differentiator is that a deal with Ditto Music is nonexclusive and is only for distribution and promotion. In other words, Koopa retains the rights, whereas with a "traditional" music label this is usually not the case.

The real example of getting control by giving it up is in fact Ditto Music, not the band. Ditto managed to get their second "rights-retained artist" (which they spin as being an "unsigned" artist) onto the Top 40 charts in the UK in March 2007 when Midas entered the UK charts at number three. According to its Web site, Ditto Music has hundreds of artists as clients. It has these clients because it doesn't want to "own" the rights for its clients' music. Sure, it is not a multibillion-dollar giant like EMI or Sony Music, but it is certainly a sign of a new business model emerging—a model that seeks to get control (distribution of new music) by giving it up (rights).[6]

As television-executive-turned-Internet-entrepreneur Jordan Levin puts it, "Ultimately these big media companies are all wrestling with the same thing—the power is being taken out of their hands. This is an industry that for its entire history has imposed its model on consumers. They've always said, 'We'll tell you when you'll watch our TV show or see our movie.' But that's fundamentally changing. The whole structure of people who control content is being supplanted by the content users themselves." And according to MGM executive Harry Sloan, "We've got to get the creativity to stand against user-created content, because that's what people are watching at my house." In an interview with *Variety,* Sloan described his seventeen-year-old son with the television on behind him and "two screens in front of him, one connected to friends and one to play World of Warcraft."[7]

CURRENT TRENDS DO NOT SPELL THE END OF THE MOVIE AND TELEVISION COMPANIES BUT THE END OF THE MOVIE AND TELEVISION COMPANIES CONTROLLING WHEN AND WHERE WE WATCH WHAT WE WATCH.

I don't think these current trends mean the end of the movie and television companies, just the end of movie and television companies controlling when and where we watch what we watch.

YouTube has become an arena for copyright-protected content uploaded without the copyright holders' consent as well as user-generated content. Thanks to its $1.65 billion purchase of YouTube, Google can demonstrate that advertising tailored to each individual user's viewing habits, like the advertising on Google Search that changes depending on what you're looking for, can adequately compensate copyright holders for the loss of control over their intellectual property. Google will also be making advertising-based payments to the amateur video makers who upload their own creations to the YouTube site. No doubt the most successful of these amateurs, who may as well be film school graduates as self-taught, will face the same quandary as the rock band Koopa: if there is profit without them, who needs the movie and television companies or the record companies? Personally I think YouTube is a holding yard for a collection of garbage. Organized garbage, for sure, but I am astounded daily at not only how many clips get loaded up, but worse, how many hours people spend downloading them.

I am more of a fan of the News Corporation, AOL, and NBC deal that will use the Internet to distribute quality content.

This shows that despite some who suggest the "end of TV," the industry will not die. They just need to flip some of the paradigms that until recently had made them phenomenally successful. It is time to take some action, and have the courage to explore new business models. For example, despite the movie studios' and cinema owners' resistance, simultaneous release of movies in cinemas, on pay-per-view television, and on DVD is coming soon, because that's what the public wants. The maverick entrepreneur Mark Cuban, who coincidentally owns the NBA team the Dallas Mavericks, has staked out an early lead in this arena.

Together with fellow entrepreneur Todd Wagner, Cuban has assembled a vertically integrated media and entertainment company that includes a movie-production studio, a movie-distribution company, a chain of cinemas, and an all-high-definition digital television network "to experiment with a 'day-and-date' model in which films will be released simultaneously across theatrical, television and home video platforms, thus collapsing the traditional release windows and giving consumers a choice of how, when, and where they wish to see a movie." The model received its trial runs with two high-profile projects: George Clooney's *Good Night, and Good Luck* and the documentary *Enron: The Smartest Guys in the Room*.

> AUTHORS WHO HAVE MADE THEIR WORK AVAILABLE AS FREE DOWNLOADS HAVE SEEN INCREASES IN THE SALES OF THEIR PRINTED BOOKS.

Book publishers have also struggled to come to terms with the world of downloads. Interestingly, book publishing shows perhaps the clearest benefits to both companies and intellectual-property creators in making downloads freely available. Listening to a downloaded file is no different from listening to a CD (sorry to all of you audiophiles). The best file formats can offer CD-quality sound, and video and film downloads will surely follow the same upward quality curve. But reading a book on a computer is not as enjoyable and practical as reading an actual printed book. Dedicated electronic-book readers have not yet bridged the gap in terms of the quality of the reading experience. At the same time, making part or all of a book available for free downloading as an electronic file can actually trigger hefty sales of a conventional print edition, a business model that the marketing commentator Seth Godin and the science-fiction writer Cory Doctorow have successfully pioneered.

For example, Seth Godin made chapters of his most successful book, *Permission Marketing,* available for free online before it was available for sale in bookstores. And Cory Doctorow says on his Web site:

> I've been giving away my books ever since my first
> novel came out, and boy has it ever made me a
> bunch of money. . . . I believe that we live in an era
> where anything that can be expressed as bits will
> be. . . . Me, I'm looking to find ways to use copying
> to make more money and it's working: enlisting my
> readers as evangelists for my work and giving them
> free ebooks to distribute sells more books."[8]

PROFITS THROUGH PARTICIPATION AND PERSONALIZATION

I opened this chapter by saying that your brand does not belong to you. While this is not news to anyone, it is worthwhile to remind ourselves. The identity of your brand and its fate in the marketplace ultimately rest with your customers. Likewise, the identity of your employment brand and its fate in the marketplace ultimately rest with your staff.

What customers think and say about your products and services determines whether they will be profitable and for how long. And what staff think and say about your business determines whether you can attract, develop, and retain the best and brightest to work for and with you. For good or ill, nothing beats word of mouth.

> NOTHING BEATS WORD OF MOUTH.

Nothing beats it, that is, except for word of *mouse*. What has changed with the Internet and the new communications technologies is the speed with which perceptions of your brand can spread, infecting people with an enthusiasm for contact with your brand, or with a disdain for it, as the case may be. The old rule of thumb was that whereas a satisfied customer might tell one person what was great about a product or service, a dissatisfied customer would tell eight people what was wrong with it. Nowadays, there is no limit to how many people dissatisfied customers can communicate with via the Internet. And although there may also be no limit to how many people satisfied customers can communicate with, human nature still means that people get more fired up about complaining when something goes wrong than they do about enthusing when something goes right. A 2006 *BRW* article cites research by Melbourne Business School that notes a correlation between

> NOTHING BEATS IT, THAT IS, EXCEPT FOR WORD OF *MOUSE*.

negative word-of-mouse publicity and a company's perfor-mance. There was no corresponding correlation between good feedback and good performance.

Let's briefly consider Dom in my office, who not so long ago had an ordinary experience with his local Australian bank. It was so bad it prompted him to start a user group on Face-book.com called "I hate banks: let's start a consumer rebellion." Within one hour sixty-four of his "friends" had joined. Inter-estingly enough, a significant number of them were from West-ern Europe, so obviously we are talking global here. Now assume that like Dom, his first-degree connections also have dozens of "friends" independent of Dom's network. Before you know it hundreds of people, many of whom Dom has never met, have joined his club. They then ask their contacts, who in turn ask their contacts, and so on until this user group has thousands of members beating up on banks.

After I gently reminded Dom that the banking sector is one of our best sources of clients, and that I like those clients very much, and so should he, given that they pay his bills, the growth of the user group was stemmed (or should I say halted) but not before showing me a live example of just how powerful word of mouse can be. By the way, my behavior here goes against the flip I am putting forth in this chapter, and had I been the bank, I would have been more interested in learning from the experience, knowing full well I had no ability to shut it down.

A search by industry on the Web site my3cents.com can show you how powerful customers banding together can be. A search about the banking sector brings up posts like "Bank

XYZ—Not like all the other banks—WORSE!" And another that reads "Big banks need to follow rules too."

According to their Web site:

> My3cents.com is a leading source of real consumer advice. Visitors come to learn, interact and voice opinions regarding companies, products and services in our open community. Learn from other consumer experiences, and help others learn from your own personal consumer experiences. Join the revolution today and start being heard!

My 3 Cents is not a popular Web site, but there are hundreds just like it, such as CRM Lowdown, which I mentioned earlier and which gives people an opportunity to vent, or more important, find out about certain products and companies. Such sites are evidence of the increasing transparency and accountability I have been talking about, which is one of the four forces of change we face daily. The result of this and other changes discussed in this book is that the power of customers and staff is increasing, and the power of companies is decreasing.

The explosion of the blogosphere, platforms like MySpace, and user-generated content are all part of the consumer voice. It is easier to be heard (or read) now than ever before, and search technology makes it simpler to find specific information on what you need. More and more consumers, and businesses, too, will do a search before making a purchase or a decision about a product or service. Job applicants are Googling a company to see what other people say about it, and giving more weight to these opinions than to what may be on the company's own Web site.

I say embrace the transparency. Once upon a time it would have cost hundreds of thousands of dollars to truly find out how a customer felt about your product. Now it takes a few hours and a free Internet search. If I was a senior leader at Bank XYZ, I would want to know what happened to the customer who is now complaining on My 3 Cents. If I was the Aussie bank that Dom was complaining about, I would want to read his post, and all the comments, to find out how I could improve my services.

I personally know this experience well. Every week I present to an audience, sometimes as small as six and other times as big as a few thousand. As disturbing as feedback can be, you need to suck it up and learn from it, both the good and the bad. When I first launched my blog, I disabled comments. Why? Because it was integrated into my commercial Web site, and the last thing I wanted was a client engaging me in their business to read things other people say about my ideas and work. Talk about missing the point. It is about dialogue these days, not monologue. It is a conversation, not a dictation. I have since changed my approach, and with the exception of the odd inappropriate link (I think you know what I mean) that some computer-generated program has placed there, my blog leaves all comments exactly as they were posted.

Such public dialogue is not to be taken lightly. General Motors ran a competition where anyone could make their own commercial for the new Chevy Tahoe (in conjunction with Donald Trump's television show *The Apprentice*). It should be noted that GM's foray into viral marketing was incomplete from the outset. Even within the strictures set out by GM (they provided the videos that users could use, so it wasn't *entirely* user-generated), there was still a significant consumer backlash that made the whole campaign backfire.

A slew of global-warming-based criticisms and anti-SUV vitriol whipped around the Web in the wake of the appeals for consumer involvement. GM was quick to remove some of the "offensive" content, although to their credit they left some of the more benign critical ads online.

GM had an opportunity to learn from this backlash. As one blogger noted:

> . . . instead of chocking this up to a bad marketing decision, they could really use the information here. There are a growing number of people who believe that the proliferation of SUVs is getting ridiculous. Does everyone need a vehicle that can climb snowy mountains?
>
> But what SUVs solve for many is a "cooler" alternative to minivans for their growing families.[9]

Their decision to remove some of the ads just created more controversy and backlash. As another blogger responded:

> Dear Chevy: Gas mileage, the environment, and big cars are not exactly a new issue. Hell, I got 32 miles per gallon on a 1967 Toyota Corolla! What did you think the public would do, given the chance?[10]

The company, however, claims they left all the ads online with the exception of vulgar and offensive content. Chevy general manager Ed Peper commented on his company blog:

> We at GM are not culturally unaware; we realize
> that there are people who would never purchase an
> SUV. That's why we make more vehicles that get
> over 30 miles-per-gallon than any other
> manufacturer. . . .
> Anyway, it sure got people talking about the
> Tahoe. Which was the whole idea, after all.[11]

That's a cracking response from GM. So they got it about half right, I guess.

What is most exciting about this trend is that while Web 2.0 is about user-generated content, Web 3.0 and Web 4.0 are going to add *huge* value to the consumer, and companies that do a great job will tap into the power of word of mouse for the positive. The word will spread loud and clear who the market leaders are, and what products and services you should buy, and what the best companies are to work for.

Here is what the evolution might look like. The idea comes from Tim Berners-Lee, who is widely credited with inventing the World Wide Web, and is what he had envisioned it to be like all along. Web 3.0 is often referred to as the *semantic Web*. This means it is about ordering and giving meaning to content and information. It refers to both a *philosophy* for future Web design (that the content should be readily searchable and understood by software programs so it can be ordered, sorted, and accessed), and also to *technologies* that enable that philosophy to be borne out.

At the moment you can use the Internet for a whole lot of

useful stuff, but you can't just have a machine do that for you because Web pages are designed to be read by people, not machines. Semantic Web is a vision of an Internet where computers can understand the information. Talk about fast, good, cheap, and easy. Web 3.0 is not really here yet, but a visit to the Media Labs at MIT would suggest it will come.

While Web 2.0 is user-generated and social-networking content, and Web 3.0 is about a learning, semantic Web, some already look ahead to what Web 4.0 might be. Marketing guru Seth Godin, whom I mentioned in connection with book downloads, has an interesting idea that is worth sharing: that Web 4.0 could be about forming an intense and close connection with a small and intimate group of people.

Consider this flash into the future, found on his blog:

> I'm typing an email to someone, and we're brainstorming about doing a business development deal with Apple. A little window pops up and lets me know that David over in our Tucson office is already having a similar conversation with Apple and perhaps we should coordinate.
>
> Google watches what I search. It watches what other people like me search. Every day, it shows me things I ought to be searching for that I'm not. And it introduces me to people who are searching for what I'm searching for.
>
> I'm late for a dinner. My GPS phone knows this (because it has my calendar, my location, and the traffic status). So, it tells me, and then it alerts the people who are waiting for me.

I visit a blog for the first time. My browser knows what sort of stories I am interested in and shows me highlights of the new blog based on that history.

Web4 is about smaller, far more intense connections with trusted colleagues and their activities. It's a tribe.[12]

According to Godin, for Web 4.0 to become a reality we need:

- an e-mail client who is smart about what I'm doing and what my opted in colleagues are doing. Once that gains traction, plenty of vendors will work to integrate with it.
- a cell phone and cell-phone provider that is not just a phone.
- a word processor that knows about everything I've written and what's on the Web that's related to what I'm writing now.
- moves by Google and Yahoo and others to make it easy for us to become nonanonymous, all the time, everywhere we go.

People, companies, and governments are watching what we do; our computers and the search engines we use are recording what we look for, read, and say—why not use this information in a way that adds power to our lives? It might scare some people, but it sure excites me.

LET THE CUSTOMER PLAY

Sony stumbled over this chapter's flip when players of its Ever-Quest game created a black economy via PayPal to buy and sell virtual assets in the game's alternative online world. Sony told its customers in effect, "We want to control those game features and any income stream associated with them." Sony's customers said, "Back off. It's our game now." In the face of massive customer resentment, Sony had to back down. In doing so, it allowed Ever-Quest to continue to grow and continue to pour profits into its corporate coffers. Sony won loyalty and profits by losing control.

In a different version of events, some companies have set out deliberately to exploit the desire of the audience to partici-pate. Reality TV is one example, and its most successful fran-chise is the *Idol* format, which includes *American Idol,* the original British show *Pop Idol,* and more than thirty other *Idol* franchises around the world. The *Idol* shows not only star am-ateur talent, but the results of the shows' talent competitions are determined by viewers' votes. Viewer choice also deter-mines the outcome on other extremely successful reality shows such as *Survivor* and *Big Brother.* It is no longer enough to pas-sively watch a TV show. We want to participate, and vote off the people we don't like.

> IT IS NO LONGER ENOUGH TO PASSIVELY WATCH TV. WE WANT TO PARTICIPATE.

In addition to making the actual shows extremely popu-lar, driving up traditional advertising revenue, the reality for-mat lends itself perfectly to be more integrated with other mediums—such as the Internet—which also drive revenues. Perhaps the most significant spin-off, however, is the revenue from voting itself. It is no secret that the voting format is a big cash spinner.

To get some idea of the power of this new format, which perfectly taps the desire of the audience to have more control over their experience, check out this information that appeared in a white paper on the *Idol* format by interactive marketing specialist SMLXL:

- Over 3.2 billion viewers over the past six years.
- The thirty-plus countries the format is aired in cover a total of 560 million TV households.
- There were 215 million unique viewers for the final alone.
- During the past five years *Pop Idol* viewers have generated 1.9 billion votes.
- The fourth *American Idol,* which took place in 2004, was at the time the biggest texting event in the world, with 41 million text-message-based votes cast. What is more remarkable is that 30 percent of those texters had never sent a text message before.
- The 2004 figures turned out to be only the beginning, with the spring 2006 run of *American Idol* attracting 64.5 million text-message votes.
- New applications include trivia, sweepstakes entries, text message chats, a fan club, a vote number reminder, and downloadable ring tones.
- It just keeps getting bigger—680 million votes in the last twelve months.
- So for a typical twenty-episode run, viewers have about fourteen opportunities to vote.

Consider that there is about a 90 percent profit margin on each text message, and you see what an enviable income stream this generates for *Idol*'s producers and broadcasters, not to mention the cell-phone service providers.

Today's viewers want to customize their viewing experience in the same way Starbucks has been customizing their lattes. It is no longer enough to just watch people on television; I need to have some control over who stays on what shows and for how long.

Interestingly, it took a music-industry man (not a TV producer) and a gutsy UK television station to try the reality format, and to add voting capability to it. This was not a "crowd" initiative. Fremantle Media and 19 Management are reaping the rewards of a concept popularly credited to one man: Simon Fuller. Fuller, who founded 19 Management in 1985, was previously known as the manager of the Spice Girls and S Club 7. The Spice Girls and the birth of the *Idol* format in 1998 make Simon Fuller a world-class flipstar.

The most interesting iteration of the *Idol* format came in 2005 in Finland. They ran side by side with the show proper an uncontrolled version called *Jokamiehen Idol* (People's Idol), where anyone and everyone could post to the Internet their own "Idol" performance, whether they had auditioned for the show or not, and whether they were any good or not; 150,000 Finns listened to the amateur posts and cast 1,950,000 votes, which according to SMLXL was more than actually voted for the televised "real" *Idol*.

Australia did the same with Telstra in 2006 when they allowed users to become the "Street Idol," winning $20,000 and a trip to the Grammy Awards.

What is cool about this is that it is an example of the producer giving up even more control. It is the same participation that has driven the phenomenal success of YouTube.

CUSTOMERS WANT WHAT THEY WANT

Formerly of the MIT Media Lab, Nicholas Negroponte famously said, "Customers don't want choice; they want what they want." This is true, but a limited and controlled set of choices allowed companies to spread their risk, and get closer to giving what the customer wanted to a broad enough market to maintain some level of economies of scale so they could deliver their "almost right" products to the market in a cost-effective way. Digital distribution, global markets, better production technology, and more sophisticated supply chains giving the ability to the customer to more easily declare what he or she wants are just some of the things allowing companies to move from "almost right" product choices to "mass personalization."

But customers generally don't really know what they want. Many times a product is successful when the company is able to convince consumers that they want something that previously they did not. Or in extreme cases like the motorcar, the digital watch, or more recently the digital music player, convince them they want something that previously did not even exist. Remember there is no wisdom in crowds when it comes to innovation, and maverick innovators are still required here.

A SUCCESSFUL PRODUCT OFTEN CONVINCES CONSUMERS THAT THEY WANT SOMETHING THAT PREVIOUSLY THEY DID NOT.

Let's revisit the Scion. Although the brand story and styling of the Scion are closely controlled by Toyota, part of that story is that this can be "your" ride. Just how you want it. The Scion has been designed to be personalized. This is the trend to which I am referring—the desire of the consumer to have products that they feel are uniquely theirs. This is why the story is so powerful, as talked about in detail in chapter 4, "Absolutely, Positively Sweat the Small Stuff."

James Farley, vice president of the Scion at the time of its unveiling, said, "Scion is about personalization. It's about providing buyers with a personalized dealership experience, a personalized ordering process, and personalized vehicles."

The Scion team developed almost forty accessories that could be fitted to the car to meet the customer's exact wants, including fog lamps, rear spoilers, and auxiliary interior light kits. Scions also lend themselves to postpurchase customization, with a whole market of non-Toyota companies springing up to offer various types of customization, including additions of subwoofers, superchargers, and non-Scion decals.

The Scion FUSE concept car unveiled at the New York Auto Show in 2006 takes this to the next level. Another Scion executive, Mark Templin, says that as well as pushing the "creative envelope," the Scion also demonstrates "boundless personalization" potential. Everything from the customizable multicolored headlights to the different options for in-car WiFi access, video consoles, and televisions can be tailored down to the individual tastes of the customer.

Your company's version of personalization does not need to be as sophisticated as the Toyota Scion. Perhaps you could start giving your customers a wider choice of payment methods. Maybe it is more delivery options. I am willing to bet you have all had that experience of buying a piece of furniture only to then be told it will need to be delivered in a few days (or months) as they keep no stock on hand. I can understand this, knowing how expensive retail space is, but your average consumer gets annoyed by it. If this is not bad enough, the retailer then tells you on what day it will be delivered, regardless of whether this suits you or not. So you arrange to duck out of work for an hour or so, but then they only give you a window of about six hours. And then they refuse to call you one hour before they are due to arrive. So basically you, the customer, end up having to take half a day off work, which probably costs you more in lost productivity than your initial purchase was worth, and you sit around waiting for the delivery truck to arrive. And if they are anything like the furniture store I referred to earlier, they will leave their rubbish in your living room because "this is not their responsibility."

Probably the best and most successful example of "mass personalization" around the world is IKEA. The build-it-yourself model that IKEA embraces comes from their desire to constantly drive down prices but not sacrifice quality. They sell all their furniture disassembled (in "flat packs") so costs of construction are passed on to the consumer (that is, the price in dollars is lower but the cost in effort is higher). The idea came from when Gilles Lundgren was struggling to fit a table into the back of his car, so he took the legs off it and it fit "in five easy pieces."

People like this model for a few reasons. First, they understand that their personal effort is part of the model that secures them lower prices, so they are happy to have a hand in building what they buy. But second, the fact that many IKEA products can be assembled in a number of ways (desks that can have a few hutches or be just plain, cupboards that can have multiple shelves in a few different layouts) means that consumers can assemble furniture that they feel is uniquely theirs—both *functionally* and *aesthetically*.

It obviously works. IKEA had an annual revenue of over $26 billion to March 2007 and employs over one hundred thousand people. Profit margins are 18 percent and sales have trebled in the last decade, while at the same time prices have come down.

I have to be honest here. Until very recently I hated IKEA with a passion. The thought of having to take down my own order, load my own shopping cart and car, and then not be able to pay on my preferred credit card was enough to drive me nuts. Paying extra for some service is what I prefer. (Perhaps IKEA should consider a premium service, where you get a one-to-one service for sixty minutes for a percentage of your purchase, with a minimum charge). However, a recent experience has changed my view, and now I am a huge fan. Redoing a playroom for my kids, I found that IKEA's myriad furniture options can indeed create a customer experience that is fast, good, cheap, and (mass-) personalized.

The point: where possible give your customers what they want, and not what you think they want. Or worse, what you want. Do an audit of your product choices, delivery options, and payment methods. Ask yourself, are they set up to favor the customer or to favor you?

THE TRUE WISDOM OF CROWDS

It may come as a shock to you, but the smartest people in the world don't all work for you. But you still want to access their brains and creativity, and these days that is easier than ever before. Consider the Chicago T-shirt manufacturer Threadless.

It makes all its T-shirts to designs that its customers post on the company Web site and that other site visitors can rate on a scale of one to five. Each week, the company puts the most popular new designs into production, paying each winning designer $1,500 in cash and $500 in merchandise. On an average day well over one hundred designs are posted on Threadless.com, and in an average month the company sells almost eighty thousand T-shirts at $15 a piece.

According to Threadless cofounder Jacob DeHart, "We've got four rules we follow. We let the [Threadless customer] community create the content. We let the community build itself— no advertising. We let the community help with the business; we add features based on user feedback. And we reward members of the community for participating."

Similarly, on the Web site of Seattle's fast-growing Jones Soda, which sells a range of organic teas and carbonated beverages, customers post and vote on photographs, with the winning images becoming incorporated into Jones Soda bottle labels. Founder and CEO Peter van Stolk says, "We founded this company with the philosophy that the world does not need another soda. That forced us to look at things differently: How could we create a new kind of connection with customers, let them play with the brand, let them take ownership of it? Everything at this company is about sharing ownership of the brand with our customers. This is not *my* brand. This is not *our* soda. It belongs to our customers."[13]

Although the crowd itself is not wise in the sense of generating new ideas, it contains a lot of insanely talented individuals who have the ability and desire to contribute in meaningful ways to the brands they love. Sometimes, purely for love. Consider LEGO.

In 2003 LEGO suffered its greatest-ever loss, $238 million. In response, rather than keep doing the same thing and hoping it would turn around, LEGO adapted. It tore up the detailed blueprint it had been operating under and became a flipstar. The company folded its game-software division, as well as a number of underperforming LEGO kit designs, and decreased the number of unique LEGO blocks and other kit pieces it manufactures by over 10 percent. Fairly normal cost cutting, but a bold, aggressive start to the action.

LEGO followed that by deciding to reinvigorate one of its most successful recent products, the Mindstorms LEGO robot kit. Costing $199 at retail, the Mindstorms kit had sold almost one million units since its 1998 release and by 2003 was still selling around 40,000 units a year with no advertising. But the kit was too complex for most children, and many of its most loyal fans were adults.

After some initial development of a new Mindstorms prototype in-house, LEGO decided to invite a small group of adult users who had become celebrities in the Mindstorms world to become a top-secret Mindstorms Users Panel (MUP): Steve Hassenpflug, an Indiana software engineer; John Barnes, the owner of a firm that manufactures ultrasonic sensors; Ralph Hempel, author of several Mindstorms how-to books; and David Schilling, cofounder of a Mindstorms user group called SMART, the Seattle Mindstorms and Robotic Techies. Working on the project without pay, and even buying their own plane tickets when they visited LEGO headquarters in Denmark, the

MUP-sters influenced every aspect of what became Mindstorms NXT, a runaway bestseller for LEGO at a list price of $249 since its 2006 release, in exchange for a few free prototypes and other LEGO kits and the satisfaction of being involved in creating the new version of something they loved.

Assembling the MUP was a huge break from LEGO's tradition of controlling every aspect of design and production in-house. A flip, no less. To create Mindstorms NXT, they also went outside for software programming. But they had already taken an important step in this direction with the original Mindstorms product release in 1996. When a hacker quickly broke and published the Mindstorms software code, LEGO considered legal action against him. Then it took a far more profitable legal action: it altered the Mindstorms software license

THE OPEN-SOURCE MODEL IS THE ANTITHESIS OF THE COMMAND-AND-CONTROL SYSTEMS OF PRODUCT DEVELOPMENT AND EXPLOI-TATION THAT MOST BUSI-NESSES STILL FOLLOW.

by inserting a "right to hack." As fanatical Mindstorms hackers published code for making Mindstorms robots that could deal blackjack, scrub toilets, and perform a host of other functions that had never been in LEGO's product brief, LEGO became one of the first manufacturers to reap the benefits of open source software development.

On a final note, LEGO has started to use this same approach to sell more of its innovative new products. LEGO enlisted influential consumers as online evangelists. After a new locomotive was shown to the 250 most hard-core LEGO train fans, their word of mouse helped the first ten thousand units sell out in ten days with no other marketing.[14]

The open-source model is the antithesis of the command-and-control systems of product development and exploitation that most businesses still follow. In command-and-control R&D, the company sets all the parameters at the top executive level, and a research-and-development department attempts to execute accordingly. Attempts, because as history amply demonstrates, there is no effective way to dictate specific discoveries in a closed system.

In chapter 6, "Mass-Market Success: Find It on the Fringe," I argued that you have to be open to ideas from the fringe. You have to be open to serendipity, to finding things you weren't looking for. The open-source model does this by recruiting the energies not of a proprietary research force whose resources

THE TRUE WISDOM OF VIRTUAL CROWDS IS THAT PEOPLE WILL BE PURSUING THEIR INDIVIDUAL INTERESTS AND AGENDAS.

are limited even in the largest companies and organizations, or even the most powerful governments, but of anyone who cares to contribute. This ensures that new ideas will have a chance to emerge from the fringe. It's a flip on a flip. The true wisdom of virtual crowds is that people will be pursuing their individual interests and agendas. It's a market of the mind, which winnows out the good ideas from the bad in an uncontrolled, Darwinian survival of the fittest.

Nothing demonstrates the potential of the open-source model better than Linux. Linux itself grew out of two earlier open-source projects—Unix and GNU. Unix, the free-time pet project of two AT&T Bell Labs researchers in the late 1960s, was initially quite successful but quickly became heavily protected by patents. In response, GNU emerged as an open-source alternative to Unix, but lacked some central elements of a viable operating system, such as a microkernel.

Linux, the brainchild of Finnish computer scientist Linus Torvalds, was released in the public domain in 1991 and was a fully viable operating system. Today, only 2 percent of the Linux kernel is directly attributable to Torvalds, but it is hugely successful. IBM and other large IT corporations use Linux as an operating system, and there are numerous small companies selling Linux software, which runs on everything from supercomputers and servers to cell phones and PDAs. In 2008 the global Linux market is projected to reach $35.7 billion a year.

None of this could have happened cost-effectively within a single company's, or even a consortium's, proprietary software development. To compile the millions of lines of source code in Red Hat Linux—the first important, but now discontinued, product from Red Hat—would have taken eight thousand man-years of conventional development time at a cost of $1.08 billion a year.

I said that open-source development represents the true wisdom of crowds, and the thousands of contributors to Linux certainly make up quite a crowd. But this does not actually contradict chapter 6's flip, "Mass Market Success: Find It on the Fringe." To appreciate this, we have to look closely at the development of the Linux kernel, and consider the concept of *lead users,* people who innovate solutions for their own specific needs in advance of similar general needs.

Only 2 percent of the current Linux kernel is directly attributable to Linus Torvalds. But by functioning as, in effect, a lead user of the Unix-like concepts and designs that he drew on to create the first Linux kernel, Torvalds put his stamp on everything Linux and pointed out a direction for its future development. By the same token, LEGO's Mindstorms User Panel, recruited from outside the company, served as lead users for the development of Mindstorms NXT. By definition, lead users are to be found on the fringe of mainstream activities, or to put it another way, on the cutting edge of developing trends.

Lead users act like fringe ideas that attract mainstream interest. As I said in chapter 6, innovations don't begin in crowds, but crowds are necessary to validate innovations. There may be no *I* in team, but there definitely is an *I* in innovation, and lead users put it there. An article published in *BRW* in Australia cited research that looked at fifty projects in development over about a year and found that products using the lead-user approach generated returns that were on average eight times greater than products developed the conventional way.

In his book *Democratizing Innovation,* Eric von Hippel shows that lead-user product development can be a far more effective means of innovation than conventional product development within a closed system. At 3M, for example, von Hippel and his research associates found that lead-user development resulted in five major new product lines with annual sales forecasts of $146 million per product, whereas conventional solely-within-the-company product development over the study period resulted in forty-one incremental improvements to existing product lines and only one major new product line with an annual sales forecast of $18 million.

Von Hippel says of the lead-user product-development

process, "This is not traditional market research—asking customers what they want. This is identifying what your most advanced users are already doing and understanding what their innovations mean for the future of your business."

Some of the most interesting new branding and product development today are being done by companies that use social-networking technology to solicit contributions from lead users and/or validate innovations by seeing what customers in general think of them. For example, the Vancouver-based shoe designer John Fluevog, who has boutiques in several North American cities and distributes his shoes to high-end shops around the world, practices what he calls "open-source footwear" by inviting design submissions from customers. Although only a few such submissions become actual products, Fluevog says, "Even submissions we can't make add to the stimulation. Our customers get more involved, and we get insight into who they are and what they're doing."[15]

Consider a slightly different version of the same phenomenon—the InnovationXChange. I had the pleasure of meeting John Wolpert, who developed the BRIDGE methodology, which instructs professional intermediaries on how to help companies share intent and find mutually beneficial collaborations without exposing secrets to one another directly.

Less than four weeks later, I serendipitously bumped into InnovationXChange CEO Grant Kearney boarding a plane to Sydney. IXC is a not-for-profit company that helps companies, universities, and research institutions collaborate and share information in the midst of highly complex international intellectual-property laws, while still maintaining strict confidentiality. In their own words, they "make it safe for organizations to 'talk.'"

They have dealt with everyone from food makers to pharmaceutical companies, in one case linking a pharmaceutical company working on a partially finished drug with a nanotechnology company on the other side of the world that revolutionized the drug's delivery mechanism. IXC is a fascinating organization, adding real value to Australian businesses.

This trend is global. For instance, the Web site Global Ideas Bank (www.globalideasbank.org) is a public forum where people can post ideas at any level of development (from a small inkling of an idea to a well-progressed idea that is hitting stumbling blocks), and people can comment and contribute.

SOME THINGS ARE (STILL) WORTH PAYING FOR

Innovation contests and prize challenges are another means of open-source, lead-user product development. The pharmaceutical giant Eli Lilly recently incubated and spun off a new enterprise called InnoCentive, which acts as a middleman between companies ("seekers") and a worldwide community of self-selected scientists and engineers ("solvers"). Seeker companies post challenges with cash awards for the development of new products and processes, and scientist solvers post their suggested solutions by a specified deadline.

In early 2007 InnoCentive's open challenges included an award of $140,000 for new pressure-sensitive adhesives and $1 million for a biomarker for the disease ALS (amyotrophic lateral sclerosis), as well as $15,000 each for a novel method of dust control and packaging to limit the breakage of snack chips. Successful solutions to past challenges include a new method for assessing the risk of breast cancer, UV-resistant coatings, and a next-generation paper binder, among several dozen others for companies such as Procter & Gamble, Du-Pont, Dow, and Boeing.

Contests to spark innovation have a long history. In the eighteenth century the British government offered twenty thousand pounds, an enormous sum then, for the solution to finding an accurate longitude on the open sea. The solution didn't come from highly credentialed astronomers of the day, but instead from a clockmaker. His name was John Harrison. After four different incarnations of his longitudinal chronometers, he finally built one that worked extremely accurately. For petty political reasons he was initially refused his reward! He fought it and fought it, and finally—*years* after building a properly working device that could measure longitude at sea—he was granted his reward. He was *eighty,* and lived only three more years. Thankfully, today such contests are known for paying up more reliably. Can you imagine the PR and word-of-mouse frenzy that would follow if they did not?

CONTESTS TO SPARK INNOVATION HAVE A LONG HISTORY. IN THE EIGHTEENTH CENTURY THE BRITISH GOVERNMENT OFFERED TWENTY THOUSAND POUNDS, AN ENORMOUS SUM THEN, FOR THE SOLUTION TO FINDING AN ACCURATE LONGITUDE ON THE OPEN SEA.

Netflix, the DVDs-by-mail service, which operates on a lending-library subscription model, asks customers to rate the movies they watch and then uses an algorithm to make recommendations of new movies based on over 1.6 billion total customer ratings. As with similar algorithms that lie behind Amazon .com's customer recommendations and Google's search-tailored advertising, the more accurate Netflix can make its "Cinematch" algorithm, the more customer satisfaction and loyalty it will engender and the better positioned it will be against competitors, including coming digital download services.

In October 2006 Netflix announced a $1 million prize and a five-year submission deadline for an algorithm that would increase the accuracy of Cinematch in predicting customer responses to movies by at least 10 percent. According to the *New York Times,* the company, which did not expect quick results, "underestimated the power of an open competition." Less than six months after the start of the competition, the model of the leading contenders, a team of Hungarian scientists, was "already 6.75 percent better than Cinematch."

Incidentally, showing that it knows how to "Think AND, Not OR," in January of 2007 Netflix introduced a new feature to its DVD-rental subscription plans. For no additional charge, Netflix customers got the capability to watch streaming movies from Netflix, as well as continue to receive and return DVDs through the mail.

On a side note, Netflix sounds a little bit like a nascent kind of Web 3.0, but it probably falls a bit short. It is certainly an example of organizing information. But to truly be a part of the semantic Web, the data would have to be ordered and stored in such a way that machines could understand what the data *meant* so they could parse it for useful information. It would need what information specialists call a *metalanguage*.

Netflix really is just a good database with a good search algorithm. If it was really Web 3.0, you should be able to ask the database, "I want to see a funny film, but I want it to be dark humor, not slapstick, and I want it to have a serious edge." The only way of doing that with Netflix would be to wade through all the user comments. In Web 3.0, the computer could understand qualitative judgments and comments.

Summing up the benefits of innovation contests at Netflix and elsewhere, the *New York Times* noted the "two essential features of prizes. They pay for nothing but performance, and they ensure that anyone with a good idea—not just the usual experts—can take a crack at a tough problem."[16]

Procter & Gamble adopted a strategy called Connect and Develop as part of its fight back following its market slump in 2001. It employed a network of scouts to look for (connect) new ideas, and then to develop them. Proctor & Gamble has an internal R&D operation with a nearly $2-billion-a-year budget, but they are smart enough to still "Think AND, Not OR" and look outside, too, by using InnoCentive's eighty thousand independent, self-selected "solvers." It's like an R&D version of e-lance.com, where the company posts R&D projects and people can contribute. Procter & Gamble's decision to move away from just traditional in-house R&D and combine it with this new outsourced variety has led to a 60 percent increase in R&D productivity and the launch of more than one hundred successful products, including the hit Olay Regenerist.

Cisco Systems has an outstanding internal culture and empowered staff, but another vital factor in their long-standing success is an "acquire and develop" strategy: they buy up small, innovative companies and use their resources and market clout to realize their potential. Rather than create an open competition, Cisco will let fringe entities develop an idea on their own time and with their own money until that innovation looks ready to be integrated into existing technologies or commercialized in its own right. Cisco will then buy the company, ideas and talent both, and take it to market.

Or for another favorite, Intel has built "lablets" in strategic universities to work closely with researchers and graduate students to publish research that Intel will not own. The goal is to build the amount of innovative activity in the areas that Intel works in. In the future they could adopt a connect-and-develop or even an acquire-and-develop approach to some of the innovations created from these relationships.

These strategies make more sense when you consider data published by the National Science Foundation showing that between 1981 and 2001 the percentage of total R&D expenditure that can be attributed to large businesses has shrunk from 70.6 percent to 39.4 percent. This is especially telling because the percentage that can be attributed to businesses with fewer than one thousand people has grown from 4.4 percent to 24.7 percent. It makes sense to connect, acquire, and develop when smaller firms are where a significant chunk of the action is.

THE FASTER YOU MOVE THE LESS LIKELY YOU ARE TO TAKE THE TIME TO FIND AND EVALUATE EXTERNAL OPTIONS. YOU NEED TO GET OVER THIS NATURAL FOCUS, AND GET OUT AND SEE WHAT IS HAPPENING.

The other advantage of these approaches by companies such as Procter & Gamble, Cisco, and Intel is that these smaller entrepreneurial ventures are putting everything on the line and have already taken the risk. They are generally more nimble organizations, able to move quickly when the market changes, and in reality their success depends on getting it right. If a large company fails to innovate in one product area, there are a few thousand others to absorb the loss. If one of these smaller companies fails to stay on trend, it will find itself out of business fast.

As a side note, of the four forces of change, compression of time is the most likely to prevent you from embracing the opportunities that open-source and lead-user development present. As you try to keep up with change, you are forced to move more quickly. The faster you move the less likely you are to take the time to find and evaluate external options. You need to get over this natural focus, and get out and see what is happening.

The whole idea of open source, like rewards for ideas, is not a new one. Although the term *open source* is alleged to have come out of a strategy session held at Palo Alto in response to Netscape's decision in January 1998 to release its source code (think "open" source code), its origins are older, and the concept as it is being presented here is far more broad in application than simply source code.

Let me give you a great Australian example. CAMBIA is a not-for-profit research center founded in 1992 by Richard Jefferson in Canberra. Focusing on plants and biotechnology, CAMBIA was an acronym for the Center for the Application of Molecular Biology to International Agriculture.

These days CAMBIA has evolved into more broad life-science applications but retains the name CAMBIA, which in Italy and Spain means "change." CAMBIA develops new methodologies and technologies for plant improvement and then provides that information to the wider market. In effect CAMBIA is an open-source biotechnology research center and a driver for much innovation.

The key goal of CAMBIA is to promote open-source innovation in the wider arena of biotechnology. They see open source as an enabler of innovation, and have developed a suite of technologies, patents, and licenses that in their view will give innovators greater free-dom to develop and market new biotechnologies. CAM-BIA makes specific reference to helping smaller businesses and ventures who, without as-sistance in what can be both scientifically and legally com-plex space, would not gain suf-ficient access to the market. In

TAP INTO THE BRILLIANCE THAT INDIVIDUALS WHO DON'T WORK FOR YOU HAVE, WHETHER THEY BE CUSTOM-ERS, BORED SCIENTISTS IN ACADEMIA, OR TEENAGE KIDS WITH AN IDEA ABOUT HOW TO ADVERTISE YOUR PRODUCT BETTER.

a way, CAMBIA helps to create and spread innovation by re-moving control from the hands of the wealthiest companies.

I am not suggesting you give everything away for free, nor am I suggesting you expose your competitive advantage. This would make no sense unless it made your business more profit-able. You should, however, tap into the brilliance that individu-als who don't work for you have, whether they be customers, bored scientists in academia, or teenage kids with an idea about how to advertise your product better.

INVISIBLE PROFITS: SUCCESS THROUGH CO-OPETITION AND PARTNERSHIP

I had the pleasure of sitting next to a senior engineer from Maunsell, Australia, on a flight from Perth to Brisbane. His name is Richard Jackson. Richard now claims that one of his great achievements in life is to sit next to me for more than four hours in a confined space and survive. It was the start of a rela-tionship with Maunsell, and now AECOM globally, a company that ranks among the most exciting I have ever worked with. In that conversation one thing stood out and has influenced my thinking greatly.

In the development of huge infrastructure projects, competitive firms are often engaged simultaneously to get the project finished on spec and on time. Apparently these "partnerships" have traditionally been hostile ones. When I asked Richard to describe the biggest change facing his industry, he said it was a move toward win-win relationships with suppliers and traditional competitors, away from the older-style hostile partnerships.

Recently I was listening to a podcast of a presentation by Bill Clinton, at the annual TED conference in Monterey, California, in which he used the word *interdependent* repeatedly. This idea of teaming and partnership even with traditional competitors seems to be gathering real momentum.

A 2000 Booz Allen Hamilton Consulting report notes that more than 50 percent of business alliances in the "new wave" of business collaborations (which refers to the massive increase of business partnerships since around the year 2000, facilitated by technology and spurred on by an increasingly crowded market) are between *competitors,* and lead on average to a 7 percent higher return than traditional collaborations.

Toyota realizes that sharing their hybrid technology has real benefits for them. They already have technology-sharing agreements with Nissan and Ford (who use Toyota technology in the popular Ford Escape). Toyota not only enjoys licensing fees and royalties from other companies using their technology, but the more people who use their technology the higher their output volume, so the lower their per-unit production costs become. It's win-win.

Apple recognized that the best way to achieve their goals lay in cooperating with other established companies in partnerships where they could ride each other's innovations. In 2006 they partnered with Intel to deliver Mac computers running off Intel chips. The companies' faith in the power of collaboration is clear when you look at the way they described the venture in a joint press release on June 6, 2005.

> JUST BECAUSE YOU HAVEN'T FIGURED OUT A WAY TO MAKE AN IDEA WORK, IT DOES NOT MEAN SOMEONE ELSE CAN'T.

Apple CEO Steve Jobs said, "Our goal is to provide our customers with the best personal computers in the world, and looking ahead Intel has the strongest processor roadmap by far . . . we think Intel's technology will help us create the best personal computers for the next ten years."

Intel CEO Paul Otellini said, "We are thrilled to have the world's most innovative personal computer company as a customer. Apple helped found the PC industry and throughout the years has been known for fresh ideas and new approaches. We look forward to providing advanced chip technologies, and to collaborating on new initiatives, to help Apple continue to deliver innovative products for years to come."

Just because you haven't figured out a way to make an idea work, it does not mean someone else can't. Sell it, license it, partner on it. Stop keeping it hidden, especially if you are getting no return on the asset. You wouldn't leave a huge piece of real estate or a factory doing nothing. You would work that and your other physical assets as much as possible. It is time we started thinking of our intellectual assets the same way.

According to Ron Sampson, secretary of the not-for-profit National Institute for Strategic Technology Acquisition and Commercialization in Manhattan, 90 to 95 percent of all patents are idle. Proctor & Gamble only uses about seven thousand of their thirty-six thousand patents, according to company spokesman Jeff LeRoy.

Companies will patent any meaningful advance but only work on the ones that fit into their immediate R&D. The worst for this are pharmaceutical companies. The phenomenon is called "the tragedy of the anticommons"—that is, patents keep knowledge in the hands of just a few people (monopoly on knowledge), which is bad for innovation. In fact, in the European Union, a report by a group of Italian academics found that as many as 18 percent of patents were "blocking" patents, which are patents taken out not so you can develop them into something, but to make sure someone else doesn't. That's not cool, and that is not progressive enough for the flip world.

THE ULTIMATE POWER TRIP: LEVERAGING POWER BY SHARING IT

The antiquated 1950s view of business is the boss walking around trying to squeeze every last bit of work out of employees, worried they are ordering Christmas presents online. I actually worry about the opposite, getting my people to take all their vacation time. We have a system that notifies me when someone is about to lose vaca-

tion days because they haven't SMART PEOPLE WORK
taken them in time. Every one of FOR YOU. USE THEM!
my direct reports is in danger.

> —Jonathan Schwartz, President/COO,
> Sun Microsystems

Smart people work for you. Use them! One of the biggest complaints that my own research, and that of companies such as Gallup, has found is that one of the most frustrating things about work is when your company doesn't fully utilize your skills.

Technology can help

The same technology that is putting the power back in the hands of the consumer can also help you to generate collaboration and knowledge sharing within your own workforce. Consider the following example that appeared in an *Australian Financial Review* article in 2007.

Geek Squad, the computer-repair company started by University of Minnesota IT graduate Robert Stephens, uses a whole suite of electronic collaborative software to keep teams all across the United States working in sync with one another. Geek Squad wikis are accessible and editable by all of their almost twelve thousand employees, so they provide *useful* and *up-to-date* information all the time. Some evidence suggested that appropriately deployed in-house wikis could cut company meeting times "in half" and decrease e-mail traffic volume by 75 percent.

While some are skeptical about those figures, the direction is right. It's interesting to note that the *AFR*'s source for the Geek Squad story, a book called *Wikinomics: How Mass Collaboration*

Changes Everything, by Don Tapscott and Anthony D. Williams, is supplemented by a wiki that readers can contribute to and edit online. Recognizing that the book's information could quickly become obsolete, the authors say a wiki is the only way to keep it relevant.

The *AFR* article also cited Xerox, where CTO Sophie Vandebroek flipped the whole R&D strategy on its head: instead of the strategy emanating from the boardroom and filtering downward, she started a company wiki where researchers in the R&D teams could collaboratively define the entire R&D strategy.

Perhaps the most extreme example is IBM's InnovationJam in 2006. The event brought together over one hundred thousand IBM employees in a moderated series of online discussions over seventy-two hours, where they brainstormed potential new products. Nothing was out-of-bounds for discussion. Such is IBM's commitment to the project that chief executive Sam Palmisano has committed $100 million to develop ideas with the most social and economic potential.[17]

IBM's first jam took place in 1998 and was a single, full-day collaboration between R&D labs. It was so successful that it was turned into a three-day extravaganza and eventually put online. An enormously detailed Web site for the 2007 jam includes "tips for successful jammers" that lead off with, "Don't be shy—you're the expert." This encapsulates the philosophy of "To Get Control, Give it Up."

Also, the jam has now gone way beyond just innovating in products (traditional R&D). Now every one of the 330,000 IBM employees has been explicitly given license to offer suggestions and "riff" on everything from new products and services ideas to new business processes and business models. It really is an innovation free-for-all.

ON TOYOTA'S ASSEMBLY LINES, THE LOWEST-LEVEL WORKER CAN SHUT DOWN THE LINE FOR ANY REASON, EVEN IF IT'S JUST THE SUSPICION OF A PROBLEM.

I feel like I am overdoing the Toyota examples, but here I go again. The truth is the company is a flipstar in many ways. Toyota became admired first and foremost for its quality manufacturing system. Ford, GM, and Chrysler have all sent high-level staff to observe Toyota's assembly lines, and in recent years the Detroit Three have greatly improved the quality of their own vehicles, in large part by copying the Japanese companies that once copied them.

But one feature of the Toyota manufacturing system seems to be too tough for Detroit to emulate. On Toyota's assembly lines, the lowest-level worker can shut down the line for any reason, even if it's just the suspicion of a problem. In fact, Toyota's managers start to worry if the line doesn't shut down

IF YOUR STAFF IS SO BORED THEY ARE SPENDING HOURS DOWNLOADING JUNK FROM THE INTERNET, THEN YOU SHOULD LOOK AT YOUR STAFF SELECTION PROCESS, AND MORE IMPORTANT AT THE WORK YOU ARE ENGAGING IN AND THE CULTURE THAT EXISTS. BUILD A MORE EXCITING PLACE TO WORK!

frequently, because it probably means quality problems are slipping through without being fixed. Yet despite all these stops and starts, Toyota produces vehicles in a shorter time than any other car manufacturer in the world.

What a flip! Empowering the lowest-level worker to shut down an entire assembly line lifts the entire workforce's dedication, performance, and morale. Meanwhile, the Detroit Three, trapped in the bunker of a command-and-control mentality, recoil from distributing such power down to the factory floor, even as their market shares erode and their losses mount. With Ford reporting its biggest-ever loss of $12.7 billion in 2006, it's no wonder CEO Alan Mulally went to Japan in December of that year to discuss a possible partnership with Toyota.

I would like to give you some simple day-to-day examples of managers trying to hold on to power, and it having a potentially adverse impact on results. Even though these examples seem quite insignificant, they represent a much deeper challenge in businesses today.

I confessed earlier to not being a fan of YouTube, and mentioned how I can't understand why people spend so much time downloading what I consider garbage from this and other sites—but remember, regardless of whether you or I do it, millions of others do it every day. And they are both your customers and your staff. It is this kind of attitude that leads to companies trying to ban sites like YouTube from being viewed by their staff at work. And herein lies one of the lessons of "To Get Control, Give It Up." If your staff is so bored they are spending hours downloading junk from the Internet, then you should look at your staff selection processes, and more important at the work you are engaging in and the culture that exists. Build a more exciting place to work!

Or consider the insurance company that doesn't give the majority of their call-center staff access to the Internet or even e-mail. Imagine not having e-mail at work. I know for some of you this sounds more like heaven, but seriously e-mail is a ubiquitous business tool and one way that we keep in touch. Oh, and they also banned cell phones. Imagine doing this in a workplace filled with young people, who can't go to the toilet without phoning five of their friends. The result was that the staff continued to use their phones, they just had to hide them. And staff attrition rose to very expensive levels. This is when I was called in.

My favorite example comes from the airline industry. A senior safety engineer at a leading airline had a problem with his trainee safety-inspection officers (or whatever their official title was). They were listening to their iPods while doing routine safety checks on the planes. On first hearing this, being someone who boards literally hundreds of planes each year, even I was disturbed.

When I said I could understand why it would be a problem if customers saw this, the manager looked at me blankly. "I had not thought of that."

"What are you worried about, then?"

"Customer perception is not my area, I was simply worried that mistakes would be made."

Fair enough, I thought, and asked, "So what did you do?"

"I banned iPods."

I asked him how that went. Badly was the answer. The young trainees started to put their iPods in their pockets and run the headphones under their uniforms, up and out of their collars, and into their noise-protection earmuffs, leaving only a small indication of the recognizable white iPod headphones.

"What did you do next?" I asked, grinning.

He employed an independent consultant to come in and do random accuracy checks on the trainees. This is not an uncommon activity, as the airline does things like this to benchmark its safety performance anyway, and has been doing so for a long time. What he found was that there was no correlation between listening to the iPods, albeit hidden, and inaccuracy.

"What happened next?" I asked.

"They bought me an iPod for Christmas."

I laughed. Just because he couldn't work effectively while listening to music, it doesn't mean someone else couldn't. In life we perceive the world through the lens of our own experience. Then anything that does not match our personal experience we think is wrong, and attempt to bring that person or thing back into alignment. The problem with this is that we usually attempt to do it by "controlling" it and using our positional authority. Are you married, or in a relationship? If so, let me ask you, have you ever tried to change your partner by controlling them? How did that go for you?

I thought so.

This experience reminded me of the financial adviser I met who was struggling with some of his staff, again younger staff, who refused to wear ties. He asked that they do, but they simply ignored him. Over time it got worse, to the point where some stopped wearing suits altogether. When one of his team came to work in jeans and flip-flops, he completely lost it.

He conducted a survey of his key customers about staff dress and the impact it had on their experience. He had expected to see a compelling response demanding that the staff wear more formal attire. Instead he found the opposite.

MANAGERS HAVE FAILED TO GIVE SPACE, RESOURCES, AND, MOST IMPORTANT, TRUST TO THEIR STAFF TO DO THEIR JOBS.

Almost 75 percent of the respondents had not even noticed the more casual attire of these advisers and para-planners. A further 20 percent actually said they preferred it because they felt more relaxed, and only 5 percent said they preferred more formal attire. He has since changed his "rules."

These examples are superficial, but are evidence of a much deeper problem in workplaces around the world: a failure on the part of managers to give space and resources and, most important, trust to their staff to do their jobs. They micromanage everything, from whether they wear a tie, to what time they arrive, down to what they say in their e-mails. This all leads to frustration for the staff, and lost productivity and profitability as a result.

According to Gallup, there are twelve questions — each of which relates to a specific need—that indicate whether someone is fully engaged. Of the twelve, the first two relate directly to the idea of letting go of control:

> "A COMPANY WILL GET NO-WHERE IF ALL THE THINKING IS LEFT TO MANAGEMENT . . . WE INSIST THAT ALL EMPLOYEES CONTRIBUTE THEIR MINDS."
> —AKIO MORITA

- Do you have the materials and equipment you need to do your work right?
- At work, do you have the opportunity to do what you do best every day?

Akio Morita, the founder and former chairman of Sony, has this to say about listening to staff: "A company will get nowhere if all the thinking is left to management. Everybody in the company must contribute and for the lower level employees their contribution must be more than just manual labor. We insist that all our employees contribute their minds."

START HERE

The old mantra of a job for life and the "loyalty" companies want so desperately from their employees also bespeak trying to keep control. Now, don't get me wrong, retention is extremely important. Crunching some numbers recently with a law-firm client of mine, the net return per person more than triples between the second year and the third year for professional legal staff.

However, expecting them to stay for very long periods these days is probably unrealistic. Some will, but most will not. This is why Ernst & Young has adopted a "start here" approach to its graduate recruitment. Instead of sweating bullets trying to keep the best and brightest from leaving, the company wishes them well and gives them every support in moving out to move up. It has built its graduate-based employment brand on the tagline "start here." The company does this for a few reasons. One is the departing staff might come back, assuming of course they had a pleasant exit experience. Secondly, they may move into non-accounting-firm employment and later become clients. But most of all because the company believes that if employees want to leave, let them. There is no point having someone unhappy at work.

Rather than try to keep the employees, the company strives to keep a relationship with them as they move to other companies and advance in their careers. When the best and brightest of these young workers have become senior executives and CEOs, they'll remember Ernst & Young as the great place where they got their start and they'll be inclined to send their consulting work to their old friends rather than to one of Ernst & Young's competitors.

More to the essence of "To Get Control, Give It Up" is the example of a "tailored taste" being implemented by the Australian Defence Force. In research I helped conduct for the ADF, we found that the younger generation shied away from the forces not just because of the work, the risk of war, or even some ideological opposition, but because they would be "locked in" for a set period of time. They were not at all keen on being controlled in this way. One of the recommendations made to combat this was the gap-year program. A "gap year" is becoming a sort of rite of passage for young Australians after they finish high school. They head to the UK, the States, or somewhere else in the world for twelve months as a gap between high school and university. Some, however, don't have the cash to head overseas, and would love to take a year off but want to grow and learn at the same time. Enter the ADF gap year.

Defence believes they have a superb product. They believe that if people at least try it they will enjoy their time in Defence and join longer term. Defence is putting in place a two-year program as a gap-year option. It's "try before you buy." That is, students finishing year twelve can enroll in one of the forces for two years only, get trained, develop some leadership skills, stay fit, and travel overseas on at least one operation while there for the year. They can then leave and pursue their other interests, and at worst the army or navy or air force had a committed soldier for a couple of years. Or in the best case, they could leave and spread the word on how awesome their time was in the forces. Some of these gappers will love the experience so

GET OUT OF THEIR WAY, AND LET SMART PEOPLE DO WHAT SMART PEOPLE DO.

much that they will enroll for full service, attend the Australian Defence Force Academy or something similar, and become fully fledged members of the ADF. This is at least the theory. Markets call this a tailored taste.

The way I look at it, particularly the Ernst & Young "start here" positioning, is that if your employees know you have no illusions about them staying for life, they will converse with you actively about their plans because they don't feel threatened. This helps you better prepare, better retain knowledge, and better orient their replacement.

Empowering your staff works when you have hired good staff. Once you have them, you really need to use them. Get out of their way, and let smart people do what smart people do.

FIVE THINGS TO DO NOW

1. Ask yourself if your current business model is under threat, or if it could be if technology continues to develop rapidly. Start now to consider how you would exploit such forced change so you are not on the defensive when it happens.

2. Come up with ten ways you could tap into a broader population to stimulate ideas for you and your business.

3. Start a structured program within your company that brings your best and brightest people together regularly to work on new ideas or existing problems. Give it a cool name and start using it to attract and retain talented people.

4. Come up with five ways you could collaborate with your existing competitors to bring a product to market, enhance an existing product or service, or even crack open an entirely new market.

5. Does your business have IP that is very valuable to the marketplace, that you are not benefiting from because it is hidden or overprotected? How could you better exploit this asset? In other words, how could you win through losing control?

conclusion:
GET MOVING!

Getting inspired while reading a book is the easy part. Talking to yourself and getting yourself psyched about a new idea for your career and company take little courage. Sure you should be commended for picking the book up in the first place, and you should be celebrated for making it to the end. However, this is the catalyst, not the action. You have to get stuff done.

When you put this book down and head back to work, you will be met by people who have not yet been *flipped*. They won't see the world through the same ideas as you now do. They will be entrenched in out-of-date and inefficient practices, which have made them successful to date.

You should not reject, ridicule, or try to "fix" these people. At least, not right away. Nor should you develop a sense of superiority because you now "get it." You will need these people in order to flip your organization.

What you must do now is exercise finesse, generate real and impressive results, and develop an inspiring view of your business's future.

Finesse because people will want to reject you and your ideas. New can often scare people. Your job is to "play the game" and win these people over. Going head-to-head is usually not the best approach. Although, if you get no love—go hard!!! You might as well go down fighting. Then start your own company and kick your old organization's butt in the marketplace.

Results because at the end of the day in business this is what matters. No point having cool ideas, as flipped as they may be, if they don't get results. Being different for different's sake is no different from change for change's sake. If all it does is keep you interested and add some variety to your life, get a hobby. Play some, take some risks, too. Just remember business is about getting results.

And finally *an inspiring view of the future* because people can and will change if they believe it is for the better. They want to know you have their best interests at heart, and deep down I think they just want to be inspired by the possibility of a better future.

My hope is that *Flip* has given you an inspiring view of the future. I wanted to provide some clarity about the changes we are dealing with daily, and most of all I wanted to give you an empowering mind-set for meeting those changes head-on.

Get off the sidelines and into the ring. Throw some mud at the wall and see what sticks. No one really knows the answers. Not today, and certainly not tomorrow. But this is no excuse not to get in the game. Play a hand or two in the new world of business and see how you fare. Or in closing, consider these words of a flipstar from days gone by:

> It is not the critic who counts: not the man who points out how the strong man stumbles or where the doer of deeds could have done better. The credit belongs to the man who is actually in the arena, whose face is marred by dust and sweat and blood, who strives valiantly, who errs and comes up short again and again, because there is no effort without error or shortcoming, but who knows the great enthusiasms, the great devotions, who spends himself for a worthy cause; who, at the best, knows, in the end, the triumph of high achievement, and who, at the worst, if he fails, at least he fails while daring greatly, so that his place shall never be with those cold and timid souls who knew neither victory nor defeat.
> —Theodore Roosevelt, "Citizenship in a Republic,"
> a speech given at the Sorbonne, Paris, April 23, 1910

Or put more simply:
Get up off your butt and do something!!!

acknowledgments

Thanks first to my wife, Sharon, for not killing me when I wake up in the middle of the night to write down a new idea and most of all for being a great mum to our beautiful kids as I traipse around the world from hotel room to hotel room.

Second, to Dom Thurbon in my office. Dude, you're a flip-star no doubt. This book would not have been written without your help. And to Hilary Hinzmann in New York, this book would be nothing more than ordinary if it had not been for your wisdom, prose, and basically keeping it real.

Also to my agent, Mary Cunnane, who persevered for literally eighteen months as *Flip* was, well, flipped again and again and again. We made it.

To Henry Ferris and the team at William Morrow, thanks for having the vision to see this was a book the market needed. It may have been a little ahead of its time when you first saw it, but now the world is ready. Thanks for your support, and for the guidance and for trusting me back then.

Finally, to all of my clients. Without you, why write the book? Thanks for your support and I look forward to hanging out some more in the future.

notes

Chapter 1. The Four Forces of Change

1. Pankaj Ghemawat, "The Myth of Globalisation," *Australian Financial Review,* "Review" section, March 16, 2007.

2. The *Time* magazine article where I found this information is no longer available online, but the main findings of the original report it references, produced by technology research firm IDC, are summarized by the Pakistan Software Export Board in its March 2007 online bulletin, available at: www.pscb.org.pk/bulletin/mar07/Bullet_details_march07.htm#test2.

3. Pew Research Center, "Luxury or Necessity? Things We Can't Live Without: The List Has Grown in the Last Decade," December 14, 2006, available at: pewresearch.org/pubs/323/luxury-or-necessity.

Chapter 2. Action Creates Clarity

1. Spencer Reiss, "His Space," *Wired,* July 14, 2006, available at: www.wired.com/wired/archive/14.07/murdoch.html.

2. T. Gilovich, and V. H. Medvec, "The Experience of Regret: What, When, and Why," *Psychological Review* 102 (1995), 379–395.

3. David Gray, "Wanted: Chief Ignorance Officer," *Harvard Business Review,* November 2003, 22–24.

4. Bain & Co., "Change or Die," *Fast Company,* May 2005, 60.

5. Daniel Gilbert, *Stumbling on Happiness* (New York: Alfred A. Knopf, 2006).

Chapter 3. Fast, Good, Cheap: Pick Three—Then Add Something Extra

1. Michael Arndt, "McDonald's 24/7," *BusinessWeek,* cover story, February 5, 2007.

2. Kasra Ferdows, Michael A. Lewis, and Jose A. D. Machuca, "Rapid Fire Fulfillment," *Harvard Business Review* 82, no. 11 (November 2004).

3. Noel Capon, *The Marketing Mavens* (New York: Crown Business, 2007), 177–179.

4. Saul Hansell, "EBay Profit Rises 52 Percent," *New York Times,* April 19, 2007.

Chapter 4. Absolutely, Positively Sweat the Small Stuff

1. Marc Gunther, "The Green Machine," *Fortune,* August 7, 2006; Michael Barbaro, "Home Depot to Display an Environmental Label," *New York Times*, April 17, 2007.

2. L'Oréal 2005 Sustainable Development Report, available at: www.loreal.com/_en/_ww/group_new/pdf/LOREAL_RDD_GB.pdf.

3. Howard Schultz's original memo can be viewed online at: starbucksgossip.typepad.com/_/2007/02/starbucks_chair_2.html.

4. Jeff Vrabel, "Macartney to Anchor New Starbucks Label," www.billboard.com, March 21, 2007, available at: www.billboard .com/bbcom/news/article_display.jsp?vnu_content_id=1003561098.
5. Gina Chon, "A Way Cool Strategy: Toyota's Scion Plans to Sell Fewer Cars," *Wall Street Journal,* November 10, 2006.

Chapter 5. Business Is Personal

1. Ben Stein, "The Hard Rain That's Falling on Capitalism," *New York Times,* January 28, 2007.
2. Yankelovich Partners, "A Crisis of Confidence: Rebuilding the Bonds of Trust," 2004, available at: www.compad.com.au/cms/ prinfluences/workstation/upFiles/955316.State_of_Consumer_ Trust_Report_-_Final_for_Distribution.pdf.
3. "Building Trust at Home in a Global World," *Australian Financial Review* 19 (March 2007).
4. www.neojobmeeting.com.
5. Finextra Research, "First Direct Hails Text Banking Milestone," February 11, 2005, available at: www.finextra.com/fullstory.asp?id= 13225.
6. Australian Interactive Media Association, "Campaign Success for Mission Impossible 3 and Expansion of Hypertag Solution into New Zealand for AURA and Adshel," available at: www.aimia .com.au/i-cms?page=2263.
7. Laurie J. Flynn, "IBM to Introduce Workers' Networking Software," *New York Times,* January 22, 2007.
8. Meredith Levine, "Tell the Doctor All Your Problems, but Keep It to Less Than a Minute," *New York Times,* June 1, 2004; Wendy Levinson, et al., "Surgeons' tone of voice: a clue to malpractice history," *Surgery*, 132, no. 1 (July 2002), 5–9; Wendy Levinson, "Doctor-patient communication and medical malpractice:

implications for pediatricians," *Pediatric Annals*, 26, no. 3 (May 1997), 186–193.

Chapter 6. Mass-Market Success: Find It on the Fringe

1. W. C. Kim, and R. Mauborgne, *Blue Ocean Strategy: How to Create Uncontested Market Space and Make Competition Irrelevant* (Cambridge, MA: Harvard Business School Press, 2005), 7–8.
2. "BMW 750hL: The Ultimate Clean Machine," available at: www .bmwworld.com/models/750hl.htm.
3. Stephen Totilo, "No Ruler Required for Xbox 360's Cost-Effective Geometry: Trial Version of Game Has Been Downloaded More than 200,000 Times," January 12, 2006, available at: www.mtv.com/news/articles/1520696/20060112/index.jhtml ?headlines=true; Tom Bramwell, "Reshaping the Past," February 2, 2006, available at: www.eurogamer.net/article.php?article_id= 62715; David Kushner, "The Infinite Arcade," *Wired,* August 2006, available at: www.wired.com/wired/archive/14.08/nintendo .html.
4. "Success of Nintendo's Wii Hinges on Games, Not Hardware," FoxNews.com, October 12, 2006, available at: www.foxnews.com/ story/0,2933,220299,00.html.
5. Joseph B. Treaster, "One-Stop Car Insurance Service: Body Work Included," *New York Times,* May 26, 2007.

Chapter 7. To Get Control, Give It Up

1. Kathy Rebello, "Inside Microsoft (Part 2): The Untold Story of How the Internet Forced Bill Gates to Reverse Course," *Business-Week,* July 15, 1996, available at: www.businessweek.com/1996/29/ b34842.htm.
2. "Intellectual Piracy in China," *News Hour with Jim Lehrer,*

October 13, 2005, available at: www.pbs.org/newshour/bb/asia/july-dec05/china_10-13.html.

3. Bloomberg News, "Piracy in China of Music, Software Draws Ire of U.S.," *Seattle Times,* April 10, 2007.

4. Charles Pillar, "How Piracy Opens Doors for Windows," *Los Angeles Times,* April 9, 2006.

5. M. A. Anastasi, "Sony Exec: We Will Beat Napster," New Yorkers For Fair Use, August 17, 2000, available at: www.nyfairuse.org/sony.xhtml.

6. Andrew Dubber, "The Real Reason Koopa Is Important," February 27, 2007, New Music Strategies, available at: www.newmusicstrategies.com/2007/02/27/the-real-story-of-koopa.

7. Patrick Goldstein, "Hollywood Is Seeing Fans Pull a Power Play," *Los Angeles Times,* January 23, 2007.

8. For more information on Cory Doctorow's work, see his Web site: www.craphound.com.

9. Tara Hunt, "Chevy Tahoe's First Mistake," HorsePigCow (blog), April 3, 2006, available at: www.horsepigcow.com/2006/04/chevy-tahoe-first-mistake.html.

10. B. J. Ochman, "Won't It Be Funny When Chevy Tahoe Sends Cease & Desist Letters to Bloggers!," whatsnextblog.com, April 2006, available at: www.whatsnextblog.com/archives/2006/04/post_49.asp.

11. Ed Peper, "Now That We've Got Your Attention," GM Fast-Lane (blog), April 6, 2006, available at: fastlane.gmblogs.com/archives/2006/04/now_that_weve_g_l.html.

12. Seth Godin, "Web4," Seth Godin's Blog, January 19, 2007, available at: sethgodin.typepad.com/seths_blog/2007/01/web4.html.

13. William C. Taylor, "To Charge Up Customers, Put Customers in Charge," *New York Times*, June 18, 2006.

14. Brendan I. Koerner, "Geeks in Toyland," *Wired*, February 2006.

15. Taylor, "To Charge Up Customers, Put Customers in Charge," *New York Times*.

16. David Leonhardt, "You Want Innovation? Offer a Prize," *New York Times,* January 31, 2007.

17. "All Profit in the Wiki Workplace," *Australian Financial Review,* March 29, 2007, 57.

index